Tube Substitution Handbook

Complete Handbook to Replacements for Vacuum Tubes and Picture Tubes

William Smith WB9TUI
&
Barry Buchanan

Second Edition, 1998
PROMPT® Publications is an imprint of Sams Technical Publishing, 5436 W. 78th St., Indianapolis, IN 46268.

International Standard Book Number: 0-7906-1148-1

Technical Advisors: Barry Buchanan, William Smith WB9TUI
Research: Barry Buchanan, William Smith
Text: William Smith, Sara Wright
Technical Drawings: Jim Young
Layout Design: Sara Wright
Cover Design: Kelly Ternet

PRINTED IN THE UNITED STATES OF AMERICA

9 8 7 6 5

Tube Substitution Handbook

CONTENTS

INTRODUCTION

For many years, vacuum tubes were standard components in all electronic equipment. Now that times have changed and the industry has advanced, tubes are considered by many to be obsolete. Although the days of running to any corner store for a needed tube are gone, many technicians and hobbyists still repair and use tube-based equipment. *Tube Substitution Handbook* is written for this group.

Tube substitution is one of the only feasible methods to repair or restore original tube equipment, but it should not be performed haphazardly. This handbook, itemizing all known vacuum tubes that have been or are still being manufactured, along with their replacements, will make tube substitution not only possible, but relatively straight-forward and efficient.

This guide is divided into three parts:

Section 1: Vacuum Tube Substitution, listing all known vacuum tubes (excluding picture tubes) in numeric-alphabetic order, with reference to appropriate substitutions and diagrams detailing base pin arrangement.

Section 2: Picture Tube Substitution, listing all known picture tubes in numeric-alphabetic order, with reference to appropriate substitutions and diagrams detailing base pin arrangement.

Section 3: Base Diagrams, illustrating nearly 900 pin arrangements which detail each of the tubes listed in the handbook.

Vacuum Tube Substitutes

This section lists, in numeric-alphabetic order, all known receiving tubes. To use this listing, look in the first column for the tube type you want to replace, then find a suitable substitute in the third column. The substitutes are only those tubes which have similar or improved characteristics and can be inserted directly into the socket. No rewiring or mechanical changes are needed, although the associated service controls may have to be readjusted. The symbols which may follow a substitute, refer to special characteristics of the tube and are shown in the key below. **When no substitutes are shown, none are recommended.** The second column contains the basing code. To use this number, turn to Section 3 to reference your tube substitution selection against the diagrammed pin connections.

You should not replace a tube in the third column with one from the first column as this may result in a less than satisfactory match. For example, where a high heater-to-cathode potential exists, a 6AX4GTA will hold up, but a 6W4 will not. You will find 6AX4GTA listed as a substitute for 6W4, but not vice versa.

In some instances of tube replacement, the space available should be taken into consideration. For example, although electrically identical to a 6L6GB, a 6L6G is more than one inch taller and a half inch larger in diameter, making the substitution physically impossible in some compact equipment.

The X-radiation levels emitted from all television receivers produced after January 15, 1970 must meet rigid government standards. These radiations are primarily produced by high-voltage rectifier tubes, shunt-regulator tubes, and picture tubes. The radiation, however, is carefully controlled in tubes of later design. Whenever possible, only tubes produced after 1970 are suggested for substitution. Although other tubes may appear to be acceptable substitutes, they have not been listed because of the possibility of radiation hazard.

Key to Symbols:

- * Appropriate for parallel-filament circuits.
- ! Appropriate for series circuits not requiring controlled warm-up.
- # May not work in all circuits.

Tube Type	Basing	Replacement
01A	4D	12A, 12A/112A, 112A
045	-	-
0A2	5BO	0A2WA, 150C2, 150C4, 6073, 6073/0A2, 6626, 6626/0A2WA, M8223, STV150/30
0A2A	5BO	0A2#, 0A2WA, 6073#, 6073/0A2#, 6626, 6626/0A2WA
0A2WA	5BO	0A2#, 6073#, 6073/0A2#, 6626, 6626/0A2WA
0A3	4AJ	0A3/VR75, 0A3A, KD21, VR75
0A3/VR75	4AJ	0A3, 0A3A, KD21, VR75
0A3A	4AJ	0A3#, 0A3/VR75#, 0A3A, VR75
0A4	4V	0A4/PL1267, 0A4G, 1267, 1267/0A4, PL1267, WTT132, Z300T
0A4/PL1267	4V	0A4, 0A4G, 1267, 1267/0A4, PL1267, WTT132, Z300T
0A4A	4V	0A4/PL1267, 0A4G, 1267, 1267/0A4, PL1267, WTT132, Z300T
0A4G	4V	0A4/PL1267, 1267, 1267/0A4, PL1267, WTT132, Z300T
0A5	-	-
00A	4D	-
0B2	5BO	0B2WA, 108C1, 6074, 6074/0B2, 6627, 6627/0B2WA, M8224, STV108/30
0B2W	5BO	0B2#, 0B2W, 0B2WA, 6074#, 6074/0B2#, 6627, 6627/0B2WA
0B2WA	5BO	0B2#, 0B2W#, 6074#, 6074/0B2#, 6627, 6627/0B2WA
0B3	4AJ	0B3/VR90, 0B3A, VR90
0B3/VR90	4AJ	0B3, 0B3A, VR90
0B3A	4AJ	0B3, 0B3/VR90, VR90
0C2	5BO	-
0C3	4AJ	0C3/VR105, 0C3A, 0C3W, KD24, VR105, VR105W, WT269
0C3/VR105	4AJ	0C3, 0C3A, 0C3W, KD24, VR105, VR105W, WT269
0C3A	4AJ	0C3#, 0C3/VR105, 0C3W, KD24, VR105, VR105W, WT269
0C3W	4AJ	0C3#, 0C3/VR105#, 0C3A#, KD24#, VR105#, VR105W, WT269
0D3	4AJ	0D3/VR150, 0D3A, 0D3W, 150C3, KD25, VR150, VR150/0D3, VR150W, WT294
0D3/VR150	4AJ	0D3, 0D3A, OD3W, 150C3, KD25, VR150, VR150/0D3, VR150W, WT294
0D3A	4AJ	0D3#, 0D3/VR150, 0D3W, 150C3, KD25, VR150, VR150/0D3
0D3W	4AJ	0D3#, 0D3/VR150#, 0D3A#, 150C3#, KD25#, VR150#, VR150/0D3, VR150W, WT294
0E3	-	0E3/85A1, 85A1
0E3/85A1	-	0E3, 85A1
0G3	-	0G3/85A2, 5TV85-10, cont'd
0G3, cont'd		85A2, 5651A
0G3/85A2	-	0G3, 5TV85-10, 85A2, 5651A
0Y4	-	0Y4G
0Y4G	-	0Y4
0Z3	-	-
0Z4	4R	0Z4A, 0Z4A/0Z4, 0Z4G (may require shield), 1003, WTT114
0Z4A	4R	0Z4, 0Z4A/0Z4, 0Z4G (may require shield), 1003
0Z4A/0Z4	4R	0Z4, 0Z4A, 0Z4G (may require shield), 1003
0Z4G	4R	0Z4, 0Z4A, 0Z4A/0Z4
1	-	-
1A3	5AP	1A3/DA90/1D13, 1D13, DA90
1A3/DA90/1D13	5AP	1A3, 1D13, DA90
1A4	4K	1A4P, 1A4T, 1B4, 1B4G, 1B4P, 1B4P/951, 1B4T, 1C4*, 1K4*, 32, 34, 951
1A4P	4M	1A4, 1A4T, 1B4, 1B4G, 1B4P, 1B4P/951, 1B4T, 1C4*, 1K4*, 32, 34, 951
1A4T	4K	1A4, 1A4P, 1B4, 1B4G, 1B4P, 1B4P/951, 1B4T, 1C4*, 1K4*, 32, 34, 951
1A5	6X	1A5G, 1A5GT, 1A5GT/G, 1T5GT, DL31
1A5G	6X	1A5GT, 1A5GT/G, 1T5GT, DL31
1A5GT	6X	1A5G, 1A5GT/G, 1T5GT, DL31
1A5GT/G	6X	1A5G, 1A5GT, 1T5GT, DL31
1A6	6L	1C6*
1A7	7Z	1A7G, 1A7GT, 1A7GT/G, 1B7G*, 1B7GT*, 1B7GT/G*, DK32, X14
1A7G	7Z	1A7GT, 1A7GT/G, 1B7G*, 1B7GT*, 1B7GT/G*, DK32, X14
1A7GT	7Z	1A7G, 1A7GT/G, 1B7G*, 1B7GT*, 1B7GT/G*, DK32, X14
1A7GT/G	7Z	1A7G, 1A7GT, 1B7G*, 1B7GT, 1B7GT/G, DK32, X14
1AB5	5BF	-
1AB6	7DH	1AB6/DK96, 1AC6*, 1AC6/DK92*, 1C3, 1C3/DK96/1AB6, 1H35, DK92*, DK96, X25
1AB6/DK96	7DH	1AB6, 1AC6*, 1AC6/DK92*, 1C3, 1C3/DK96/1AB6, 1H35, DK96, X25
1AC5	8CP	-
1AC6	7DH	1AB6*, 1AB6/DK96*, 1AC6/DK92, 1C3, 1C3/DK96/1AB6*, DK92, DK96*, X18, X20
1AC6/DK92	7DH	1AB6*, 1AB6/DK96*, 1AC6, 1C3, 1C3/DK96/1AB6*, DK92, DK96*, X18, X20
1AD2	12GV	1AD2/1BY2A, 1AD2A
1AD2/1BY2A	12GV	1AD2, 1BY2A
1AD2A	12GV	1AD2, 1AD2/1BY2A
1AD4	-	1AD4/DF62, DF62
1AD4/DF62	-	1AD4, DF62
1AD5	8CP	-
1AE4	6AR	1L4#, 1L4/DF92#, 1T4#
1AF4	6AR	1AE4*, 1AF4/1AJ4, cont'd

Tube Type	Basing	Replacement
1AF4, cont'd		1AM4, 1F1, 1T4*, 1T4SF, 1U4*, DF96, W25
1AF4/1AJ4	6AR	1AE4, 1AF4, 1AJ4, 1AM4, 1F1, 1T4*, 1T4SF, 1U4*, DF96, W25
1AF5	6AU	1AF5/1AH5, 1AH5, 1AR5, 1FD1, 1S5*, 1S5/DAF91, DAF91*, DAF96, ZD25
1AF5/1AH5	6AU	1AF5, 1AH5, 1AR5, 1FD1, 1S5*, 1S5/DAF91*, DAF91*, DAF96, ZD25
1AF33	-	-
1AF34	-	-
1AG4	-	-
1AG5	1AG5	-
1AH4	-	-
1AH5	7DJ	1AF5, 1AF5/1AH5, 1AR5, 1FD1, 1S5*1S5/DAF91, DAF91*, DAF96, ZD25
1AJ2	12EL	-
1AJ4	7DP	1AE4*, 1AF4, 1AF4/1AJ4, 1AM4, 1F1, 1T4*, 1T4SF, DF96, W25
1AJ5	-	-
1AK4	1AK4	-
1AK5	-	-
1AM4	6AR	1AE4*, 1AF4, 1AF4/1AJ4, 1AJ4, 1F1, 1T4*, 1T4SF, DF96, W25
1AN5	7ES	1AN5/DF97, DF97
1AN5/DF97	7ES	1AN5, DF97
1AQ5	7AT	1R5*, 1R5/DK91*, 1R5SF*, 1H33, DK91*
1AR5	6AU	1AF5, 1AF5/1AH5, 1AH5, 1FD1, 1S5*, 1S5/DAF91*, 1U5, DAF91*, DAF96, ZD95
1AS5	6BW	1U5SF, 1DN5*, 1U5*, DAF92*
1AU2	9U	-
1AU3	3C	1N2, 1N2A, 1N2A/1AU3
1AX2	9Y	-
1AY2	1AY2	-
1AY2A	1AY2	-
1B3	3C	1AU3, 1B3/8016, 1B3GT, 1B3GT/1G3GTA, 1G3GT, 1G3GT/1B3GT, 1G3GTA, 1G3GTA/1B3GT, 1J3, 1J3/1K3A, 1J3A, 1K3, 1K3/1J3, 1K3A, 1K3A/1J3, 1N2, 1N2A, 1N2A/1AU3, 8016, DY30, U41
1B3/8016	3C	1AU3, 1B3, 1B3GT, 1B3GT/1G3GTA, 1G3GT, 1G3GT/1B3GT, 1G3GTA, 1G3GTA/1B3GT, 1J3, 1J3/1K3A, 1J3A, 1K3, 1K3/1J3, 1K3A, 1K3A/1J3, 1N2, 1N2A, 1N2A/1AU3, 1N2A, 8016, DY30, U41
1B3GT	3C	1AU3, 1B3GT/1G3GTA, 1B3GTA, 1B3GTA/1B3GT, 1G3GT, 1G3GT/1B3GT, 1G3GTA, 1G3GTA/1B3GT, 1J3, 1J3/1K3A, 1J3A, 1K3, 1K3/1J3, 1K3A, 1K3A/1J3, 1N2, 1N2A, 1N2A/1AU3, 8016, DY30, U41
1B3GT/1G3GTA	3C	1AU3,1B3GT,cont'd

Tube Type	Basing	Replacement
1B3GT/1G3GTA, cont'd		1B3GTA, 1B3GTA/1B3GT, 1G3GT, 1G3GT/1B3GT, 1G3GTA, 1G3GTA/1B3GT, 1J3, 1J3/1K3A, 1J3A, 1K3, 1K3/1J3, 1K3A, 1K3A/1J3, 1N2, 1N2A, 1N2A/1AU3, 8016, DY30, U41
1B3GTA	3C	1AU3, 1B3GT, 1B3GT/1G3GTA, 1B3GTA/1B3GT, 1G3GT, 1G3GT/1B3GT, 1G3GTA, 1J3, 1J3/1K3A, 1J3A, 1K3, 1K3/1J3, 1K3A, 1K3A/1J3, 1N2, 1N2A, 1N2A/1AU3, 8016, DY30, U41
1B3GTA/1B3GT	3C	1AU3, 1B3GT, 1B3GT/1G3GTA, 1G3GT, 1G3GT/1B3GT, 1G3GTA, 1J3, 1J3/1K3A, 1J3A, 1K3, 1K3/1J3, 1K3A, 1K3A/1J3, 1N2, 1N2A, 1N2A/1AU3, 8016, DY30, U41
1B4	4M	1A4, 1A4P, 1A4T, 1B4G, 1B4P, 1B4P/951, 1B4T, 1C4*, 1K4*, 32, 34, 951
1B4G	4M	1A4, 1A4P, 1A4T, 1B4, 1B4P, 1B4P/951, 1B4T, 1C4*, 1K4*, 32, 34, 951
1B4P	4M	1A4, 1A4P, 1A4T, 1B4, 1B4P/951, 1B4T, 1C4*, 1K4*, 32, 34, 951
1B4P/951	4M	1A4, 1A4P, 1A4T, 1B4, 1B4P, 1B4T, 1C4*, 1K4*, 32, 34, 951
1B4T	4K	1A4, 1A4P, 1A4T, 1B4, 1B4P, 1B4P/951, 1B4T, 1C4*, 1K4*, 32, 34, 951
1B5	6M	1B5/25S, 25, 25S
1B5/25S	6M	1B5, 25, 25S
1B6	7AV	-
1B7	7Z	-
1B7G	7Z	1A7G*, 1A7GT*, 1A7GT/G*, 1B7GT, 1B7GT/G
1B7GT	7Z	1A7G*, 1A7GT*, 1A7GT/G*, 1B7G, 1B7GT/G
1B7GT/G	7Z	1A7G*, 1A7GT*, 1A7GT/G*, 1B7G, 1B7GT
1B8	8AJ	-
1B8GT	8AJ	1D8GT#
1B22	-	-
1B23	-	-
1B27	-	-
1B29	-	-
1B35A	-	-
1B44	-	-
1B48	-	5517, 5517/CX1013, CK1013
1B49	-	-
1B56	-	-
1B87	-	-
1B88	-	-
1BC2	9RG	1BC2A, 1BC2B
1BC2A	9RG	1BC2B
1BC2B	9RG	-
1BC2A	9RG	1BC2B
1BC2B	9RG	-
1BG2	-	DY51
1BH2	9RG	1BH2A

VACUUM TUBES

Tube Type	Basing	Replacement
1BH2A	9RG	-
1BK2	9Y	1BK2/1RK41, 1RK41, 1S2
1BK2/1RK41	9Y	1BK2, 1RK41, 1S2
1BL2	-	-
1BQ2	-	1BQ2/DY802, DY802
1BQ2/DY802	-	1BQ2, DY802
1BX2	9Y	1X2A, 1X2A/1X2B, 1X2B, 1X2B/1X2A, 1X2C, 1X2C/1BX2
1BY2	12HZ	1AD2, 1AD2/1BY2A, 1BY2A
1BY2A	12HZ	1AD2, 1AD2/1BY2A, 1AD2A#
1C1	7AT	1R5, 1R5/DK91, DK91, X17
1C2	-	1AC6, 1AC6/DK92, DK92
1C3	7DH	1AB6, 1AB6/DK96, 1C3/DK96/1AB6, DK96
1C3/DK96/1AB6	7DH	1AB6, 1AB6/DK96, 1C3, DK96
1C4	4K	1A4*, 1A4P*, 1A4T*, 1B4*, 1B4G*, 1B4P*, 1B4P/951*, 1B4T*, 1K4, 32*, 34*, 951*
1C5	6X	1C5/DL35/N14, DL35, N14
1C5/DL35/N14	6X	1C5, DL35, N14
1C5G	6X	1C5GT, 1C5GT/G, 1Q5G, 1Q5GT, 1Q5GT/G, 1T5GT*
1C5GT	6X	1C5G, 1C5GT/G, 1Q5G, 1Q5GT, 1Q5GT/G, 1T5GT*, DL35, DL36, N14
1C5GT/G	6X	1C5G, 1C5GT, 1Q5G, 1Q5GT, 1Q5GT/G, 1T5GT*
1C6	6L	1A6*
1C7	7Z	1C7G, 1D7G*
1C7G	7Z	1D7G*
1C8	-	-
1C21	4V	5823
1D3	-	-
1D4G	5B	1D4GT
1D4GT	5B	1D4G
1D5	5R	1D5G, 1D5GP, 1D5GT, 1E5G, 1E5GP, 1E5GT, 1K5G*, 1M5G*
1D5G	5R	1D5GP, 1D5GT, 1E5G, 1E5GP, 1E5GT, 1K5G*, 1M5G*
1D5GP	5Y	1D5G, 1D5GT, 1E5G, 1E5GP, 1E5GT, 1K5G*, 1M5G*
1D5GT	5R	1D5G, 1D5GT, 1E5G, 1E5GP, 1E5GT, 1K5G*, 1M5G*
1D7	7Z	1D7G, 1C7G*
1D7G	7Z	1C7G*
1D8	8AJ	1D8GT, 1B8GT#
1D8GT	8AJ	1B8GT#
1D13	5AP	1A3, 1A3/DA90/1D13, 1D13, DA90
1D21	-	1D21/SN4, 631P1, 631P1/SN4, SN4
1D21/SN4	-	1D21, 631P1, 631P1/SN4, SN4
1DG3	8ND	1DG3A
1DG3A	8ND	1D3G#
1DK1	-	-
1DK27	-	-
1DK29	-	-
1DN5	6BW	1AS5*, 1U5, 1U5SF*, DAF92
1DY4	7DK	1DY4A
1DY4A	7DK	1DY4
1E3	9BG	DC80
1E4	5S	1E4G, 1E4GT, 1G4G, 1G4GT, 1G4GT/G

Tube Type	Basing	Replacement
1E4G	5S	1E4GT, 1G4G, 1G4GT, 1G4GT/G
1E4GT	5S	1E4G, 1G5G, 1G5GT, 1G5GT/G
1E5	5R	1D5G, 1D5GP, 1D5GT, 1E5G, 1E5GP, 1E5GT, 1K5G*, 1M5G*
1E5G	5R	1D5G, 1D5GP, 1D5GT, 1E5GP, 1E5GT, 1K5G*, 1M5G*
1E5GP	5Y	1D5G, 1D5GP, 1D5GT, 1E5G, 1E5GT, 1K5G*, 1M5G*
1E5GT	5R	1D5G, 1D5GP, 1D5GT, 1E5G, 1E5GP, 1K5G*, 1M5G*
1E7	8C	1E7G
1E7G	8C	1E7GT
1E7GT	8C	1E7G
1E8	8CN	-
1F1	-	1AF4, 1AF4/1AJ4, 1AJ4
1F2	6AR	1L4, 1L4/DF92, DF92
1F3	6AR	1T4, DF91, W17
1F4	5K	2101
1F5	6x	1F5G
1F5G	6X	1L5G*, KL35*
1F6	6W	-
1F7	7AD	1F7G, 1F7GH, 1F7GT, 1F7GV, 1F7GY
1F7G	7AD	1F7GH, 1F7GT, 1F7GV, 1F7GY
1F7GH	7AD	1F7G, 1F7GT, 1F7GV, 1F7GY
1F7GT	7AD	1F7G, 1F7GH, 1F7GV, 1F7GY
1F7GV	7AF	1F7G, 1F7GH, 1F7GT, 1F7GY
1F7GY	7AD	1F7G, 1F7GH, 1F7GT, 1F7GV
1F33	-	-
1F34	-	-
1FD1	-	1AF5, 1AF5/1AH5, 1AH5
1FD9	6AU	1S5, 1S5/DAF91, DAF91, ZD17
1G3	3C	1B3GT, 1B3GT/1G3GTA, 1G3GT, 1G3GT/1B3GT, 1G3GTA, 1G3GTA/1B3GT
1G3GT	3C	1B3GT, 1B3GT/1G3GTA, 1G3GT/1B3GT, 1G3GTA, 1G3GTA/1B3GT, 1J3/1K3A, 1K3A, 1K3A/1J3, DY30, U41
1G3GT/1B3GT	3C	1B3GT, 1B3GT/1G3GTA, 1G3GT, 1G3GTA, 1G3GTA/1B3GT, 1J3/1K3A, 1K3A, 1K3A/1J3, DY30, U41
1G3GTA	3C	1B3GT, 1B3GT/1G3GTA, 1G3GTA/1B3GT, 1J3/1K3A, 1K3A, 1K3A/1J3
1G3GTA/1B3GT	3C	1B3GT, 1B3GT/1G3GTA, 1G3GTA, 1J3/1K3A, 1K3A, 1K3A/1J3
1G4	5S	1E4G, 1E4GT, 1G4G, 1G4GT, 1G4GT/G
1G4G	5S	1E4G, 1E4GT, 1G4GT, 1G4GT/G
1G4GT	5S	1E4G, 1E4GT, 1G4G, 1G4GT/G
1G4GT/G	5S	1E4G, 1E4GT, 1G4G, 1G4GT
1G5	6X	1G5G, 1G5GT, 1G5GT/G, 1J5G, 1J5GT
1G5G	6X	1G5GT, 1G5GT/G, 1J5G, 1J5GT
1G5GT	6X	1G5G, 1G5GT/G, 1J5G, 1J5GT

Tube Type	Basing	Replacement
1G5GT/G	6X	1G5G, 1G5GT, 1J5G, 1J5GT
1G6	7AB	1G6G, 1G6GT, 1G6GT/G
1G6G	7AB	1G6GT, 1G6GT/G
1G6GT	7AB	1G6G, 1G6GT/G
1G6GT/G	7AB	1G6G, 1G6GT
1G50	-	2050#, 2050A#
1G84	-	6Q5#, 884#, 884/6Q5#, 884W#
1H2	9DT	1S2
1H4	5S	1H4G, 1H4GT
1H4G	5S	1H4GT
1H4GT	5S	1H4G
1H5	5Z	1H5/HD14/DAC32, 1H5G, 1H5GT, 1H5GT/G, DAC32, HD14
1H5/HD14/DAC32	5Z	1H5, 1H5G, 1H5GT, 1H5GT/G, DAC32, HD14
1H5G	5Z	1H5, 1H5/HD14/DAC32, 1H5GT, 1H5GT/G, DAC32, HD14
1H5GT	5Z	1H5, 1H5/HD14/DAC32, 1H5G, 1H5GT/G, DAC32, HD14
1H5GT/G	5Z	1H5, 1H5/HD14/DAC32, 1H5G, 1H5GT
1H6	7AA	1H6, 1H6GT
1H6G	7AA	1H6GT
1H6GT	7AA	1H6G
1H33	7AT	1AQ5, 1R5*, 1R5/DK91*, 1R5SF, DK91*
1H34	-	-
1H35	7DH	1AB6, 1AB6/DK96, 1C3, 1C3/DK96/1AB6, DK96
1J3	3C	1G3GTA, 1G3GTA/1B3GT, 1J3/1K3A, 1K3, 1K3/1J3, 1K3A, 1K3A/1J3, DY30, U41
1J3/1K3A	3C	1G3GTA, 1G3GTA/1B3GT, 1J3, 1K3, 1K3/1J3, 1K3A, 1K3A/1J3, DY30, U41
1J3A	3C	1G3GTA, 1G3GTA/1B3GT, 1J3/1K3A, 1K3A, 1K3A/1J3
1J5	6X	1G5G, 1G5GT, 1G5GT/G, 1J5G, 1J5GT
1J5G	6X	1G5G, 1G5GT, 1G5GT/G, 1J5GT
1J5GT	6X	1G5G, 1G5GT, 1G5GT/G, 1J5G
1J6	7AB	1J6G, 1J6GT, 1J6GX
1J6G	7AB	1J6GT, 1J6GX
1J6GT	7AB	1J6G, 1J6GX
1J6GX	7AB	1J6G, 1J6GT
1K3	3C	1G3GTA, 1G3GTA/1B3GT, 1J3, 1J3/1K3A, 1K3/1J3, 1K3A, 1K3A/1J3, DY30, U41
1K3/1J3	3C	1G3GTA, 1G3GTA/1B3GT, 1J3, 1J3/1K3A, 1K3, 1K3A, 1K3A/1J3, DY30, U41
1K3A	3C	1G3GTA, 1G3GTA/1B3GT, 1J3, 1J3/1K3A, 1K3/1J3, 1K3A/1J3, DY30, U41
1K3A/1J3	3C	1G3GTA, 1B3GTA/1B3GT, 1J3, 1J3/1K3A, 1K3/1J3, 1K3A, 1K3A/1J3, DY30, U41
1K4	4M	1A4*, 1A4P*, 1A4T*, 1B4*, 1B4G*, 1B4P*, cont'd
1K4, cont'd		1B4P/951, 1B4T*, 1C4, 32*, 34*, 951*
1K5	5Y	1D5G*, 1D5GP*, 1D5GT*, 1E5G*, 1E5GP*, 1E5GT*, 1K5G, 1M5G
1K5G	5Y	1D5G*, 1D5GP*, 1D5GT*, 1E5G*, 1E5GP*, 1E5GT*, 1M5G
1K7	7AD	1K7G
1K7G	7AD	1K7#
1L4	6AR	1AE4*, 1F2, 1L4/DF92, DF92
1L4/DF92	6AR	1F2, 1L4, DF92
1L5	6X	1F5G*, 1L5G
1L5G	6X	1F5G*, 1K5#
1L6	7DC	1U6*
1L33	-	-
1L34	-	-
1LA4	5AD	1LA4E
1LA4E	5AD	1LA4
1LA6	7AK	1LC6
1LB4	5AD	-
1LB6	8AX	-
1LC5	7AO	1LN5
1LC6	7AK	1LA6
1LD5	6AX	-
1LE3	4AA	1LF3
1LF3	4AA	1LE3
1LG5	7AO	-
1LH4	5AG	-
1LK5	-	-
1LN5	7AO	1LC5
1M1	-	1M3, 1M3/DM70, 1N3, 1N3/DM70/1M3, 1N3/DM71, DM70, DM70/1N3, DM71, Y25
1M3	8EM	1M3/DM70, 1N3, 1N3/DM70/1M3, 1N3/DM71, DM70, DM70/1N3, DM71, Y25
1M3/DM70	8EM	1M3, 1N3, 1N3/DM70/1M3, DM70, DM70/1N3, DM71, Y25
1M5	5Y	1D5G*, 1D5GP*, 1D5GT*, 1E5G*, 1E5GP*, 1E5GT*, 1K5G, 1M5G
1M5G	5Y	1D5G*, 1D5GP*, 1D5GT*, 1E5G*, 1E5GP*, 1E5GT*, 1K5G
1N2	3C	1AU3, 1B3GT, 1B3GT/1G3GTA, 1G3GTA, 1G3GTA/1B3GT, 1J3, 1J3/1K3A, 1K3, 1K3/1J3, 1K3A, 1K3A/1J3, 1N2A, 1N2A/1AU3
1N2A	3C	1AU3, 1B3GT, 1B3GT/1G3GTA, 1G3GTA, 1G3GTA/1B3GT, 1J3, 1J3/1K3A, 1K3, 1K3/1J3, 1K3A, 1K3A/1J3, 1N2A/1AU3
1N2A/1AU3	3C	1AU3, 1B3GT, 1B3GT/1G3GTA, 1G3GTA, 1G3GTA/1B3GT, 1J3, 1J3/1K3A, 1K3, 1K3/1J3, 1K3A, 1K3A/1J3, 1N2A
1N3	8EM	1M3, 1M3/DM70, 1N3/DM70/1M3, 1N3/DM71, DM70, DM70/1N3, cont'd

Tube Type	Basing	Replacement
1N3, cont'd		DM71, Y25
1N3/DM70/1M3	8EM	1M3, 1M3/DM70, 1N3, 1N3/DM71, DM70, DM70/1N3, DM71, Y25
1N3/DM71	8EM	1M3, 1M3/DM70, 1N3, 1N3/DM70, 1N3/DM70/1M3, DM70, DM70/1N3, DM71
1N5	5Y	1N5G, 1N5GT, 1N5GT/1P5GT, 1N5GT/G, 1P5G, 1P5GT, DF33, Z14
1N5G	5Y	1N5GT, 1N5GT/1P5GT, 1N5GT/G, 1P5G, 1P5GT, DF33, Z14
1N5GT	5Y	1N5G, 1N5GT/1P5GT, 1N5GT/G, 1P5G, 1P5GT, DF33, Z14
1N5GT/1P5GT	5Y	1N5GT, 1N5GT/G, 1P5G, 1P5GT, DF33, Z14
1N5GT/G	5Y	1N5G, 1N5GT, 1N5GT/1P5GT, 1P5G, 1P5GT, DF33, Z14
1N6	7AM	1N6G, 1N6GT
1N6G	7AM	1N6GT
1N6GT	7AM	1N6G
1P1	-	3C4, DL96, DL96
1P5	5Y	1N5G, 1N5GT, 1N5GT/1P5GT, 1N5GT/G, 1P5G, 1P5GT
1P5G	5Y	1N5G, 1N5GT, 1N5GT/1P5GT, 1N5GT/G, 1P5GT
1P5GT	5Y	1N5G, 1N5GT, 1N5GT/1P5GT, 1N5GT/G, 1P5G
1P10	7BA	3S4, 3S4/DL92, DL92, N17
1P11	6BX	3V4, 3V4/DL94, DL94
1P21	-	-
1P22	11K	-
1P28	-	-
1P29	2K	-
1P37	2K	-
1P39	-	-
1P40	-	-
1P41	2AR	1P41/924, 924
1P41/924	2AR	1P41, 924
1P42	2AT	-
1Q5	6AF	1C5G, 1C5GT, 1C5GT/G, 1Q5G, 1Q5GT, 1Q5GT/G, 1T5GT*, DL35, DL36, N14
1Q5G	6AF	1C5G, 1C5GT, 1C5GT/G, 1Q5GT, 1Q5GT/G, 1T5GT*, DL35, DL36, N14
1Q5GT	6AF	1C5G, 1C5GT, 1C5GT/G, 1Q5G, 1Q5GT/G, 1T5GT*, DL35, DL36, N14
1Q5GT/G	6AF	1C5G, 1C5GT, 1C5GT/G, 1Q5G, 1Q5GT, 1T5GT*, DL35, DL36, N14
1Q6	8CO	-
1R	-	-
1R4	4AH	1R4/1294, 1294
1R4/1294	4AH	1R4, 1294
1R5	7AT	1AQ5*, 1C1, 1H33*, 1R5/DK91, 1R5SF*, 1R5WA, DK91, X17
1R5/DK91	7AT	1AQ5*, 1C1, 1H33*, 1R5, 1R5SF*, 1R5WA, DK91, X17
1R5SF	7AT	1AQ5, 1R5*, 1R5/DK91#, 1H33, DK91#
1R5WA	7AT	1AQ5#, 1R5#*, 1R5/DK91#, 1R5SF#, 1H33#, DK91#
1RK23	9DT	1S2, 1S2A, 1S2A/DY87, 1S2A/DY87/DY86, 1S2B, DY86, DY87
1RK31	-	-
1RK41	9Y	1BK2, 1BK2/1RK41
1S2	9DT	1RK23, 1S2A, 1S2A/DY87, 1S2A/DY87/DY86, 1S2B, DY86, DY87
1S2A	9DT	1S2A/DY87, 1S2A/DY87/DY86, 1S2B, DY86, DY87
1S2A/DY87	9DT	1S2A, 1S2A/DY87/DY86, 1S2B, DY86, DY87
1S2A/DY87/DY86	9DT	1S2A, 1S2A/DY87, 1S2B, DY86, DY87
1S2B	9DT	1S2A, 1S2A/DY87, 1S2A/DY87/DY86, DY86, DY87
1S4	7AV	DL91
1S5	6AU	1AF5*, 1AF5/1AH5*, 1AH5*, 1AR5*, 1FD9, 1S5/DAF91, DAF91, ZD17
1S5/DAF91	6AU	1AF5*, 1AF5/1AH5*, 1AH5*, 1AR5*, 1FD9, 1S5, DAF91, ZD17
1S5SF	6AU	1AF5*, 1AF5/1AH5*, 1AH5*, 1AR5*, 1FD9, 1S5/DAF91, DAF91, ZD17
1S6	8DA	-
1S21	-	-
1SA6	6BD	1SA6GT
1SA6GT	6BD	1SA6G
1SB6	6BE	1SB6GT
1SB6GT	6BE	1SB6
1SX2	-	-
1T2	-	R16, U37
1T4	6AR	1AE4*, 1AF4*, 1AF4/1AJ4*, 1AJ4*, 1AM4*, 1F3, 1T4SF*, 1T4WA, DF91, W17
1T4SF	6AR	1AE4*, 1AF4, 1AF4/1AJ4, 1AJ4*, 1AM4, 1F1, 1T4*, DF96, W25
1T4WA	6AR	1AF4/1AJ4#, 1AM4#, 1T4#, 1T4SF#
1T5	6X	1A5G, 1A5GT, 1A5GT/G, 1C5*, 1C5/DL35/N14*, 1C5GT*, 1C5GT/G*, 1Q5G*, 1Q5GT*, 1Q5GT/G*, 1T5GT, DL31, DL35*, N14*
1T5GT	6X	1A5G, 1A5GT, 1A5GT/G, 1C5*, 1C5/DL35/N14*, 1C5GT*, 1C5GT/G*, 1Q5G*, 1Q5GT*, 1Q5GT/G*, DL31, DL35*, N14*
1T6	8DA	-
1U4	6AR	1AF4*, 1AF4/1AJ4*, 1AJ4*, 1U4WA, 5910
1U4WA	6AR	1U4#, 5910
1U5	6BW	1AS5*, 1DN5, 1U5SF*, 1U5WA, DAF92
1U5SF	6BW	1AS5, 1DN5*, 1U5*
1U5WA	6BW	1AS5#, 1SN5#, 1U5#, DAF92#
1U6	7DC	1L6*
1V	4G	1V/6Z3, 1V/KR1, 6G3, 6G3AD, 6Z3, KR1

Tube Type	Basing	Replacement
1V/6Z3	4G	1V, 1V/KR1, 6G3, 6G3AD, 6Z3, KR1
1V/KR1	4G	1V, 1V/6Z3, 6G3, 6G3AD, 6Z3, KR1
1V2	9U	-
1V5	8CP	-
1V6	-	DCF60
1W4	5BZ	-
1W5	8CP	-
1X2	9Y	1BX2, 1X2C 1X2C/1BX2, DY80, R19
1X2A	9Y	1BX2, 1X2A/1X2B, 1X2B, 1X2B/1X2A, 1X2C, 1X2C/1BX2, DY80, R19
1X2A/1X2B	9Y	1BX2, 1X2A, 1X2B, 1X2B/1X2A, 1X2C, 1X2C/1BX2, DY90, R19
1X2B	9Y	1BX2, 1X2A/1X2B, 1X2B/1X2A, 1X2C, 1X2C/1BX2, DY90, R19
1X2B/1X2A	9Y	1BX2, 1X2A/1X2B, 1X2B, 1X2C, 1X2C/1BX2, DY90, R19
1X2C	9Y	1BX2, 1X2C/1BX2
1Y2	4P	-
1Y32	-	-
1Z1	-	-
1Z2	7CB	-
2-01C	-	-
2-15A/15R	-	2-15A, 15R, 4B30, 8022
2A3	4D	2A3/5930, 2A3H, 2A3W, 5930, 5930/2A3W
2A3/5930	4D	2A3, 2A3W, 5930, 5930/2A3W
2A3H	4D	2A3, 2A3/5990, 2A3H, 2A3W, 5930, 5930/2A3W
2A3W	4D	2A3/5990, 5930, 5930/2A3W
2A4	5S	-
2A4G	5S	-
2A5	6B	2A5/KR65, KR25, KR65
2A5/KR65	6B	2A5, KR25, KR65
2A6	6G	1659
2A7	7C	2A7S
2A7S	7C	2A7
2AB2	9U	2AV2, 2AV2/2AB2
2AC15	-	-
2AF4	7DK	2AF4A, 2AF4A/2AF4B, 2AF4B, 2AF4B/2DZ4, 2DZ4#, 2DZ4/2AF4B, 2T4#
2AF4A	7DK	2AF4A/2AF4B, 2AF4B, 2AF4B/2DZ4, 2DZ4#, 2DZ4/2AF4B, 2T4#
2AF4A/2AF4B	7DK	2AF4A, 2AF4B, 2AF4B/2DZ4, 2DZ4#, 2DZ4/2AF4B
2AF4B	7DK	2AF4A, 2AF4A/2AF4B, 2AF4B, 2AF4B/2DZ4, 2DZ4#, 2DZ4/2AF4B
2AF4B/2DZ4	7DK	2AF4A, 2AF4A/2AF4B, 2AF4B#, 2DZ4#, 2DZ4/2AF4B
2AH2	12DG	2AS2A, 2BU2, 2BU2/2AH2, 2BU2/2AS2A/2AH2
2AS2	12EW	2AH2, 2AS2A, 2BU2, 2BU2/2AH2, 2BU2/2AS2A/2AH2
2AS2A	12EW	2AH2, 2BU2, 2BU2/2AH2, 2BU2/2AS2/2AH2
2AS15A	-	-
2AV2	9U	2AV2/2AB2, 2AB2
2AV2/2AB2	9U	2AV2, 2AB2
2AZ2	9Y	2BJ2A
2B3	8HC	-
2B4	5A	885
2B5	8DP	-
2B6	7J	-
2B7	7D	2B7S
2B7S	7D	2B7
2B21	4AB	2X2A, 879
2B25	-	-
2B35	-	2B35/EA50, 6D1, EA50, T6D
2B35/EA50	-	2B35, 6D1, EA50, T6D
2B/250A	5AW	4Y25, 5S1, 807, 807W, 807WA, 5933, 5933WA, 8018, HY61, P17A, QE06/50, QE06140, QV05-25, RK39
2BA2	9U	2AV2, 2AV2/2BA2
2BJ2	9RT	2BJ2A
2BJ2A	9RT	2BJ2#
2BN4	7EG	2BN4A
2BN4A	7EG	2BN4
2BU2	12JB	2AH2, 2AS2A, 2BU2/2AH2, 2BU2/2AS2A/2AH2
2BU2/2AH2	12JB	2AH2, 2AS2A, 2BU2, 2BU2/2AS2A/2AH2
2BU2/2AS2A/2AH2	12JB	2AH2, 2AS2A, 2BU2, 2BU2/2AH2
2C4	5AS	-
2C21	7BH	2C21/1642, 1642
2C21/1642	7BH	2C21, 1642
2C22	4AM	6J5#, 6J5GT#, 7193, 7193/2C22, 7196
2C26	-	-
2C34	-	2C34/RK34, RK34
2C34/RK34	-	2C34, RK34
2C39	-	2C39A
2C39A	-	2C39
2C40	-	2C40A, 4037A#
2C40A	-	2C40, 4037A#
2C42	-	-
2C43	-	-
2C46	-	-
2C50	8BD	-
2C51	8CJ	2C51W*, 396A, 1219, 5670*, 5670/2C51W*, 5670WA*, 6185, 6385*
2C51W	8CJ	1219, 5670, 5670/2C51W, 5670WA, 6185
2C52	8BD	-
2C53	-	-
2CN3	8MU	-
2CN3A	8MU	2CN3B
2CN3B	8MU	-
2CQ3	8MK	-
2CW4	12AQ	2CW4/2DS4, 2DS4
2CW4/2DS4	12AQ	2CW4, 2DS4
2CY5	7EW	2EA5, 2EV5
2D21	7BN	2D21/5727, 2D21/PL21, 2D21W, 20A3, 5727, 5727/2D21, 5727/2D21W, E91, EN91, EN91N, M8204, PL21
2D21/5727	7BN	2D21, 2D21/PL21, 2D21W, 20A3, 5727, 5727/2D21, 5727/2D21W, E91, cont'd

VACUUM TUBES

Tube Type	Basing	Replacement
2021/5727, cont'd		EN91, EN91N, M8204, PL21
2D21/PL21	7BN	2D21, 2D21/5727, 2D21W, 20A3, 5727, 5727/2D21, 5727/2D21W, E91, EN91, EN91N, M8204, PL21
2D21W	7BN	2D21#, 2D21/5727#, 5727#, 5727/2D21#, 5727/2D21W
2DF4	9JL	-
2DS4	12AQ	2CW4, 2CW4/2DS4
2DV4	12EA	-
2DX4	7DK	-
2DY4	7DK	2DY4A
2DY4A	7DK	2DY4
2DZ4	7DK	2AF4A#, 2AF4A/2AF4B#, 2AF4B#, 2AF4B/2DZ4, 2DZ4/2AF4B
2DZ4/2AF4B	7DK	2AF4A#, 2AF4A/2AF4B, 2AF4B, 2AF4B/2DZ4, 2DZ4
2E5	6R	-
2E22	-	-
2E24	-	-
2E25	-	-
2E26	-	6893/2E26, 6893
2E30	-	-
2E31	-	-
2E32	-	-
2E35	-	-
2E36	-	-
2E41	-	-
2E42	-	-
2EA5	7EW	2EV5, 2CY5
2EG4	12AQ	-
2EN5	7FL	-
2ER5	7FN	2ES5, 2FQ5, 2FQ5A, 2FQ5A/2GK5, 2FY5, 2GK5, 2GK5/2FQ5, 2GK5/2FQ5A, XC95, XC97
2ES5	7FN	2ER5*!, 2FQ5, 2FQ5A, 2FQ5A/2GK5, 2GK5, 2GK5/2FQ5, 2GK5/2FQ5A
2EV5	7EW	2CY5, 2EA5
2F7	-	-
2FH5	7FP	2ES5, 2FQ5, 2FQ5A, 2FQ5A/2GK5, 2GK5, 2GK5/2FQ5, 2GK5/2FQ5A
2FQ5	7FP	2ES5, 2FQ5A, 2FQ5A/2GK5, 2FY5*!, 2GK5, 2GK5/2FQ5, 2GK5/2FQ5A
2FQ5A	7FP	2ES5, 2FQ5, 2FQ5A/2GK5, 2FY5*!, 2GK5, 2GK5/2FQ5, 2GK5/2FQ5A
2FQ5A/2GK5	7FP	2ES5, 2FQ5, 2FQ5A, 2GK5, 2GK5/2FQ5, 2GK5/2FQ5A
2FS5	7GA	2GU5
2FV6	7FQ	2EA5#, 2EU5#, 2CY5
2FY5	7FP	2ER5, 2FQ5, 2FQ5A, 2FQ5A/2GK5, 2GK52GK5/2FQ5, 2GK5/2FQ5A, XC95, XC97
2G5	6R	-
2G21	-	-
2G22	-	-
2GK5	7FP	2ES5, 2FQ5, 2FQ5A, 2FQ5A/2GK5, 2GK5/2FQ5, 2GK5/2FQ5A
2GK5/2FQK	7FP	2ES5, 2FQ5, 2FQ5A, cont'd
2GK5/2FQK, cont'd		2FQ5A/2GK5, 2GK5, 2GK5/2FQ5A
2GK5/2FQ5A	7FP	2ES5, 2FQ5, 2FQ5A, 2FQ5A/2GK5, 2GK5, 2GK5/2FQ5
2GU5	7GA	2FS5
2GW5	7GK	-
2HA5	7GM	2HA5/2HM5, 2HK5, 2HM5, 2HM5/2HA5, 2HQ5, XC900
2HA5/2HM5	7GM	2HA5, 2HK5, 2HM5, 2HM5/2HA5, 2HQ5, XC900
2HK5	7GM	2HA5!, 2HA5/2HM5, 2HQ5, 2HM5, 2HM5/2HA5
2HM5	7GM	2HA5, 2HA5/2HM5, 2HK5, 2HM5/2HA5, 2HQ5
2HM5/2HA5	7GM	2HA5, 2HA5/2HM5, 2HK5, 2HM5, 2HQ5
2HQ5	7GM	2HA5, 2HA5/2HM5, 2HK5, 2HM5, 2HM5/2HA5
2HR8	9BJ	-
2J2	9DT	R20, U26, U49
2K25	-	-
2K26	-	-
2K28	-	2K28/707B, 707B
2K28/707B	-	2K28, 707B
2NH11	-	-
2NH12	-	-
2RA3	-	213, 213A
2RA5	-	205
2RA6	-	206, 619
2RA15	-	203, 215
2S	5D	2S/4S, 4S, G2
2S/4S	5D	2S, 4S, G2
2T4	7DK	2AF4#, 2AF4A#, 2AF4A/2AF4B#, 2AF4B#, 2AF4B/2DZ4#, 2DZ4#, 2DZ4/2AF4B#
2V2	8FV	3C2
2V3	4Y	-
2V3G	4Y	-
2W3	4X	2W3GT
2W3GT	4X	2W3
2X2	4AB	2B21, 2X2/879, 879
2X2/879	4AB	2B21, 2X2, 879
2X2A	4AB	2B21, 879
2X3	4Y	-
2X3G	4Y	-
2Y2	4AB	2X2, 2X2/879, 2X2A, 879
2Y3	-	-
2Z2	4B	2Z2/G84, 684, G84
2Z2/G84	4B	2Z2, 684, G84
3-400Z	-	3-400Z/8163, 8163
3-400Z/8163	-	3-400Z, 8163
3-500Z	-	-
3A2	9DT	3A2A
3A2A	9DT	3A2
3A3	8EZ	3A3/3B2, 3A3A/3B2, 3A3B, 3A3C, 3A3C/3AW3, 3AW3, 3B2, 3CU3A, 3CV3A, 3CY3, 3CY3/3DB3, 3DB3, 3DB3/3CY3, 3DB3/3DJ3, 3DJ3
3A3/3B2	8EZ	3A3, 3A3A, 3A3A/3B2, 3A3B, 3A3C, 3A3C/3AW3, 3AW3, 3B2, 3CV3A, 3CY3, 3CY3/3DB3, 3CZ3A, 3DB3, 3DB3/3CY3, 3DB3/3DJ3, 3DJ3

Tube Type	Basing	Replacement
3A3A	8EZ	3A3/3B2, 3A3A, 3A3A/3B2, 3A3B, 3A3C, 3A3C/3AW3, 3AW3, 3B2, 3CU3A, 3CV3A, 3CY3, 3CY3/3DB3, 3DB3, 3DB3/3CY3, 3DB3/3DJ3, 3DJ3
3A3A/3B2	8EZ	3A3/3B2, 3A3A, 3A3B, 3A3C, 3A3C/3AW3, 3AW3, 3B2, 3CV3A, 3CY3, 3CY3/3DB3, 3CZ3A, 3DB3, 3DB3/3CY3, 3DB3/3DJ3, 3DJ3
3A3B	8EZ	3A3C, 3A3C/3AW3, 3AW3, 3CV3A, 3CY3, 3CY3/3DB3, 3DB3, 3DB3/3CY3, 3DB3/3DJ3, 3DJ3
3A3C	8EZ	3A3C/3AW3, 3AW3, 3DB3, 3DB3/3DJ3, 3DJ3
3A3C/3AW3	8EZ	3A3C, 3AW3, 3DB3, 3DB3/3DJ3, 3DJ3
3A4	7BB	3A4/DL93, DL93
3A4/DL93	7BB	3A4, DL93
3A4T	-	3A4#, 3A4/DL93#, DL93#
3A5	7BC	3A5/DCC90, DCC90
3A5/DCC90	7BC	3A5, DCC90
3A8	8AS	3A8G, 3A8GT
3A8G	8AS	3A8GT
3A8GT	8AS	3A8G
3AF4	7DK	3AF4A, 3AF4A/3AF4B, 3AF4A/3DZ4, 3AF4B, 3DZ4, 3DZ4/3AF4B
3AF4A	7DK	3AF4A/3AF4B, 3AF4A/3DZ4, 3AF4B, 3DZ4#, 3DZ4/3AF4B
3AF4A/3AF4B	7DK	3AF4A, 3AF4A/3DZ4, 3AF4B, 3DZ4#, 3DZ4/3AF4B
3AF4A/3DZ4	7DK	3AF4A#, 3AF4A/3AF4B#, 3AF4B#, 3DZ4#, 3DZ4/3AF4B
3AF4B	7DK	3AF4A#, 3AF4A/3AF4B, 3AF4A/3DZ4, 3DZ4#, 3DZ4/3AF4B
3AJ8	9CA	XCH81
3AL5	6BT	XAA91
3AT2	12FV	3AT2A, 3BL2A, 3BM2A, 3BN2A
3AT2A	12FV	3AT2B, 3BL2A, 3BM2A, 3BN2A
3AT2B	12FV	3AT2A, 3BL2A, 3BM2A, 3BN2A
3AT6	-	-
3AU6	7BK	3BA6, 3BC5#, 3BC5/3CE5#, 3CB6#, 3CB6/3CF6#, 3CE5#, 3CE5/3BC5#, 3CF6#, XF94
3AV6	7BT	-
3AW2	12HA	3AW2A
3AW2A	12HA	3AW2
3AW3	8EZ	3A3A, 3A3A/3B2, 3A3B, 3A3C, 3A3C/3AW3, 3B2, 3CY3, 3CY3/3DB3, 3DB3, 3DB3/3CY3#, 3DB3/3DJ3, 3DJ3
3B2	8GH	3A3/3B2, 3A3A/3B2, 3A3B, 3A3C, 3A3C/3AW3, 3AW3, 3CV3A, 3CY3, 3CY3/3DB3, 3DB3, 3DB3/3CY3#, 3DB3/3DJ3, 3DJ3
3B4	7CY	3B4/DL98, 3B4WA, DL98, HD30
3B4/DL98	7CY	3B4, 3B4WA, DL98, HD30
3B4WA	7CY	3B4#, 3B4/DL98#, DL98, HD30
3B5	7AP	3B5GT, 3C5GT, 3Q5, 3Q5G, 3Q5GT, 3Q5GT/G
3B5GT	7AP	3B5, 3C5GT, 3Q5, 3Q5G, 3Q5GT, 3Q5GT/G, DL33, N15, N16
3B7	7BE	1288, 1291, 1292
3B22	-	3B22/EL1C, EL1C
3B22/EL1C	-	3B22, EL1C
3B24	-	-
3B28	4P	-
3BA6	7BK	3AU6, 3BZ6#
3BC5	7BD	3AU6#, 3BC5/3CE5, 3CB6#, 3CB6/3CF6#, 3CE5, 3CE5/3BC5, 3CF6#, 3DK6#
3BC5/3CE5	7BD	3AU6#, 3BC5, 3BZ6#, 3CB6#, 3CB6/3CF6#, 3CE5, 3CE5/3BC5, 3CF6#, 3DK6#
3BE6	7CH	3BY6, 3CS6
3BF2	12GQ	-
3BH2	-	GY501
3BL2	12HK	3AT2B, 3BL2A, 3BM2A, 3BN2A
3BL2A	12HK	3AT2B, 3BM2A, 3BN2A
3BM2	12HK	3AT2B, 3BL2A, 3BM2A
3BM2A	12HK	3AT2B, 3BL2A
3BN2	12FV	3AT2B, 3BL2A, 3BM2A
3BN2A	12FV	3AT2B, 3BL2A, 3BM2A
3BN4	7EG	3BN4A
3BN4A	7EG	3BN4
3BN6	7DF	-
3BS2	12HY	3BS2A, 3BS2B, 3BS2B/3BW2, 3BT2, 3BT2A, 3BT2/3BS2, 3BT2/3BS2B, 3BT2A, 3BW2, 3BW2/3BS2A/3BT2, 3BW2/3BS2B/3BT2A
3BS2A	12HY	3BS2B, 3BS2B/3BW2, 3BT2, 3BT2/3BS2B, 3BT2A, 3BW2, 3BW2/3BS2A/3BT2, 3BW2/3BS2B/3BT2A
3BS2B	12HY	3BS2B/3BW2, 3BT2, 3BT2/3BS2B, 3BT2A, 3BW2, 3BW2/3BS2A/3BT2, 3BW2/3BS2B/3BT2A
3BS2B/3BW2	12HY	3BS2B, 3BT2, 3BT2/3BS2B, 3BT2A, 3BW2, 3BW2/3BS2A/3BT2, 3BW2/3BS2B/3BT2A
3BT2	12HY	3BS2, 3BS2B, 3BS2B/3BW2, 3BT2/3BS2, 3BT2/3BS2B, 3BT2A, 3BW2, 3BW2/3BS2A/3BT2, 3BW2/3BS2B/3BT2A
3BT2/3BS2	12HY	3BS2, 3BS2B, 3BS2B/3BW2, 3BT2, 3BT2/3BS2B, 3BT2A, 3BW2, 3BW2/3BS2A/3BT2, 3BW2/3BS2B/3BT2A
3BT2/3BS2B	12HY	3BS2, 3BS2B, 3BS2B/3BW2, 3BT2, 3BT2/3BS2, 3BT2A, 3BW2, 3BW2/3BS2A/3BT2, 3B42/3BS2B/3BT2A
3BT2A	12HY	3BS2A, 3BS2B, 3BS2B/3BW2, 3BW2, 3BW2/3BS2A/3BT2, 3BW2/3BS2A/3BT2A
3BU8	9FG	3BU8/3GS8, 3BU8A, 3GS8, 3GS8/3BU8, 3HS8, 3KF8
3BU8/3GS8	9FG	3BU8, 3BU8A, 3GS8, 3GS8/3BU8, 3HS8, 3KF8

Tube Type	Basing	Replacement
3BU8A	9FG	3BU8, 3BU8/3GS8, 3GS8, 3GS8/3BU8, 3HS8, 3KF8
3BW2	12HY	3BS2A, 3BS2B/3BW2, 3BS2B/3BW2, 3BT2, 3BT2A, 3BT2/3BS2, 3BT2/3BS2B, 3BT2A, 3BW2/3BS2A/3BT2, 3BW2/3BS2B/3BT2A
3BW2/3BS2A/3BT2	12HY	3BS2A, 3BS2B, 3BS2B/3BW2, 3BT2, 3BT2A, 3BT2/3BS2, 3BT2/3BS2B, 3BT2A, 3BW2, 3BW2/3BS2B/3BT2A
3BW2/3BS2B/3BT2A	12HY	3BS2A, 3BS2B, 3BS2B/3BW2, 3BT2, 3BT2/3BS2B, 3BT2A, 3BW2, 3BW2/3BS2A/3BT2, 3BW2/3BS2B/3BT2A
3BX6	9AQ	3BY7, XF80, XF85
3BY6	7CH	3BE6, 3CS6
3BY7	9AQ	3BX6, XF80, XF85
3BZ6	7CM	3BA6#, 3CB6#, 3CB6/3CF6#, 3CE6#, 3CF6#, 3DK6#, 3MV7
3C2	8FV	-
3C4	6BX	1P1, 3C4/DL96, 3E5, 3V4*, 3V4/DL94*, DL94*, DL96, N25
3C4/DL96	6BX	1P1, 3C4, 3E5, DL96, N25
3C5	7AP	3C5GT
3C5GT	7AP	3B5, 3B5GT, 3Q5, 3Q5G, 3Q5GT, 3Q5GT/G, DL33, N15, N16
3C6	7BW	3C6/XXB, XXB
3C6/XXB	7BW	3C6, XXB
3C21	-	-
3C23	3G	-
3C24	3G	3C24/24G, 24G
3C24/24G	3G	3C24, 24G
3C30	-	-
3C45	4BL	3C45/6130, 6130, 6130/3C45
3C45/6130	4BL	3C45, 6130, 6130/3C45
3CA3	8MH	3CA3A
3CA3A	8MH	3CA3
3CB6	7CM	3AU6#, 3BC5#, 3BC5/3CE5#, 3BZ6#, 3CB6/3CF6, 3CF6, 3DK6, 3CE5#, 3CE5/3BC5#
3CB6/3CF6	7CM	3AU6#, 3BC5#, 3BC5/3CE5#, 3BZ6#, 3CB6, 3CE5#, 3CE5/3BC5#, 3CF6, 3DK6
3CE5	7BD	3AU6#, 3BC5, 3BC5/3CE5, 3BZ6#, 3CB6#, 3CB6/3CF6#, 3CE5/3BC5, 3CF6#, 3DK6#
3CE5/3BC5	7BD	3AU6#, 3BC5, 3BC5/3CE5, 3BZ6#, 3CB6#, 3CB6/3CF6, 3CE5, 3CF6#, 3DK6#
3CE6	7CM	3BZ6
3CF6	7CM	3AU6#, 3BC5#, 3BC5/3CE5#, 3BZ6#, 3CB6, 3CB6/3CF6, 3CE5#, 3CE5/3BC5#, 3DK6
3CN3	8MU	3CN3A, 3CN3B
3CN3A	8MU	3CN3, 3CN3B
3CN3B	8MU	3CN3, 3CN3A
3CS6	7CH	3BE6, 3BY6
3CU3	8MK	3A3C, 3AC3/3AW3, 3AW3, 3CU3A, 3CV3A
3CU3A	8MK	3A3C, 3AC3/3AW3, 3AW3, 3CV3A
3CV3	8EZ	3A3C, 3AC3/3AW3, 3AW3, 3CU3A, 3CV3A, 3CY3, 3CY3/3DB3, 3CZ3A, 3DB3, 3DB3/3CY3, 3DB3/3DJ3, 3DJ3
3CV3A	8EZ	3A3C, 3AC3/3AW3, 3AW3, 3CY3, 3CY3/3DB3, 3CZ3A, 3DB3, 3DB3/3CY3, 3DB3/3DJ3, 3DJ3
3CX3	8MT	3CX3/3DA3, 3DA3, 3DA3/3DH3, 3DF3A, 3DH3
3CX3/3DA3	8MT	3CX3, 3DA3, 3DA3/3DH3, 3DF3A, 3DH3
3CX100A5	-	-
3CY3	8MX	3CY3/3DB3, 3DB3, 3DB3/3CY3, 3DB3/3DJ3, 3DJ3
3CY3/3DB3	8MX	3CY3, 3DB3, 3DB3/3CY3, 3DB3/3DJ3, 3DJ3
3CY5	7EW	3EA5, 3EV5
3CZ3	8EZ	3A3C, 3AC3/3AW3, 3AW3, 3CZ3A
3CZ3A	8EZ	3A3C, 3AC3/3AW3, 3AW3
3D6	6BA	3D6/1299, 3LE4*, 3LF4*, 1299, DL29
3D6/1299	6BA	3D6, 1299, DL29
3D21	-	-
3D22	7BV	3D22A
3D22A	7BV	3D22
3DA3	8MY	3CX3, 3CX3/3DA3, 3DA3/3DH3, 3DH3
3DA3/3DH3	8MY	3CX3, 3CX3/3DA3, 3DA3, 3DH3
3DB3	8MX	3CY3, 3CY3/3DB3, 3DB3/3CY3, 3DB3/3DJ3, 3DJ3
3DB3/3CY3	8MX	3CY3, 3CY3/3DB3, 3DB3, 3DB3/3DJ3, 3DJ3
3DB3/3DJ3	8MX	3CY3, 3CY3/3DB3, 3DB3, 3DB3/3CY3, 3DJ3
3DC3	8MZ	-
3DE6	-	3BZ6
3DF3	8MT	3DF3A
3DF3A	8MT	3DF3
3DG4	5DE	Z2669
3DH3	8NG	3CX3, 3CX3/3DA3, 3DA3, 3DA3/3DH3
3DHH12	-	-
3DHH13	-	-
3DJ3	8MX	3CY3, 3CY3/3DB3, 3DB3, 3DB3/3CY3, 3DB3/3DJ3
3DK6	7CM	3BC5#, 3BC5/3CE5#, 3BZ6#, 3CB6, 3CB6/3CF6, 3CE5#, 3CE5/3BC5#, 3CF6, 3MR24
3DR3	8NL	-
3DS3	8NL	3DR3
3DT6	7EN	3DT6A
3DT6A	7EN	3DT6
3DX4	7DK	-
3DY4	7DK	3DY4A
3DY4A	7DK	3DY4
3DZ4	7DK	3AF4A, 3AF4A/3AF4B#, 3AF4A/3DZ4, 3AF4B#, 3DZ4/3AF4B
3DZ4/3AF4B	7DK	3AF4A, 3AF4A/3AF4B, 3AF4A/3DZ4, 3AF4B, 3DZ4, 3DZ4/3AF4B
3E5	6BX	1P1, 3C4, 3C4/DL96, 3V4*, 3V4/DL94*, DL94*, cont'd

Tube Type	Basing	Replacement
3E5, cont'd		DL96, N25
3E6	7CJ	-
3E22	-	-
3E29	-	3E29/829A, 829A
3E29/829A	-	3E29, 829A
3EA5	7EW	3CY5, 3EV5
3EH7	9AQ	3EH7/XF183, 3EJ7, 3EJ7/XF184, 3HM6#, 3HT6#, 3JC6#, 3JC6A#, 3JD6#, XF183, XF184
3EH7/XF183	9AQ	3EH7, 3EJ7, 3EJ7/XF184, XF183, XF184
3EJ7	9AQ	3EH7, 3EH7/XF183, 3EJ7/XF184, 3HM6#, 3HT6#, 3JC6#, 3JC6A#, 3JD6#, XF183, XF184
3EJ7/XF184	9AQ	3EH7, 3EH7/XF183, 3EJ7, XF183, XF184
3ER5	7FN	3ER5/YC95, 3FQ5, 3FQ5A, 3FY5, 3GK5, 3GK5/3FQ5, YC95
3ER5/YC95	7FN	3ER5, 3FQ5, 3FQ5A, 3FY5, 3GK5, 3GK5/3FQ5, YC95
3ES5	7FN	-
3EV5	7EW	3EA5, 3CY5
3FH5	7FP	3ER5#, 3ER5/YC95#, 3ES5, 3FQ5, 3FQ5A, 3GK5, 3GK5/3FQ5, YC95#
3FQ5	7FP	3ER5#, 3ER5/YC95#, 3ES5, 3FQ5A, 3GK5, 3GK5/3FQ5, YC95#
3FQ5A	7FP	3ER5, 3ER5/YC95, 3FQ5, 3GK5, 3GK5/3FQ5, YC95
3FS5	7GA	3GU5
3FX7	-	3DHH13
3FY5	7FP	3ER5, 3ER5/YC95, 3FQ5, 3GK5, 3GK5/3FQ5, YC95, YC97
3GK5	7FP	3FQ5, 3FQ5A, 3GK5/3FQ5
3GK5/3FQ5	7FP	3GK5, 3FQ5, 3FQ5A
3GS8	9LW	3BU8, 3BU8/3GS8, 3BU8A, 3GS8/3BU8, 3HS8, 3KF8
3GS8/3BU8	9LW	3BU8, 3BU8/3GS8, 3BU8A, 3GS8, 3HS8, 3KF8
3GU5	7GA	3FS5#
3GW5	7GK	-
3HA5	7GM	3HA5/3HM5, 3HK5, 3HK5/3HA5, 3HM5, 3HM5/3HA5, 3HQ5
3HA5/3HM5	7GM	3HA5, 3HK5, 3HK5/3HA5, 3HM5, 3HM5/3HA5, 3HQ5
3HK5	7GM	3HA5, 3HA5/3HK5, 3HA5/3HM5, 3HK5/3HA5, 3HM5, 3HM5/3HA5, 3HQ5
3HK5/3HA5	7CM	3HA5, 3HA5/3HK5, 3HA5/3HM5, 3HK5, 3HM5, 3HM5/3HA5, 3HQ5
3HM5	7GM	3HA5, 3HA5/3HM5, 3HA5/3HM5, 3HK5, 3HK5/3HA5, 3HM5/3HA5, 3HQ5
3HM5/3HA5	7GM	3HA5, 3HA5/3HM5, 3HK5, 3HK5/3HA5, 3HM5, 3HQ5
3HM6	9PM	3HT6, 3JC6, 3JC6A, 3JD6
3HQ5	7GM	3HA5, 3HA5/3HM5, 3HK5, 3HK5/3HA5, 3HM5, 3HM5/3HA5
3HS8	9FG	3BU8, 3BU8/3GS8, cont'd
3HS8, cont'd		3BU8A, 3GS8, 3GS8/3BU8, 3KF8
3HT6	9PM	3HM6, 3JC6, 3JC6A, 3JD6
3JC6	9PM	3HM6, 3HT6, 3JC6A, 3JD6
3JC6A	9PM	3HM6, 3HT6, 3JC6, 3JD6
3JD6	9PM	3HM6, 3HT6, 3JC6, 3JC6A
3JH6	-	-
3KD1	-	-
3KF8	9FG	3BU8/3GS8, 3BU8, 3BU8A, 3GS8, 3GS8/3BU8, 3HS8
3KT6	9PM	-
3L4	-	-
3LE4	6BA	3D6*, 3D6/1299*, 3LF4#, 1299*
3LF4	6BA	3D6*, 3D6/1299*, 3LE4#, 1299*
3LH4	-	-
3MP26	-	-
3MR24	7CM	3DK6
3MV7	7CM	3BZ6
3Q4	7BA	3S4#, 3S4/DL92#, 3S4SF*#, 3W4*#, 3Z4*#, DL92#, DL95, N18
3Q5	7AP	3B5, 3C5GT, 3Q5G, 3Q5GT, 3Q5GT/G
3Q5G	7AP	3B5, 3B5GT, 3C5GT, 3Q5, 3Q5GT, 3Q5GT/G
3Q5GT	7AP	3B5, 3B5GT, 3C5GT, 3Q5, 3Q5G, 3Q5GT/G, DL33, N15, N16, WT389
3Q5GT/G	7AP	3B5, 3B5GT, 3C5GT, 3Q5, 3Q5G, 3Q5GT
3S4	7BA	1P10, 3S4/DL92, 3S4SF*, 3Q4#, 3W4*, 3Z4*, DL92, N17
3S4/DL92	7BA	1P10, 3S4, DL92, N17
3S4SF	7BA	3W4, 3Z4, 3S4*, 3S4/DL92*, 3Q4*#, DL92*
3TF7	-	-
3V4	6BX	1P11, 3C4*, 3C4/DL96*, 3E5*, 3V4/DL94, 3V4WA, DL94, DL96*, N19
3V4/DL94	6BX	1P11, 3V4, 3V4WA, DL94, N19
3V4WA	6BX	1P11#, 3V4#, DL94#, N19#
3V5	-	-
3W4	7BA	3S4SF, 3Z4, 3S4*, 3S4/DL92*, 3Q4*#, DL92*
3Z4	7BA	3S4SF, 3W4, 3S4*, 3S4/DL92*, 3Q4*#, DL92*
4-65A	-	4-65A/8165, 8165
4-65A/8165	-	4-65A, 8165
4-125A	5BK	4-125A/4D21, 4D21, 4D21A, 6155
4-125A/4D21	5BK	4-125A, 4D21, 4D21A, 6155
4-250	-	4-250/5D22, 4-250A, 4-250A/5D22, 5D22, 6156
4-250/5D22	-	4-250, 4-250A, 4-250A/5D22, 5D22, 6156
4-250A	-	4-250, 4-250/5D22, 4-250A/5D22, 5D22, 6156
4-250A/5D22	-	4-250, 4-250/5D22, 4-250A, 5D22, 6156
4-400A	5BK	4-400A/8438, 8438
4-400A/8438	5BK	4-400A, 8438
4-400B	-	4-400B/7527, 7527
4-400B/7527	-	4-400B, 7527
4-400C	-	4-400C/6775, 6775

Tube Type	Basing	Replacement
4-400C/6775	-	4-400C, 6775
4A6	8L	4A6G
4A6G	8L	4A6
4AF4	-	-
4AU6	7BK	4AU6/4BA6, 4BA6, 4BC5#, 4BC5/4CE5#, 4BZ6#, 4BZ6/4JH6#, 4CB6#, 4CB6/4DE6#, 4CE5#, 4CE5/4BC5, 4CF6#, 4DE6#, 4JH6#
4AU6/4BA6	7BK	4AU6, 4BA6, 4BC5#, 4BC5/4CE5#, 4BZ6#, 4BZ6/4JH6#, 4CB6/4DE6#, 4CE5#, 4CE5/4BC5, 4CF6#, 4DE6#, 4JH6
4AV6	7BT	-
4B24	-	EL3C, EL3C/4B24, 6484
4B30	-	15R, 8022
4B31	-	-
4B32	-	-
4B368	-	-
4BA6	7BK	4AU6, 4AU6/4BA6, 4BZ6#, 4BZ6/4JH6#, 4JH6#
4BC5	7BD	4AU6#, 4AU6/4BA6#, 4BA6#, 4BC5/4CE5, 4CB6#, 4CB6/4DE6#, 4CE5, 4CE5/4BC5, 4CF6#, 4DE6#, 4DK6#
4BC5/4CE5	7BD	4BC5, 4CE5, 4CE5/4BC5
4BC8	9AJ	4BC8/4RHH2, 4BQ7A, 4BQ7A/4BZ7, 4BS8, 4BS8/4BZ8, 4BX8, 4BZ7, 4BZ8, 4RHH2, 5BK7A
4BC8/4RHH2	9AJ	4BC8, 4BQ7A, 4BQ7A/4BZ7, 4BS8, 4BS8/4BZ8, 4BX8, 4BZ7, 4BZ8, 4RHH2, 5BK7A
4BE6	7CH	4CS6
4BL8	9DC	4BL8/XCF80, 5EA8, 5EA8/5GH8A, 5CQ8#, 5GH8A, 5RHP1, 5U8, XCF80
4BL8/XCF80	9DC	4BL8, 5EA8, 5EA8/5GH8A, 5GH8A, 5RHP1, 5U8, XCF80
4BN4	7EG	-
4BN6	7DF	-
4BQ7	9AJ	4BC8, 4BC8/4RHH2, 4BQ7A, 4BQ7A/4BZ7, 4BS8, 4BS8/4BZ8, 4BX8, 4BZ7, 4BZ8, 4RHH2, 5BK7A
4BQ7A	9AJ	4BC8, 4BC8/4RHH2, 4BQ7A/4BZ7, 4BS8, 4BS8/4BZ8, 4BX8, 4BZ7, 4BZ8, 4RHH2, 5BK7A
4BQ7A/4BZ7	9AJ	4BC8, 4BC8/4RHH2, 4BQ7A, 4BS8, 4BS8/4BZ8, 4BX8, 4BZ7, 4BZ8, 4RHH2, 5BK7A
4BS4	-	PC93
4BS8	9AJ	4BC8, 4BC8/4RHH2, 4BQ7A, 4BQ7A/4BZ7, 4BZ7, 4BS8/4BZ8, 4BX8, 4BZ7, 4BZ8, 4RHH2, 5BK7A
4BS8/4BZ8	9AJ	4BC8, 4BC8/4RHH2, 4BQ7A, 4BQ7A/4BZ7, 4BZ7, 4BS8, 4BX8, 4BZ7, 4BZ8, 4RHH2, 5BK7A
4BU8	9FG	4BU8/4GS8, 4GS8, 4GS8/4BU8, 4HS8, 4HS8/4KS8, 4KE8, 4KS8, 4MK8
4BU8/4GS8	9FG	4BU8, 4GS8, 4GS8/4BU8, 4HS8, 4HS8/4KF8, 4KE8, 4KF8, 4MK8
4BU8A	9FG	4BU8/4GS8, 4GS8, 4GS8/4BU8
4BX8	9AJ	4BC8, 4BC8/4RHH2, 4BQ7A, 4BQ7A/4BZ7, 4BS8, 4BS8/4BZ8, 4BZ7, 4BZ8, 4RHH2, 5BK7A
4BY6	7CH	4CS6
4BZ6	7CM	4AU6#, 4AU6/4BA6#, 4BA6#, 4BZ6/4JH6, 4CB6#, 4CB6/4DE6#, 4CF6#, 4DE6#, 4DK6#, 4JH6
4BZ6/4JH6	7CM	4AU6#, 4AU6/4BA6#, 4BA6#, 4BZ6, 4CB6#, 4CB6/4DE6#, 4CF6#, 4DE6#, 4DK6#, 4JH6
4BZ7	9AJ	4BC8, 4BC8/4RHH2, 4BQ7A, 4BQ7A/4BZ7, 4BS8, 4BS8/4BZ8, 4BX8, 4BZ8, 4RHH2, 5BK7A
4BZ8	9AJ	4BC8, 4BC8/4RHH2, 4BQ7, 4BQ7A/4BZ7, 4BS8, 4BS8/4BZ8, 4BX8, 4BZ7, 4RHH2, 5BK7A
4C22	-	HF100, HF100/4C22
4C35A	-	-
4CB6	7CM	4AU6#, 4AU6/4BA6#, 4BA6#, 4BC5#, 4BC5/4CE5#, 4BZ6#, 4BZ6/4JH6#, 4CB6/4DE6, 4CE5#, 4CE5/4BC5#, 4CF6, 4DE6, 4DK6, 4JH6#
4CB6/4DE6	7CM	4AU6#, 4AU6/4BA6#, 4BA6#, 4BC5#, 4BC5/4CE5#, 4BZ6#, 4BZ6/4JH6#, 4CB6, 4CE5#, 4CE5/4BC5, 4CF6, 4DE6, 4DK6, 4JH6#
4CE5	7BD	4AU6#, 4AU6/4BA6#, 4BA6#, 4BC5, 4BC5/4CE5, 4BZ6#, 4BZ6/4JH6#, 4CB6#, 4CB6/4DE6#, 4CE5/4BC5, 4CF6#, 4DE6#, 4DK6#, 4JH6#
4CE5/4BC5	7BD	4AU6#, 4AU6/4BA6#, 4BA6#, 4BC5, 4BC5/4CE5, 4BZ6#, 4BZ6/4JH6#, 4CB6#, 4CB6/4DE6#, 4CE5, 4CF6#, 4DE6#, 4DK6#, 4JH6#
4CF6	7CM	4AU6#, 4AU6/4BA6#, 4BA6#, 4BC5#, 4BC5/4CE5#, 4BZ6#, 4BZ6/4JH6#, 4CB6, 4CB6/4DE6, 4CE5#, 4CE5/4BC5#, 4DE6, 4DK6, 4JH6#
4CM4	9KG	PC86
4CM6	9KG	-
4CS6	7CH	4BE6
4CX7	9FC	-
4CX250B	-	4CX250B/7203, 7203, 7203/4CX250B
4CX250B/7203	-	4CX250B, 7203, 7203/4CX250B
4CX250F	-	7204, 7204/4CX250F
4CH250FG	-	-
4CX250K	-	4CX250K/8245, 8245
4CX250K/8245	-	4CX250K, 8245
4CX250R	-	-

Tube Type	Basing	Replacement
4CX300A	-	4CX300A/8167, 8167
4CX300A/8167	-	4CX300A, 8167
4CX350A	-	4CX350A/8321, 8321
4CX350A/8321	-	4CX350A, 8321
4CX1000A	-	-
4CY5	7EW	-
4D21	-	4-125A, 4-125A/4D21, 4D21A
4D21A	-	4-125A, 4-125A/4D21
4D32	-	-
4DE6	7CM	4AU6#, 4AU6/4BA6#, 4BA6#, 4BC5#, 4BC5/4CE5#, 4BZ6#, 4BZ6/4JH6#, 4CB6, 4CB6/4DE6, 4CE5#, 4CE5/4BC5#, 4DK6, 4CE5#, 4JH6#
4DK6	7CM	4BC5#, 4BC5/4CE5#, 4BZ6#, 4BZ6/4JH6#, 4CB6, 4CB6/4DE6, 4CE5#, 4CE5/4BC5#, 4CF6, 4DE6, 4JH6#
4DL4	-	PC88
4DT6	7EN	4DT6A
4DT6A	7EN	4DT6
4E27	-	4E27/257B, 257B
4E27/257B	-	4E27, 257B
4E27A	-	4E27A/5-125B, 5-125B
4E27A/5-125B	-	4E27A, 5-125B
4EH7	9AQ	4EH7/LF183, 4EJ7, 4EJ7/LF184, 4HM6#, 4HM6/4JC6A#, 4HT6#, 4JC6#, 4JC6A#, 4JC6A/4JD6#, 4JD6#, LF183, LF184, YF183
4EH7/LF183	9AQ	4EH7, 4EJ7, 4EJ7/LF184, 4HM6#, 4HM6/4JC6A#, 4HT6#, 4JC6#, 4JC6A#, 4JC6A/4JD6#, 4JD6, LF183, LF184, YF183
4EJ7	9AQ	4EH7, 4EH7/LF183, 4EJ7/LF184, 4HM6#, 4HM6/4JC6A, 4HT6#, 4JC6#, 4JC6A#, 4JC6A/4JD6#, 4JD6#, LF183, LF184, YF184
4EJ7/LF184	9AQ	4EH7, 4EH7/LF183, 4EJ7, LF183, LF184, YF184
4ES8	9DE	4ES8/XCC189, 4KN8, 4KN8/4RHH8, 4RHH8, XCC189
4ES8/XCC189	9AJ	4ES8, 4KN8, 4KN8/4RHH8, 4RHH8, XCC189
4EW6	7CM	4EW6/4GM6, 4GM6, 4LU6
4EW6/4GM6	7CM	4EW6, 4GM6, 4LU6
4FQ5	-	4FQ5A, 4GK5
4FQ5A	-	4GK5
4FS7	9MP	4HG8
4FY5	-	4GK5
4G280K	-	2D21, 2D21/5727, 2D21/PL21, 5727, 5727/2D21, PL21
4GJ7	9QA	4GJ7/4GX7, 4GJ7/XCF801, 4GX7, XCF801
4GJ7/4GX7	9QA	4GJ7, 4GX7, 4GJ7/XCF801, XCF801
4GJ7/XCF801	9QA	4GJ7, 4GJ7/4GX7, 4GX7, XCF801
4GK5	7FP	PC95
4GM6	7CM	4EW6, 4EW6/4GM6, 4LU6
4GS7	9GF	4LJ8#, 5FG7#
4GS8	9LW	4BU8, 4BU8/4GS8, 4GS8/4BU8, 4HS8, cont'd
4GS8, cont'd		4HS8/4KF8, 4KF8, 4MK8
4GS8/4BU8	9LW	4BU8, 4BU8/4GS8, 4GS8, 4HS8, 4HS8/4KF8, 4KE8, 4KF8, 4MK8
4GW5	7GK	-
4GX7	9QA	4GJ7, 4GJ7/4GX7, 4GJ7/XCF801, 4GX7, XCF801
4GZ5	7CV	-
4H4	-	-
4HA5	7GM	4HA5/4HM5, 4HA5/PC900, 4HK5!, 4HM5, 4HM5/4HA5, 4HQ5, PC900
4HA5/4HM5	7GM	4HA5, 4HA5/PC900, 4HM5, 4HM5/4HA5, 4HQ5, PC900
4HA5/PC900	7GM	4HA5, 4HA5/4HM5, 4HM5, 4HQ5, PC900
4HA7	12FQ	4HA7/4HC7, 4HC7#
4HA7/4HC7	12FR	4HA7#, 4HC7#
4HC7	12FR	4HA7#, 4HA7/4HC7
4HG8	9MP	-
4HK5	7GM	4HA5, 4HA5/4HM5, 4HA5/PC900, 4HM5, 4HM5/4HA5, 4HQ5, PC900
4HM5	7GM	4HA5, 4HA5/4HM5, 4HA5/PC900, 4HK5!, 4HM5/4HA5, 4HQ5, PC900
4HM5/4HM5	7GM	4HA5, 4HA5/PC900, 4HK5!, 4HM5, 4HM5/4HA5, 4HQ5, PC900
4HM6	9PM	4HM6/4JC6A, 4HT6, 4JC6, 4JC6A, 4JC6A/4JD6, 4JD6
4HM6/4JC6A	9PM	4HM6, 4HT6, 4JC6, 4JC6A, 4JC6A/4JD6, 4JD6
4HQ5	7GM	4HA5, 4HA5/4HM5, 4HA5/PC900, 4HK5!, 4HM5, 4HM5/4HA5, 4HQ5, PC900
4HR8	9BJ	4HR8/PF86, PF86
4HR8/PF86	9BJ	4HR8, PF86
4HS8	9FG	4BU8, 4BU8/4GS8, 4GS8, 4GS8/4BU8, 4HS8/4KF8, 4KF8, 4MK8
4HS8/4KF8	9FG	4BU8, 4BU8/4GS8, 4GS8, 4GS8/4BU8, 4HS8, 4KF8, 4MK8
4HT6	9PM	4HM6, 4HM6/4JC6A, 4JC6, 4JC6A, 4JC6A/4JD6, 4JD6
4JC6	9PM	4HM6, 4HM6/4JC6A, 4HT6, 4JC6A, 4JC6A/4JD6, 4JD6
4JC6A	9PM	4HM6, 4HM6/4JC6A, 4HT6, 4JC6, 4JC6A/4JD6, 4JD6
4JC6A/4JD6	9PM	4HM6, 4HM6/4JC6A, 4HT6, 4JC6, 4JC6A, 4JD6
4JD6	9PM	4HM6, 4HM6/4JC6A, 4HT6, 4JC6, 4JC6A, 4JC6A/4JD6
4JH6	7CM	4BZ6, 4BZ6/4JH6
4JK6	7CM	4EW6, 4EW6/4GM6, 4GM6, 4JL6, 4LU6
4JL6	7CM	4EW6, 4EW6/4GM6, 4GM6, 4JK6, 4LU6
4JW8	9DC	5JW8
4KE8	9DC	-
4KF8	9FG	4BU8, 4BU8/4GS8, 4GS8, 4GS8/4BU8, 4HS8, 4HS8/4KF8, 4MK8
4KN8	9AJ	4KN8/4RHH8, 4RHH8
4KN8/4RHH8	9AJ	4KN8, 4RHH8
4KT6	9PM	-

Tube Type	Basing	Replacement
4LJ8	9GF	-
4LU6	7CM	-
4MK8	9FG	4BU8, 4BU8/4GS8, 4GS8, 4GS8/4BU8, 4HS8, 4HS8/4KF8, 4KF8
4MP	-	-
4MP12	-	-
4MP26	-	-
4PR60A	-	-
4PR65A	-	4PR65A/8187, 8187
4PR65A/8187	-	4PR65A, 8187
4RHH2	9AJ	4BC8, 4BC8/4RHH2, 4BQ7A, 4BQ7A/4BZ7, 4BS8, 4BS8/4BZ8, 4BZ7, 4BZ8
4RHH8	9DE	4KN8, 4KN8/4RHH8
4S	5D	2S, 2S/4S, G2
4X150A	-	4X150A/7034, 7034, 7034/4X150A
4X150A/7034	-	4X150A, 7034, 7034/4X150A
4X150D	-	7035, 7035/4X150D, 7609
4X250B	-	7203#
4X250F	-	7204#
4X500A	-	-
4Y25	-	5S1, 807, 807W, 807WA, 5933, 5933WA, 8018, HY61, P17A, QE06/50, QE06140, QV05-25, RK39
5-125B	-	4E27A, 4E27A/5-125B
5A/160H	-	6AM6, 6AM6/EF91, EF91
5A6	9L	-
5AF4	7DK	-
5AF4A	7DK	-
5AM8	9CY	-
5AN8	9DA	-
5AQ4	-	5AQ4/GZ32, GZ32
5AQ4/GZ32	-	5AQ4, GZ32
5AQ5	7BZ	-
5AR4	5DA	5AR4/GZ34, 52KU, 53KU, 54KU, GZ30, GZ32, GZ33, GZ34, GZ34/5AR4, GZ37, R52, U54, U77
5AR4/GZ34	5DA	5AR4, GZ34, 52KU, 53KU, 54KU, GZ30, GZ32, GZ33, GZ34, GZ34/5AR4, GZ37, R52, U54, U77
5AS4	5T	5AS4A, 5AS4A/5U4GB, 5AU4, 5AU4/5V3A, 5DB4, 5U4GB, 5U4GB/5AS4A, 5V3, 5V3A, 5V3A/5AU4
5AS4A	5T	5AS4, 5AS4A/5U4GB, 5AU4, 5AU4/5V3A, 5DB4, 5U4GB, 5U4GB/5AS4A, 5V3, 5V3A, 5V3A/5AU4
5AS4A/5U4GB	5T	5AS4, 5AS4A, 5AU4, 5AU4/5V3A, 5DB4, 5U4GB, 5U4GB/5AS4A, 5V3, 5V3A, 5V3A/5AU4
5AS8	9DS	-
5AT4	5L	-
5AT8	9DW	5BE8#, 5BR8#, 5BR8/5FV8#, 5CG8#, 5CL8#, 5CL8A#, 5DH8#, 5FG7#, 5FV8#, 5FV8/5BR8/5CL8A#
5AU4	5T	5AS4A, 5AS4A/5U4GB, 5AU4G, 5AU4/5V3A, 5U4GB, 5U4GB/5AS4A, 5V3, 5V3A, cont'd
5AU4, cont'd		5V3A/5AU4, 5U4GB
5AU4/5V3A	5T	5AS4A, 5AS4A/5U4GB, 5AU4, 5AU4G, 5U4GB, 5U4GB/5AS4A, 5V3, 5V3A, 5V3A/5AU4
5AU4G	5T	5AS4A, 5AS4A/5U4GB, 5AU4, 5AU4/5V3A, 5U4GB, 5U4GB/5AS4A, 5V3, 5V3A, 5V3A/5AU4, 5U4GB
5AV8	9DZ	5B8#
5AW4	5T	5AS4, 5AS4A, 5AS4A/5U4GB, 5AU4, 5AU4/5V3A, 5DB4, 5R4G, 5R4GTY, 5R4GY, 5R4GYA, 5R4GYB, 5U4GA, 5U4GB, 5U4GB/5AS4A, 5V3, 5V3A, 5V3A/5AU4, U54
5AX4	5T	5AR4, 5AR4/GZ34, 5AS4, 5AS4A, 5AS4A/5U4GB, 5AU4, 5AU4/5V3A, 5AX4GT, 5R4G, 5R4GTY, 5R4GY, 5R4GYA, 5R4GYB, 5T4, 5U4G, 5U4GA, 5U4GB, 5U4GB/5AS4A, 5V3A, 5V3A/5AU4, 5V4G, 5V4GA, 5V4GY, GZ34, GZ34/5AR4, U54
5AX4GT	5T	5AR4, 5AR4/GZ34, 5AS4, 5AS4A, 5AS4A/5U4GB, 5AU4, 5AU4/5V3A, 5R4G, 5R4GTY, 5R4GY, 5R4GYA, 5R4GYB, 5T4, 5U4G, 5U4GA, 5U4GB, 5U4GB/5AS4A, 5V3A, 5V3A/5AU4, 5V4G, 5V4GA, 5V4GY, GZ34, GZ34/5AR4, U54
5AZ3	12BR	-
5AZ4	5T	-
5B/250A	-	4Y25, 5S1, 807, 807W, 807WA, 5933, 5933WA, 8018, HY61, P17A, QE06/50, QE06140, QV05-25, RK39
5B/255M	-	-
5B8	9EC	5AV8#
5BC3	9QJ	5BC3A
5BC3A	9QJ	5BC3
5BC8	-	5BQ7A, 5BQ7A/5BZ7, 5BZ7
5BE8	9EG	5BR8#, 5BR8/5FV8#, 5CL8#, 5CL8A#, 5FG7#, 5FV8#, 5FV8/5BR8/5CL8A#
5BK7	9AJ	4BC8, 4BC8/4RHH2, 4BQ7A, 4BQ7A/4BZ7, 4BS8, 4BS8/4BZ8, 4BX8, 4BZ7, 4BZ8, 4RHH2, 5BK7A
5BK7A	9AJ	4BC8, 4BC8/4RHH2, 4BQ7A, 4BQ7A/4BZ7, 4BS8, 4BS8/4BZ8, 4BX8, 4BZ7, 4BZ8, 4RHH2
5BM8	-	-
5BQ5	-	-
5BQ7	9AJ	5BQ7A, 5BQ7A/5BZ7, 5BS8, 5BZ7
5BQ7A	9AJ	5BQ7A/5BZ7, 5BS8, 5BZ7
5BQ7A/5BZ7	9AJ	5BQ7A, 5BS8, 5BZ7
5BR8	9FA	5BE8#, 5BR8/5FV8, 5CL8#, 5CL8A#, cont'd

Tube Type	Basing	Replacement
5BRB, cont'd		5CL8A/5BR8, 5FV8, 5FV8/5BR8/5CL8A
5BR8/5FV8	9FA	5BE8#, 5BR8, 5CL8#, 5CL8A#, 5CL8A/5BR8, 5FV8, 5FV8/5BR8/5CL8A
5BS8	9AJ	5BQ7A, 5BQ7A/5BZ7, 5BZ7
5BT8	9FE	-
5BW8	9HK	-
5BZ7	9AJ	5BQ7A, 5BQ7A/5BZ7, 5BS8
5C21	-	C6J, C6J/5C21
5C22	-	-
5CG4	5L	5AR4, 5AR4/GZ34, 5V4G, 5V4GA, 5V4GY, 5Z4, 5Z4G, 5Z4GT, 5Z4GT/G, 5Z4MG, GZ34, GZ34/5AR4, OSW3107
5CG8	9GF	5AT8#, 5BE8#, 5BR8#, 5BR8/5FV8#, 5CL8#, 5CL8A#, 5EA8, 5EA8/5GH8A, 5FG7, 5FV8#, 5FV8/5BR8/5CL8A, 5GH8A
5CL8	9FX	5BE8#, 5BR8#, 5BR8/5FV8#, 5CL8A, 5CL8A/5BR8, 5FV8#, 5FV8/5BR8/5CL8A#
5CL8A	9FX	5BE8#, 5BR8#, 5BR8/5FV8#, 5CL8, 5CL8A/5BR8, 5FV8#, 5FV8/5BR8/5CL8A#
5CL8A/5BR8	9FX	5BE8#, 5BR8#, 5BR8/5FV8#, 5CL8, 5CL8A, 5FV8#, 5FV8/5BR8/5CL8A#
5CM6	9CK	-
5CM8	9FZ	5CR8#, 5KZ8
5CQ8	9GE	5EA8#, 5EA8/5GH8A#, 5GH8A, 5MQ8#, 5U8#
5CR8	9GJ	5CM8#, 5KZ8#
5CU4	8KD	-
5CZ5	9HN	-
5D22	-	4-250, 4-250/5D22, 4-250A, 4-250A/5D22, 6156
5DB4	5T	5AS4, 5AS4A, 5AS4A/5U4GB, 5U4GB, 5U4GB/5AS4A
5DH8	9EG	5BR8#, 5BR8/5FV8#, 5CL8#, 5CL8A#, 5FG7#, 5FV8#, 5FV8/5BR8/5CL8A#
5DJ4	8KS	5DN4
5DN4	8KS	-
5EA8	9AE	5EA8/5GH8A, 5GH8, 5GH8A, 5GQ8#, 5U8, XCF80
5EA8/5GH8A	9AE	5EA8, 5GH8A, 5U8, XCF80
5EH8	9JG	5X8#
5ES8	9DE	5ES8/YCC189, YCC189
5ES8/YCC189	9AJ	5ES8, YCC189
5EU8	9JF	-
5EW6	7CM	5EW6/5GM6, 5GM6
5EW6/5GM6	7CM	5EW6, 5GM6
5FG7	9GF	5BE8#, 5BR8#, 5BR8/5FV8#, 5FV8#, 5FV8/5BR8/5CL8A#
5FV8	9FA	5BE8#, 5BR8, 5BR8/5FV8, 5CL8#, 5CL8A#, 5CL8A/5BR8, 5DH8#, 5FV8/5BR8/5CL8A
5FV8/5BR8/5CL8A	9FA	5BR8, 5BR8/5FV8, 5CL8A, 5CL8A/5BR8, 5FV8
5GH8	9AE	5EA8, 5EA8/5GH8A, 5GH8A, 5GQ8#, 5U8
5GH8A	9AE	5EA8, 5EA8/5GH8A, 5GH8, 5GQ8#, 5MQ8#

Tube Type	Basing	Replacement
5GJ7	9QA	5GJ7/LCF801, 5GX7, LCF801
5GJ7/LCF801	9QA	5GJ7, 5GX7, LCF801
5GM6	7CM	5EW6, 5EW6/5GM6
5GQ8	9AE	5EA8, 5EA8/5GH8A, 5GH8, 5GH8A, 5U8, XCF80
5GS7	9GF	5LJ8#
5GX6	7EN	5GX6/5HZ6, 5HZ6
5GX6/5HZ6	7EN	5GX6, 5HZ6
5GX7	9QA	5GJ7, 5GJ7/LCF801, LCF801
5HA7	12FQ	-
5HB7	9QA	-
5HC7	12FR	-
5HG8	9MP	5HG8/LCF86, LCF86
5HG8/LCF86	9MP	5HG8, LCF86
5HZ6	7EN	5GX6, 5GX6/5HZ6
5J2	-	KY80
5J6	7BF	5J6/5MHH3, 5MHH3
5J6/5MHH3	7BF	5J6, 5MHH3
5JK6	7CM	5EW6, 5EW6/5GM6, 5GM6, 5JL6#
5JL6	7CM	5EW6, 5EW6/5GM6, 5GM6, 5JK6#
5JW8	9DC	4JW8
5KD8	9AE	-
5KE8	9DC	-
5KZ8	9FZ	5CR8#
5LJ8	9GF	5GS7!#
5MB8	9FA	-
5MHH3	7BF	5J6, 5J6/5MHH3
5MK9	-	-
5MQ8	9AE	5CQ8#, 5EA8, 5EA8/5GH8A, 5GH8#, 5GH8A, 6ML8, 6MU8
5P29	-	6CN6
5R4	5T	5AS4A#, 5AS4A/5U4GB#, 5R4, 5R4GB, 5R4GTY, 5R4GY, 5R4GYA, 5R4GYB, 5U4GB#, 5U4GB/5AS4A#, 2076, 2076/5R4GB, 2076/5R4GYB
5R4G	5T	5AS4A#, 5AS4A/5U4GB#, 5R4GB, 5R4GTY, 5R4GY, 5R4GYA, 5R4GYB, 5U4GB#, 5U4GB/5AS4A#, 2076, 2076/5R4GB, 2076/5R4GYB
5R4GA	5T	5AS4A#, 5AS4A/5U4GB#, 5R4GB, 5R4GTY, 5R4GY, 5R4GYA, 5R4GBY, 5U4GB#, 5U4GB/5AS4A#, 2076, 2076/5R4GB, 2076/5R4GYB
5R4GB	5T	5AS4A#, 5AS4A/5U4GB#, 5R4GYB, 5U4GB#, 5U4GB/5AS4A#, 2076, 2076/5R4GB, 2076/5R4GYB
5R4GTY	5T	5AS4A#, 5AS4A/5U4GB#, 5R4G, 5R4GB, 5R4GY, 5R4GYA, 5R4GYB, 5U4GB#, 5U4GB/5AS4A#, 2076, 2076/5R4GB, 2076/GYB
5R4GY	5T	5AS4A#, 5AS4A/5U4GB#, 5R4G, 5R4GB, 5R4GTY, 5R4GYA, 5R4GYB, 5R4WGA, 5R4WGB, 5R4WGY, 5R4WGYA, 5U4GB#, cont'd

Tube Type	Basing	Replacement
5R4GY, cont'd		5U4GB/5AS4A#, 274B, 2076, 2076/5R4GB, 2076/5R4GYB
5R4GYA	5T	5AS4A#, 5AS4A/5U4GB#, 5R4G, 5R4GB, 5R4GTY, 5R4GY, 5R4GYB, 5U4GB#, 5U4GB/5AS4A#, 2076, 2076/5R4GB, 2076/5R4GYB
5R4GYB	5T	5AS4A#, 5AS4A/5U4GB#, 5R4G, 5R4GB, 5R4GTY, 5R4GY, 5R4GYA, 5U4GB#, 5U4GB/5AS4A#, 2076, 2076/5R4GB, 2076/5R4GYB
5R4WGA	5T	5R4WGB, 5R4WGY, 5R4WGYA
5R4WGB	5T	5R4WGYA
5R4WGY	5T	5R4WGA, 5R4WGB, 5R4WGYA
5R4WGYA	5T	5R4WGB
5RDDH1	-	-
5RHH2	-	5BQ7A, 5BQ7A/5BZ7, 5BZ7
5RHP1	-	4BL8, 4BL8/XCF80
5RHR1		
5RK16	-	-
5S1	-	4Y25, 807, 807W, 807WA, 5933, 5933WA, 8018, HY61, P17A, QE06/50, QE06140, QV05-25, RK39
5T4	5T	5AR4, 5AR4/GZ34, 5CG4, 5R4G, 5R4GB, 5R4GTY, 5R4GY, 5R4GYA, 5R4GYB, 5T4, 5T4G, 5W4GT, 5W4GT/G, 5Y3, 5Y3/6087, 2076, 2076/5R4GB, 6087, GZ31, GZ34, GZ34/5AR4, U52
5T4G	5T	5AR4, 5AR4/GZ34, 5CG4, 5R4G, 5R4GB, 5R4GTY, 5R4GY, 5R4GYA, 5R4GYB, 5T4, 5T4GA, 5T4GY, 5W4, 5W4GT, 5W4GT/G, 5Y3, 5Y3/6087, 5Y3G, 5Y3G/GT, 5Y3GA, 5Y3GT, 5Y3GT/G, 5Z4, 5Z4G, 5Z4GT, 5Z4GT/G, 5Z4MG, 2076, 2076/5R4GB, 2076/5R4GYB, 6087, GZ34, GZ34/5AR4
5T4GA	5T	5AR4, 5AR4/GZ34, 5CG4, 5R4G, 5R4GB, 5R4GTY, 5R4GY, 5R4GYA, 5R4GYB, 5T4, 5T4G, 5T4GY, 5W4, 5W4GT, 5W4GT/G, 5Y3, 5Y3/6087, 5Y3G, 5Y3G/GT, 5Y3GA, 5Y3GT, 5Y3GT/G, 5Z4, 5Z4G, 5Z4GT, 5Z4GT/G, 5Z4MG, 2076, 2076/5R4GB, 2076/5R4GYB, 6087, GZ34, GZ34/5AR4
5T4GY	5T	5AR4, 5AR4/GZ34, 5CG4, 5R4G, 5R4GB, 5R4GTY, 5R4GY, 5R4GYA, 5R4GYB, 5T4, 5T4G, 5T4GA, 5W4, 5W4GT, cont'd

Tube Type	Basing	Replacement
5T4GY, cont'd		5W4GT/G, 5Y3, 5Y3/6087, 5Y3G, 5Y3G/GT, 5Y3GA, 5Y3GT, 5Y3GT/G, 5Z4, 5Z4G, 5Z4GT, 5Z4GT/G, 5Z4MG, 2076, 2076/5R4GB, 2076/5R4GYB, 6087, GZ34, GZ34/5AR4
5T8	9E	-
5TV85-10	-	0G3, 0G3/85A2, 85A2
5U4	5T	5AR4, 5AR4/GZ34, 5AS4, 5AS4A, 5AS4A/5U4GB, 5AU4, 5AU4/5V3A, 5DB4, 5R4G, 5R4GTY, 5R4GY, 5R4GYA, 5R4GYB, 5T4, 5U4G, 5U4GA, 5U4GB, 5U4GB/5AS4A, 5U4WG, 5U4WGB, 5V3, 5V3A, 5V3A/5AU4, 5Z10, 5931, GZ32, GZ34, GZ34/5AR4, U52, WTT135
5U4G	5T	5AR4, 5AR4/GZ34, 5AS4, 5AS4A, 5AS4A/5U4GB, 5AU4, 5AU4/5V3A, 5DB4, 5R4G, 5R4GTY, 5R4GY, 5R4GYA, 5R4GYB, 5T4, 5U4GA, 5U4GB, 5U4GB/5AS4A, 5U4WG, 5U4WGB, 5V3, 5V3A, 5V3A/5AU4, 5Z10, 5931, GZ32, GZ34, GZ34/5AR4, U52, WTT135
5U4GA	5T	5AS4, 5AS4A, 5AS4A/5U4GB, 5AU4, 5AU4/5V3A, 5DB4, 5R4G, 5R4GTY, 5R4GY, 5R4GYA, 5R4GYB, 5U4GB, 5U4GB/5AS4A, 5V3, 5V3A, 5V3A/5AU4
5U4GB	5T	5AS4, 5AS4A, 5AS4A/5U4GB, 5AU4, 5AU4/5V3A, 5DB4, 5U4GB/5AS4A, 5V3, 5V3A, 5V3A/5AU4
5U4GB/5AS4A	5T	5AS4A, 5AS4A/5U4GB, 5AU4, 5AU4/5V3A, 5DB4, 5U4GB, 5V3, 5V3A, 5V3A/5AU4
5U4GT	5T	5AS4, 5AS4A, 5AS4A/5U4GB, 5AU4, 5AU4/5V3A, 5DB4, 5U4GB, 5U4GB/5AS4A, 5V3, 5V3A, 5V3A/5AU4
5U4WG	5T	5U4WGB, 5931, GZ31, U52
5U4WGB	5T	5931, GZ31, U52
5U8	9AE	5EA8, 5EA8/5GH8A, 5GH8, 5GH8A, 5GQ8#, XCF80
5U9	10K	5U9/LCF201, LCF201
5U9/LCF201	10K	5U9, LCF201
5V3	5T	5AU4, 5AU4/5V3A, 5V3A, 5V3A/5AU4
5V3A	5T	5AU4, 5AU4/5V3A, 5V3A/5AU4
5V3A/5AU4	5T	5AU4, 5AU4/5V3A, 5V3A
5V4	5L	5V4G, 5V4GA, 5V4GY, 5AR4, 5AR4/GZ34, 52KU, 53KU, 54KU, 274, GZ30, GZ32, GZ33, GZ34, GZ34/5AR4, OSW3107, R52, U54
5V4G	5L	5V4GA, 5V4GY, 5AR4, 5AR4/GZ34, 52KU, 53KU, 54KU, 274, GZ30, GZ32, GZ33, GZ34, cont'd

Tube Type	Basing	Replacement
5V4G, cont'd		GZ34/5AR4, OSW3107, R52, U54
5V4GA	5L	5AR4, 5AR4/GZ34, 5V4G, 5V4GY, GZ34, GZ34/5AR4
5V4GT	5L	5AR4, 5AR4/GZ34, 5V4G, 5V4GY, GZ34, GZ34/5AR4
5V4GY	5L	5AR4, 5AR4/GZ34, 5V4G, 5V4GT, GZ34, GZ34/5AR4
5V6	7S	5V6GT
5V6GT	7S	5V6
5V9	-	LCH200
5W4	5T	5AR4, 5AR4/GZ34, 5CG4, 5R4G, 5R4GTY, 5R4GY, 5R4GYA, 5R4GYB, 5T4, 5W4G, 5W4GT, 5W4GT/G, 5V4G, 5V4GA, 5V4GY, 5Y3, 5Y4/6087, 5Y3G, 5Y3G/GT, 5Y3GA, 5Y3GT, 5Y3GT/G, 5Z4, 5Z4G, 5Z4GT, 5Z4GT/G, 5Z4MG, 6087, GZ34, GZ34/5AR4, U50, U51
5W4G	5T	5AR4, 5AR4/GZ34, 5CG4, 5R4G, 5R4GTY, 5R4GY, 5R4GYA, 5R4GYB, 5T4, 5T4G, 5T4GA, 5T4GY, 5W4, 5W4GT, 5W4GT/G, 5Y3, 5Y3/6087, 5Y3G, 5Y3G/GT, 5Y3GA, 5Y3GT, 5Y3GT/G, 5Z4, 5Z4G, 5Z4GT, 5Z4GT/G, 5Z4MG, 6087, GZ34, GZ34/5AR4
5W4GT	5T	5AR4, 5AR4/GZ34, 5CG4, 5R4G, 5R4GTY, 5R4GY, 5R4GYA, 5R4GYB, 5T4, 5V4G, 5V4GA, 5V4GY, 5W4, 5W4G, 5W4GT/G, 5Y3, 5Y3/6087, 5Y3G, 5Y3G/GT, 5Y3GA, 5Y3GT, 5Y3GT/G, 5Z4, 5Z4G, 5Z4GT, 5Z4GT/G, 5Z4MG, 6087, GZ34, GZ34/5AR4
5W4GT/G	5T	5AR4, 5AR4/GZ34, 5CG4, 5R4G, 5R4GTY, 5R4GY, 5R4GYA, 5R4GYB, 5T4, 5V4G, 5V4GA, 5V4GY, 5W4, 5W4G, 5W4GT, 5Y3, 5Y3/6087, 5Y3G, 5Y3G/GT, 5Y3GA, 5Y3GT, 5Y3GT/G, 5Z4, 5Z4G, 5Z4GT, 5Z4GT/G, 5Z4MG, 6087, GZ34, GZ34/5AR4
5X3	4C	13, 80, 83V, 88
5X4	-	5X4G, 5X4GT
5X4G	5Q	5X4GA
5X4GA	5Q	5X4G
5X8	9AK	5EH8#
5X9	-	LCF200
5Y3	5T	5AR4, 5AR4/GZ34, 5AX4GT, 5CG4, 5R4G, 5R4GTY, 5R4GY, 5R4GYA, 5R4GYB, 5T4, 5V4, 5V4GA, 5V4GY, 5Y3/6087, 5Y3G/GT, 5Y3GA, 5Y3GT, 5Y3GT/G, 5Z4, 5Z4G, 5Z4GT, 5Z4GT/G, 5Z4MG, 6087, 6106, 6106/5Y3WGT, GZ34, GZ34/5AR4
5Y3/6087	5T	5AR4, 5AR4/GZ34, 5AX4GT, 5CG4, 5R4G, 5R4GTY, 5R4GY, 5R4GYA, 5R4GYB, 5T4, 5V4, 5V4GA, 5V4GY, 5Y3, 5Y3G/GT, 5Y3GA, 5Y3GT, 5Y3GT/G, 5Z4, 5Z4G, 5Z4GT, 5Z4GT/G, 5Z4MG, 6087, 6106, 6106/5Y3WGT, GZ34, GZ34/5AR4
5Y3G	5T	5AR4, 5AR4/GZ34, 5AX4GT, 5CG4, 5R4G, 5R4GTY, 5R4GY, 5R4GYA, 5R4GYB, 5T4, 5V4, 5V4GA, 5V4GY, 5Y3, 5Y3/6087, 5Y3G/GT, 5Y3GA, 5Y3GT, 5Y3GT/G, 5Z4, 5Z4G, 5Z4GT, 5Z4GT/G, 5Z4MG, 6087, 6106, 6106/5Y3WGT, GZ34, GZ34/5AR4
5Y3G/GT	5T	5AR4, 5AR4/GZ34, 5AX4GT, 5CG4, 5R4G, 5R4GTY, 5R4GY, 5R4GYA, 5R4GYB, 5T4, 5V4G, 5V4GA, 5V4GY, 5Y3, 5Y3/5087, 5Y3G, 5Y3GA, 5Y3GT, 5Z4, 5Z4G, 5Z4GT, 5Z4GT/G, 5Z4MG, 6087, 6106, 6106/5Y3WGT, GZ34, GZ34/5AR4
5Y3GA	5T	5AR4, 5AR4/GZ34, 5AX4GT, 5CG4, 5R4G, 5R4GTY, 5R4GY, 5R4GYA, 5R4GYB, 5T4, 5V4, 5V4GA, 5V4GY, 5Y3, 5Y3/8087, 5Y3G, 5Y3G/GT, 5Y3GT, 5Y3GT/G, 5Z4, 5Z4G, 5Z4GT, 5Z4GT/G, 5Z4MG, 6087, 6106, 6106/5Y3WGT, GZ34, GZ34/5AR4
5Y3GT	5T	5AR4, 5AR4/GZ34, 5AX4GT, 5CG4, 5R4G, 5R4GTY, 5R4GY, 5R4GYA, 5R4GYB, 5T4, 5V4, 5V4GA, 5V4GY, 5Y3, 5Y3/6087, 5Y3G, 5Y3G/GT, 5Y3GA, 5Y3GT/G, 5Y3WGT, 5Y3WGTA, 5Y3WGTB, 5Z4, 5Z4G, 5Z4GT, 5Z4GT/G, 5Z4MG, 6087, 6087/5Y3WGTB, 6106, 6106/5Y3WGT, 6853, GZ34, GZ34/5AR4, U50, U52, WTT202
5Y3GT/G	5T	5AR4, 5AR4/GZ34, 5AX4GT, 5CG4, 5R4G, 5R4GTY, 5R4GY, 5R4GYA, 5R4GYB, 5T4, 5V4G, 5V4GA, 5V4GY, 5Y3, 5Y3/6087, 5Y3G, 5Y3G/GT, 5Y3GA, 5Y3GT, 5Z4, 5Z4G, 5Z4GT, 5Z4GT/G, 5Z4MG, 6087, 6106, 6106/5Y3WGT, GZ34, GZ34/5AR4
5Y3WGT	5T	5Y3, 5Y3/6087, 5Y3WGT, 5Y3WGTA, 5Y3WGTB, 6087, cont'd

Tube Type	Basing	Replacement
5Y3WGT, cont'd		6087/5Y3WGTB, 6106, 6106/5Y3WGT, 6853
5Y3WGTA	5T	5Y3, 5Y3/6087, 5Y3WGT, 5Y3WGTB, 6087, 6087/5Y3WGTB, 6106, 6106/5Y3WGT, 6853
5Y3WGTB	5T	5Y3, 5Y3/6087, 5Y3WGT, 5Y3WGTA, 6087, 6087/5Y3WGTB, 6106, 6106/5Y3WGT, 6853
5Y4	5Q	5AU4, 5Y4G, 5Y4GA, 5Y4GT
5Y4G	5Q	5AU4, 5Y4GA, 5Y4GT
5Y4GA	5Q	5AU4, 5Y4G, 5Y4GT
5Y4GT	5Q	5AU4, 5Y4G, 5Y4GA
5Z3	4C	83, 1275, WT270X
5Z4	5L	5AR4, 5AR4/GZ34, 5AS4A, 5AS4A/5U4GB, 5AU4, 5AU4/5V3A, 5CG4, 5U4GB, 5U4GB/5AS4A, 5V3A, 5V3A/5AU4, 5V4G, 5V4GA, 5V4GY, 5Y5, 5Y3/6087, 5Y3G, 5Y3G/GT, 5Y3GT, 5Y3GT/G, 5Z4G, 5Z4GT, 5Z4GT/G, 5Z4MG, 52KU, 53KU, 54KU, 6087, GZ30, GZ32, GZ33, GZ34, GZ34/5AR4, OSW3107, R52, U54, U77
5Z4G	5L	5AR4, 5AR4/GZ34, 5AS4A, 5AS4A/5U4GB, 5AU4, 5AU4/5V3A, 5CG4, 5U4GB, 5U4GB/5AS4A, 5V3A, 5V3A/5AU4, 5V4G, 5V4GA, 5V4GY, 5Y3, 5Y3/6087, 5Y3G, 5Y3G/GT, 5Y3GT, 5Y3GT/G, 5Z4, 5Z4GT, 5Z4GT/G, 5Z4MG, 52KU, 6087, GZ30, GZ34, GZ34/5AR4, R52, U77
5Z4GT	5L	5AR4, 5AR4/GZ34, 5AS4A, 5AS4A/5U4GB, 5AU4, 5AU4/5V3A, 5CG4, 5U4GB, 5U4GB/5AS4A, 5V3A, 5V3A/5AU4, 5V4G, 5V4GA, 5V4GY, 5Y3, 5Y3/6087, 5Y3G, 5Y3G/GT, 5Y3GT, 5Y3GT/G, 5Z4, 5Z4G, 5Z4GT/G, 5Z4MG, 6087, GZ34, GZ34/5AR4
5Z4GT/G	5L	5AR4, 5AR4/GZ34, 5AS4A, 5AS4A/5U4GB, 5AU4, 5AU4/5V3A, 5CG4, 5U4GB, cont'd

Tube Type	Basing	Replacement
5Z4GT/G, cont'd		5U4GB/5AS4A, 5V3A, 5V3A/5AU4, 5V4G, 5V4GA, 5V4GY, 5Y3, 5Y3/6087, 5Y3G, 5Y3G/GT, 5Y3GT, 5Y3GT/G, 5Z4, 5Z4G, 5Z4GT, 5Z4MG, 6087, GZ34, GZ34/5AR4
5Z4MG	5L	5AR4, 5AR4/GZ34, 5AS4A, 5AS4A/5U4GB, 5AU4, 5AU4/5V3A, 5CG4, 5U4GB, 5U4GB/5AS4A, 5V3A, 5V3A/5AU4, 5V4GA, 5V4GY, 5Y3, 5Y3/6087, 5Y3G, 5Y3G/GT, 5Y3GT, 5Y3GT/G, 5Z4, 5Z4G, 5Z4GT, 5Z4GT/G, 6087, GZ34, GZ34/5AR4
5Z10	-	5AS4A, 5AS4A/5U4GB, 5U4G, 5U4GB, 5U4GB/5AS4A
6/30L2	-	6GA8
6A3	4D	-
6A4	5B	6A4LA#
6A4LA	5K	6A4#
6A5	6T	6A5G
6A5G	6T	6A5
6A6	7B	6A6X, B63
6A6X	7B	6A6
6A7	7C	6A7S
6A7S	7C	6A7
6A8	8A	6A8G, 6A8GT, 6A8GTX, 6A8MG, 6D8*, 6D8G*, PH4, X63
6A8G	8A	6A8, 6A8GT, 6A8GTX, 6A8MG, 6D8*, 6D8G*
6A8GT	8A	6A8, 6A8G, 6A8GTX, 6A8MG, 6D8*, 6A8G*
6A8GTX	8A	6A8, 6A8G, 6A8GT, 6A8MG, 6D8*, 6D8G*
6A8MG	8A	6A8, 6A8G, 6A8GT, 6A8GTX, 6D8*, 6D8G*
6AB4	5CE	6AB4/EC92, 6664, 6664/6AB4, EC92, EC92/6AB4
6AB4/EC92	5CE	6AB4, 6664, 6664/6AB4, EC92, EC92/6AB4
6AB5	6R	6AB5/6N5, 6N5, 6N5G
6AB5/6N5	6R	6AB5, 6N5, 6N5G
6AB6G	7AU	6N6*, 6N6G*, 6N6MG*
6AB7	8N	6AB7/1853, 6AC7, 6AC7/1852, 6AC7/1852/6134, 6AC7/EL34, 6AC7A, 6AC7Y, 6AJ7, 6AB7Y, 6F10, 6SG7*#, 6SG7GT*#, 6SG7GT*, 6SH7*, 1852, 1853, 6134, EL34, OSW2190, OSW2600
6AB7/1853	8N	6AB7, 6AC7, 6AC7/1852, 6AC7/1852/6134, 6AC7/EL34, 6AC7A, 6AC7Y, 6AJ7, 6AB7Y, 6F10, 6SG7*#, 6SG7GT*#, 6SG7GT*, 6SH7*, 1852, 1853, 6134, EL34, OSW2190, OSW2600
6AB7Y	8N	6AB7, 6AB7/1853, 6AC7, 6AC7/1852, 6AC7/1852/6134, 6AC7/EL34, 6AC7A, 6AC7Y, 6AJ7, cont'd

Tube Type	Basing	Replacement
6AB7Y, cont'd		6SG7*#, 6SG7GT*#, 6SG7Y*#, 6SH7*#, 6SH7GT*, 1852, 1853, 6134, EL34
6AB8	9AT	6AB8/ECL80, 63TP, ECL80, LN152
6AB8/ECL80	9AT	6AB8, 63TP, ECL80, LN152
6AB9	10N	-
6AC5	6Q	6AC5G, 6AC5GT, 6AC5GT/G
6AC5G	6Q	6AC5, 6AC5GT, 6AC5GT/G
6AC5GT	6Q	6AC5, 6AC5G, 6AC5GT/G
6AC5GT/G	6Q	6AC5, 6AC5G, 6AC5GT
6AC6	7W	6AC6G, 6AC6GT
6AC6G	7W	6AC6GT
6AC6GT	7W	6AC6G
6AC7	8N	6AB7, 6AB7/1853, 6AB7Y, 6AC7/1852, 6AC7/1852/6134, 6AC7/EL34, 6AC7A, 6AC7W, 6AC7WA, 6AC7Y, 6AJ7, 6F10, 6SG7*#, 6SG7GT*#, 6SG7Y*#, 6SH7*#, 6SH7GT*#, 1852, 1853, 6134, EL34, OSW2190, OSW2600
6AC7/1852	8N	6AB7, 6AB7/1853, 6AB7Y, 6AC7, 6AC7/EL34, 6AC7/1852/6134, 6AC7/EL34, 6AC7A, 6AC7GT, 6AC7W, 6AC7WA, 6AJ7, 6F10, 6SG7*#, 6SG7GT*#, 6SG7Y*#, 6SH7*#, 6SH7GT*3, 1852, 1853, 6134, EL34, 0SW2190, 0SW2600
6AC7/1852/6134	8N	6AB7, 6AB7/1853, 6AB7Y, 6AC7, 6AC7/1852, 6AC7/EL34, 6AJ7, 6AC7A, 6AC7GT, 6AC7W, 6AC7WA, 6AC7Y, 6F10, 6SG7*#, 6SG7GT*#, 6SG7Y*#, 6SH7*#, 6SH7GT*#, 1852, 1853, 6134, EL34, OSW2190, OSW2600
6AC7A	8N	6AB7, 6AB7/1853, 6AB7Y, 6AC7, 6AC7/1852, 6AC7/1852/6134, 6AC7/EL34, 6AC7GT, 6AC7Y, 6AJ7, 6F10, 6SG7*#, 6SG7GT*#, 6SG7Y*#, 6SH7*#, 6SH7GT*#, 1852, 1853, 6134, EL34, OSW2190, OSW2600
6AC7/EL34	8N	6AB7, 6AB7/1853, 6AB7Y, 6AC7, 6AC7/1852, 6AC7/1852/6134, 6A7GT, 6AC7W, 6AC7WA, 6AC7Y, 6AJ7, 6F10, 6SG7*#, 6SG7GT*#, 6SH7*#, 6SH7GT*#, EL34, 1852, 1853, 6134, EL34, OSW2190, OSW2600
6AC7GT	8N	6AB7, 6AB7/1853, 6AB7Y, 6AC7, 6AC7/1852, 6AC7/1852/6134, 6AC7/EL34, 6AC7W, 6AC7WA, 6AC7Y, 6AJ7, 6F10, cont'd
6AC7GT, cont'd		6SG7*#, 6SG7GT*#, 6SH7*#, 6SH7GT*#, EL34, 1852, 1853, 6134, EL34, OSW2190, OSW2600
6AC7W	8N	6AC7#, 6AC7/EL34#, 6AC7/1852#, 6AC7/1852/6134, 6AC7WA, 1852#, 6134, EL34#
6AC7WA	8N	6AC7#, 6AC7/EL34#, 6AC7/1852#, 6AC7/1852/6134, 6AC7W, 1852#, 6134, EL34#
6AC7Y	8N	6AC7#, 6AC7/EL34#, 6AC7/1852#, 6AC7/1852/6134, 6AC7A, 6AB7, 6AB7Y, 6AJ7, 6SG7*#, 6SG7GT*#, 6SG7Y*#, 6SH7*#, 6SH7GT*#, 6SH7*#, 6SH7GT*#, EL34, 1852, 6134, EL34, OSW2190, OSW2600
6AC9	12GN	-
6AC10	12FE	6U10#
6AC11	-	-
6AD4	8DK	-
6AD5	6Q	6AD5G, 6AD5GT
6AD5G	6Q	6AD5GT
6AD5GT	6Q	6AD5G
6AD6	7AG	6AD6G, 6AF6G, 6AF6G/GT, 6AF6GT, 6AF6GT/6AD6G
6AD6G	7AG	6AF6G, 6AF6G/GT, 6AF6GT, 6AF6GT/6AD6G
6AD7	8AY	6AD7G
6AD7G	8AY	6AD7
6AD8	9T	6AD8/6DC8, 6DC8, 6FD12, 6N8, EBF80, EBF81, EBF89, WD709, ZD152
6AD8/6DC8	9T	6AD8, 6DC8
6AD10	12EZ	6AD10A, 6T10#
6AD10A	12EZ	6AD10
6AE5	6Q	6AE5G, 6AE5G/6AF5G, 6AE5GT, 6AE5GT/G, 6AF5G
6AE5G	6Q	6AE5, 6AE5G/6AF5G, 6AE5GT, 6AE5GT/G, 6AF5G
6AE5G/6AF5G	6Q	6AE5G, 6AE5GT, 6AE5GT/G, 6AF5G
6AE5GT	6Q	6AE5, 6AE5G, 6AE5G/6AF5G, 6AE5GT/G, 6AF5G
6AE5GT/G	6Q	6AE5, 6AE5G, 6AE5G/6AF5G, 6AE5GT, 6AF5G
6AE6	7AH	6AE6G
6AE6G	7AH	6AE6
6AE7	7AX	6AE7G, 6AE7GT
6AE7G	7AX	6AE7GT
6AE7GT	7AX	6AE7G
6AE8	8DU	-
6AF3	9CB	6AL3*, 6AL3/EY88*, 6BR3, 6BR3/6RK19, 6RK19, EY88*
6AF4	7DK	6AF4A, 6AF4A/6DZ4, 6AN4#, 6DZ4, 6DZ4/6AF4A, 6T4#, EC94
6AF4A	7DK	6AF4, 6AF4A/6DZ4, 6AN4#, 6DZ4#, 6DZ4/6AF4A, 6T4#
6AF4A/6DZ4	7DK	6AF4, 6AF4A, 6AN4#, 6DZ4, 6DZ4/6AF4A, 6T4#

Tube Type	Basing	Replacement
6AF5	6Q	6AE5, 6AE5G, 6AE5G/6AF5G, 6AE5GT, 6AE5GT/G, 6AF5G
6AF5G	6Q	6AE5, 6AE5G, 6AE5G/6AF5G, 6AE5GT, 6AE5GT/G
6AF6	7AG	6AD6G, 6AF6G, 6AF6G/GT, 6AF6GT, 6AF6GT/6AD6G
6AF6G	7AG	6AD6G, 6AF6G/GT, 6AF6GT, 6AF6GT/6AD6G
6AF6GT	7AG	6AD6G, 6AF6G, 6AF6G/GT, 6AF6GT/6AD6G
6AF7G	8AG	6CD7*, 6CD7/EM34*, 6CD7/EM34-35*, 64ME*, EM34*, EM34/6CD7*, EM34-35*, EM34-35/6CD7
6AF9	10L	-
6AF10	12GX	-
6AF11	12DP	6AS11, 6BD11
6AG4Y	8Y	-
6AG5	7BD	6AG5/EF96, 6AG5WA, 6AK5*, 6AK5/5654*, 6AK5/EF95*, 6AU6#, 6AU6/6136#, 6AU6A#, 6AU6A/EF94#, 6AW6#, 6BC5, 6BC5/6CE5, 6CB6#, 6CB6A#, 6CB6A/6CF6#, 6CE5, 6CE5/6BC5, 6CF6#, 6CY5*#, 6DC6#, 6DE6#, 6EA5*#, 6EV5*#, 5654*, 6136#, 6186, 6186/6AG5WA, 6676#, 6676/6CB6A#, 8425A#, 8425A/6AU6A#, EF94#, EF95*, EF96
6AG5/EF96	7BD	6AG5, 6AG5WA, 6AK5*, 6AK5/5654*, 6AK5/EF95*, 6AU6#, 6AU6/6136#, 6AU6A#, 6AU6A/EF94#, 6AW6#, 6BC5, 6BC5/6CE5, 6CB6#, 6CB6A#, 6CB6A/6CF6#, 6CE5, 6CE5/6BC5, 6CF6#, 6CY5*#, 6DC6#, 6DE6#, 6EA5*#, 6EV5*#, 5654, 6136#, 6186, 6186/6AG5WA, 6676#, 6676/6CB6A#, 8425A#, 8425A/6AU6A#, EF94#, EF95*, EF96
6AG5WA	7BD	6AG5#, 6AG5/EF96#, 6186, 6186/6AG5WA, EF96#
6AG6	7S	6AG6G, 6AG6G/EL33, 6M6G*, EL33
6AG6G	7S	6AG6G/EL33, 6M6G*, EL33
6AG6G/EL33	7S	6AG6G, EL33
6AG7	8Y	6AG7/6AK7, 6AG7W, 6AG7Y, 6AK7, 6L10, OSW2192, OSW2601
6AG7/6AK7	8Y	6AG7, 6AG7W, 6AK7, 6L10, OSW2192, OSW2601
6AG7W	8Y	OSW2192, OSW2601
6AG7Y	8Y	6AG7#, 6AG7/6AK7#, 6AG7W, 6AK7
6AG9	12HE	6AL9
6AG10	12GT	-
6AG11	12DA	-
6AH4	8EL	6AH4GT
6AH4GT	8EL	6AH4

Tube Type	Basing	Replacement
6AH5	6AP	6AH5G
6AH5G	6AP	6AH5
6AH6	7BK	6AH6V, 6AH6WA, 6F36, 6485, 6485/6AH6WA
6AH6V	7BK	6AH6, 6AH6WA, 6F36, 6485, 6485/6AH6WA
6AH6WA	7BK	6AH6#, 6485, 6485/6AH6WA
6AH7	8BE	6AH7GT
6AH7GT	8BE	6AH7
6AH9	12HJ	-
6AJ4	9BX	6AM4#, 6CR4*#, EC84
6AJ5	7BD	6F35, 7755
6AJ6	-	-
6AJ7	8N	6AB7, 6AB7/1853, 6AB7Y, 6AC7, 6AC7/1852, 6AC7/1852/6134, 6AC7/EL34, 6AC7A, 6AC7Y, 6F10, 6SG7*#, 6SG7GT*#, 6SG7Y*#, 6SH7*#, 6SH7GT*#, 1852, 1853, 6134, EL34, OSW2190, OSW2600
6AJ8	9CA	6AJ8/ECH81, 6C12, 20D4, ECH81, ECH81/6AJ8, X719
6AJ8/ECH81	9CA	6AJ8, 6C12, 20D4, ECH81, ECH81/6AJ8, X719
6AJ9	-	ECF202
6AK4	8DK	-
6AK5	7BD	6AG5*, 6AG5/EF96#, 6AK5/5654, 6AK5/EF95, 6AK5W, 6AK5WA, 6AK5WB, 6BC5*, 6BC5/6CE5*, 6CE5*, 6CE5/6BC5*, 6CY5*#, 6EA5*#, 6EV5*#, 6F32, 403A, 1220, 5654, 5591*, 5654, 5654/6AK5W, 6096, 6096/6AK5W, 6968, DP61, E95#, E95F, EF95, EF96#, EF905, M8100#, PM05#
6AK5/5654	7BD	6AG5*, 6AG5/EF96#, 6AK5, 6AK5/EF95, 6AK5W, 6AK5WA, 6AK5WB, 6BC5*, 6BC5/6CE5*, 6CE5*, 6CE5/6BC5*, 6CY5*#, 6EA5*#, 6EV5*#, 6F32, 403A, 1220, 5654, 5591*, 5654, 5654/6AK5W, 6096, 6096/6AK5W, 6968, DP61, E95#, E95F, EF95, EF96#, EF905, M8100#, PM05#
6AK5/EF95	7BD	6AG5*, 6AG5/EF96#, 6AK5, 6AK5/5654, 6AK5W, 6AK5WA, 6AK5WB, 403A, 1220, 5654, 5654/6AK5W, 6096, 6096/6AK5W, 6968, 6F32, EF95, E95F, EF905, DP61
6AK5W	7BD	6AK5/5654, 6AK5/EF95#, 6AK5WB, 5654, 5654/6AK5W, 6096, 6096/6AK5W, EF95#
6AK5WA	7BD	6AK5#, 6AK5/5654, 6AK5/EF95#, 6AK5WB, 5654, 5654/6AK5W, 6096, 6096/6AK5W, EF95#
6AK5WB	7BD	6AK5/5654, 6AK5/EF95#, 6AK5WA, cont'd

Tube Type	Basing	Replacement
6AK5WB, cont'd		5654, 5654/6AK5W, 6096, 6096/6AK5W, EF95#
6AK6	7BK	-
6AK7	8Y	6AG7, 6AG7/6AK7, 6AG7W, 6AG7Y, 6L10, OSW2192, OSW2601
6AK8	9E	6AK8/EABC80, 6LD12, 6T8, 6T8A, DH719, EABC80
6AK8/EABC80	9E	6AK8, 6LD12, 6T8A, DH719, EABC80
6AK9	12GZ	-
6AK10	12FE	-
6AL3	9CB	6AL3/EY88, EY88
6AL3/EY88	9CB	6AL3, EY88
6AL5	6BT	6AL5/6EB5, 6AL5/5726/6663, 6AL5/EAA91, 6AL5W, 6B32, 6D2, 6EB5, 5726, 5726/6AL5W, 5726/6AL5W/6097, 6058, 6097, 6663, 6663/6AL5, 7631, D2M9, D27, D77, D152, D717, DD6, EAA91, EAA91/6AL5, EB91, EAA91, EAA901, EAA901S
6AL5/6EB5	6BT	6AL5, 6AL5/5726/6663, 6AL5/EAA91, 6AL5W, 6EB5, 5726, 5726/6AL5W, 5726/6AL5W/6097, 6097, 6663, 6663/6AL5, EAA91, EAA91/6AL5
6AL5/5726/6663	6BT	6AL5, 6AL5/6EB5, 6AL5/EAA91, 6AL5W, 5726, 5726/6AL5W, 5726/6AL5W/6097, 6097, 6663, 6663/6AL5, EAA91, EAA91/6AL5
6AL5/EAA91	6BT	6AL5, 6AL5/6EB5, 6AL5/5726/6663, 6AL5W, 6B32, 6D2, 6EB5, 5726, 5726/6AL5W, 5726/6AL5W/6097, 6058, 6097, 6663, 6663/6AL5, 7631, D2M9, D27, D77, D152, D717, DD6, EAA91, EAA91/6AL5, EAA901, EAA901S, EB91
6AL5W	6BT	6AL5#, 6AL5/6EB5#, 6AL5/5726/6663, 6AL5/EAA91#, 6EB5#, 5726, 5726/6AL5W, 5726/6AL5W/6097, 6097, 6663, 6663/6AL5#, EAA91#, EAA91/6AL5#
6AL6	6AM	6AL6G
6AL6G	6AM	6AL6
6AL7	8CH	6AL7GT
6AL7GT	8CH	6AL7
6AL9	12HE	6AG9
6AL11	12BU	-
6AM4	9BX	6AJ4#
6AM5	6CH	6AM5/EL91, 6P17, 7D9, 16A, 6516, DD7, DDR7, EL91, M8082*
6AM5/EL91	6CH	6AM5, 6P17, 7D9, 16A, 6516, DD7, DDR7, EL91, EL91, M8082*
6AM6	7DB	6AM6/EF91, 6F12, 8D3, 6064, 7498, EF91, HP6, cont'd
6AM6, cont'd		M8083, PM07, QA2403, QZ77, R144, S6F12, SP6, Z77
6AM6/EF91	7BD	6AM6, 6F12, 8D3, 6064, 7498, EF91, HP6, M8083, PM07, QA2403, QZ77, R144, S6F12, SP6, Z77
6AM8	9CY	6AM8A, 6HJ8
6AM8A	9CY	6AM8*!, 6HJ8
6AN4	7DK	6AF4#, 6AF4A#, 6AF4A/6DZ4#, 6DZ4#, 6DZ4/6AF4A#, 6T4#
6AN5	7BD	6AN5WA
6AN5WA	7BD	6AN5#
6AN6	7BJ	-
6AN7	9Q	ECH80
6AN8	9DA	6AN8A
6AN8A	9DA	6AN8*!
6AQ4	7DT	6AQ4/EC91, 6L34, EC91, M8099
6AQ4/EC91	7DT	6AQ4, 6L34, EC91, M8099
6AQ5	7BZ	6AQ5/6005/6669, 6AQ5A, 6AQ5A/6HG5, 6AQ5W, 6BM5, 6HG5, 6005, 6005/6AQ5W, 6005/6AQ5W/6095, 6095, 6669, 6669/6AQ5A, 6L31, BPM04, EL90, M8245, N727
6AQ5/6005/6669	7BZ	6AQ5, 6AQ5A, 6AQ5A/6HG5, 6AQ5W, 6BM5, 6HG5, 6005, 6005/6AQ5W, 6005/6AQ5W/6095, 6095, 6669, 6669/6AQ5A, 6L31, BPM04, EL90, M8245, N727
6AQ5A	7BZ	6AG5*!, 6AG5/EF96*!, 6AQ5A/6HG5, 6AQ5/6005/6669, 6HG5, 6005, 6005/6AQ5W, 6005/6AQ5W/6095, 6095, 6669, 6669/6AQ5A, EF96*!, GBM5*!
6AQ5A/6HG5	7BZ	6AQ5A, 6AQ5A/6HG5, 6AQ5W, 6HG5, 6005, 6005/6AQ5W, 6005/6AQ5W/6095, 6095, 6669, 6669/6AQ5A
6AQ5W	7BZ	6AQ5A#, 6AQ5A/6HG5#, 6HG5#, 6005, 6005/6AQ5W, 6005/6AQ5W/6095, 6095, 6669#, 6669/6AQ5A
6AQ6	7BT	6AT6*, 6AT6/6BK6*, 6AV6*, 6AV6/EBC91*, 6BK6*, 6BT6*, 6AQ7GT, EBC91*
6AQ7	8CK	6AQ7GT
6AQ7GT	8CK	6AQ7
6AQ8	9DE	6AQ8/ECC85, 6DT8*, 6L12, B719, ECC85, ECC85/6AQ8
6AQ8/ECC85	9DE	6AQ8, 6L12, B719, ECC85, ECC85/6AQ8
6AR5	6CC	-
6AR6	6BQ	6AR6G, 6AR6WA, 6098, 6098/6AR6WA, 6384, 7756
6AR6G	6BQ	6AR6, 6AR6WA, 6098, 6098/6AR6WA, 6384, 7756

Tube Type	Basing	Replacement
6AR6WA	6BQ	6098, 6098/6AR6WA
6AR7	7DE	6AR7GT
6AR7GT	7DE	6AR7
6AR8	9DP	6AR8*
6AR11	12DM	-
6AS4	4CG	6AS4GT, 6CQ4*, 6CQ4/6DE4*, 6DA4A, 6DA4A/6DM4, 6DA4A/6DM4A, 6DE4*, 6DE4/6CQ4*, 6DM4, 6DM4A, 6DM4A/6DQ4, 6DQ4, 6DT4
6AS4GT	4CG	6CQ4*, 6CQ4/6DE4*, 6DA4A, 6DA4A/6DM4, 6DA4A/6DM4A, 6DE4*, 6DE4/6CQ4*, 6DM4, 6DM4A, 6DM4A/6DQ4, 6DQ4, 6DT4
6AS5	7CV	6CA5*, 6CA5/6EH5*, 6EH5*
6AS6	7CM	6AS6W, 6DB6*, 6F33, 409A, 5725, 5725/6AS6W, 6187*, 6187/6AS6, 7752, M8196
6AS6W	7CM	6AS6, 5725, 5725/6AS6W, 6187, 6187/6AS6, M8196
6AS7	8BD	6AS7G, 6AS7GA, 6AS7GT, 6AS7GYB, 6080, 6080W, 6080WA, 6080WB, 6520, A1834, ECC230
6AS7G	8BD	6AS7GA, 6AS7GT, 6AS7GYB, 6080, 6080W, 6080WA, 6080WB, 6520, A1834, ECC230
6AS7GA	8BD	6AS7G#, 6AS7GT, 6080, 6080W, 6080WA
6AS7GT	8BD	6AS7G, 6AS7GA, 6080, 6080W, 6080WA
6AS7GYB	8BD	6AS7G#, 6AS7GA#, 6080#, 6080W#, 6080WA#, 6080WB
6AS8	9DS	-
6AS11	12DP	6AF11, 6BD11
6AT6	7BT	6AQ6*, 6AT6/6BK6, 6AV6, 6AV6/EBC91, 6BC32, 6BK6, 6BT6, 6066, DH77, EBC90, EBC91
6AT6/6BK6	7BT	6AT6, 6AV6, 6AV6/EBC91, 6BC32, 6BK6, 6BT6, 6066, DH77, EBC90, EBC91
6AT7	-	6AT7N, 6DT8
6AT7N	-	6DT8
6AT8	9DW	6AT8A, 6BE8#, 6BE8A#, 6BR8#, 6BR8A#, 6BR8A/6FV8#, 6BR8A/6FV8A#, 6CG8#, 6CG8A#, 6CL8#, 6CL8A#, 6FG7#, 6FV8#, 6FV8A#, 6FV8A/6BR8A#
6AT8A	9DW	6AT8*!, 6BE8A#, 6BR8A#, 6BR8A/6FV8#, 6BR8A/6FV8A#, 6CG8A#, 6CL8#, 6CL8A#, 6FG7#, 6FV8#, 6FV8A#, 6FV8A/6BR8A#
6AU4	4CG	6AU4GT, 6AU4GTA, 6CQ4*, 6CQ4/6DE4*, 6DA4A*, 6DA4A/6DM4*, 6DA4A/6DM4A*, 6DE4*, 6DE4/6CQ4*, 6DM4*, 6DM4A*, 6DM4A/6DQ4*, cont'd
6AU4,cont'd		6DQ4*, 6DT4*, 6G-K17
6AU4GT	4CG	6AU4GTA, 6CQ4*, 6CQ4/6DE4*, 6DA4A*, 6DA4A/6DM4*, 6DA4A/6DM4A*, 6DE4*, 6DE4/6CQ4*, 6DM4*, 6DM4A*, 6DM4A/6DQ4*, 6DQ4*, 6DT4*, 6G-K17
6AU4GTA	4CG	6DA4A*, 6DA4A/6DM4*, 6DA4A/6DM5A*, 6DM4A*, 6DM4A/6DQ4, 6DQ4, 6DT4*
6AU5	6CK	6AU5GT, 6AV5GA*, 6AV5GT*, 6FW5
6AU5GT	6CK	6AV5GA*, 6AV5GT*, 6FW5
6AU6	7BK	6AG5#, 6AG5/EF96#, 6AU6/6136, 6AU6A, 6AU6A/EF94, 6AU6WA, 6AU6WB, 6AW6#, 6BA6, 6BA6/5749/6660, 6BA6/EF93, 6BC5#, 6BC5/6CE5#, 6CB6#, 6CB6A#, 6CB6A/6CF6#, 6CE5#, 6CE5/6BC5#, 6CF6#, 6DE6#, 6DK6#, 5749, 6136, 6136/6AU6WA, 6660, 6660/6BA6, 6676#, 6676/6CB6A, 7543, 8425A, 8425A/6AU6A, EF93, EF94, EF96#
6AU6/6136	7BK	6AG5#, 6AG5/EF96#, 6AU6, 6AU6A, 6AU6A/EF94, 6AU6WA, 6AU6WB, 6AW6#, 6BA6, 6BA6/5749/6660, 6BA6/EF93, 6BC5#, 6BC5/6CE5#, 6CB6#, 6CB6A#, 6CB6A/6CF6#, 6CE5#, 6CE5/6BC5#, 6CF6#, 6DE6#, 6DK6#, 6136, 6136/6AU6WA, 5749, 6660, 6660/6BA6, 6676#, 6676/6CB6A#, 7543, 8425A, 8425A/6AU6A, EF93, EF94, EF96#
6AU6A	7BK	6AU6*!, 6AU6/6136, 6AU6A/EF94, 6BA6*!, 6BA6/5749/6660*!, 6BA6/EF93*!, 6BC5#, 6BC5/6CE5#, 6CB6A#, 6CB6A/6CF6#, 6CE5#, 6CE5/6BC5#, 6CF6#, 5749*!, 6136, 6136/6AU6WA, 6660*!, 6660/6BA6*!, 6676#, 6676/6CB6A#, 7543*!, 8425A, 8425A/6AU6A, EF93*!, EF94
6AU6A/EF94	7BK	6AU6*!, 6AU6/6136, 6AU6A, 6BA6*!, 6BA6/5749/6660*!, 6BA6/EF93*!, 6BC5, 6BC5/6CE5, 6CB6A#, 6CB6A/6CF6#, 6CE5#, 6CE5/6BC5#, 6CF6#, 5749*!, 6136, 6136/6AU6WA, 6660*!, 6660/6BA6*!, 6676, 6676/6CB6A, 7543*!, cont'd

Tube Type	Basing	Replacement
6AU6A/EF94, cont'd		8425A, 8425A/6AU6A, EF93*#, EF94
6AU6WA	7BK	6AU6#, 6AU6/6136#, 6AU6A#, 6AU6A/EF94#, 6AU6WB, 6136#, 6136/6AU6WA, 8425A#, 8425A/6AU6A#, EF94#
6AU6WB	7BK	6AU6#, 6AU6/6136#, 6AU6A#, 6AU6A/EF94#, 6AU6WA, 6136#, 6136/6AU6WA, 8425A#, 8425A/6AU6A#, EF94#
6AU7	9A	6AU7A, 6AX7#, 7AU7
6AU7A	9A	6AU7, 7AU7
6AU8	9DX	6AU8A, 6AU8A/6BH8, 6AW8, 6AW8A, 6BA8, 6BA8A, 6BH8, 6CX8*, 6CX8/6JA8*, 6EH8*#, 6JA8*, 6X8*, 6X8A*, 2081, 2081/6AW8A
6AU8A	9DX	6AU8, 6AU8A/6BH8, 6AW8, 6AW8A, 6BA8, 6BA8A, 6BH8, 6CX8*, 6CX8/6JA8*, 6EH8*#, 6JA8*, 6X8*, 6X8A*, 2081, 2081/6AW8A
6AU8A/6BH8	9DX	6AU8, 6AU8A, 6AW8, 6AW8A, 6BA8, 6BA8A, 6BH8, 6CX8*, 6CX8/6JA8*, 6EH8*#, 6JA8*, 6X8*, 6X8A*, 2081, 2081/6AW8A
6AV3	9CB	6BR3, 6BR3/6RK19, 6RK19
6AV4	5BS	6BX4*, 6FX4*, EZ91
6AV5	6CK	6AU5GT*, 6AV5GA, 6AV5GT, 6FW5
6AV5GA	6CK	6AU5GT*, 6AV5GT, 6FW5
6AV5GT	6CK	6AU5GT*, 6AV5GA, 6FW5
6AV6	7BT	6AQ6*, 6AT6, 6AT6/6BK6, 6AV6/EBC91, 6BK6, 6BT6, 6BC32, 6066, DH77, EBC90, EBC91
6AV6/EBC91	7BT	6AQ6*, 6AT6, 6AT6/6BK6, 6AV6, 6BK6, 6BT6, 6BC32, 6066, DH77, EBC90, EBC91
6AV8	-	-
6AV11	12BY	6K11#, 6Q11#
6AW6	7CM	6AG5#, 6AG5/EF96#, 6AU6, 6AU6/6136, 6AU6A, 6AU6A/EF94, 6AU6WA, 6AU6WB, 6AW6#, 6BA6, 6BA6/5749/6660, 6BA6/EF93, 6BC5#, 6BC5/6CE5#, 6CB6#, 6CB6A#, 6CB6A/6CF6, 6CE5#, 6CE5/6BC5#, 6CF6#, 6DE6#, 6DK6#, 6136, 7543, EF94, EF96#, 6AU6#, 6AU6A#, 6BC5#, 6BC5/6CE5#, 6BH6*, 6BH6/6265/6661*, 6CB6, 6CB6A, 6CB6A/6CF6, 6CE5#, 6CE6/6BC5#, 6CF6, 6DC6, 6DE6, 5749, 6136, 6136/6AU6WA, 6660, 6660/6BA6, 6661*, 6661/6BH6*, 6676, 6676/6CB6A, 8425A, cont'd
6AW6, cont'd		8425A/6AU6A, EF93, EF94, EF96#, EF190
6AW7	8CQ	6AW7GT
6AW7GT	8CQ	6AW7
6AW8	9DX	6AU8, 6AU8A, 6AU8A/6BH8, 6AW8A, 6BA8, 6BA8A, 6BH8, 6EB8*, 6EH8*#, 6GN8*, 6GN8/6EB8*, 6JV8, 6KS8, 6LF8, 6X8*#, 6X8A*#, 2081, 2081/6AW8A
6AW8A	9DX	6AU8, 6AU8A, 6AU8A/6BH8, 6AW8, 6BA8, 6BA8A, 6BH8, 6EB8*, 6EH8*#, 6GN8*, 6GN8/6EB8*, 6JV8, 8KS8, 6LF8, 6X8*#, 6X8A*#, 2081, 2081/6AW8A
6AX2	-	-
6AX3	12BL	-
6AX4	4CG	6AS4GT, 6AU4GT*, 6AU4GTA*, 6AX4GT, 6AX4GTA, 6AX4GTB, 6CQ4*, 6CQ4/6DE4*, 6DA4, 6DA4A, 6DA4A/6DM4, 6DA4A/6DM4A, 6DE4*, 6DE4/6CQ4*, 6DM4, 6DM4A, 6DM4A/6DQ4, 6DQ4, 6DT4
6AX4GT	4CG	6AS4GT, 6AU4GT*, 6AU4GTA*, 6AX4GTA, 6AX4GTB, 6CQ4*, 6CQ4/6DE4*, 6DA4, 6DA4A, 6DA4A/6DM4, 6DA4A/6DM4A, 6DE4*, 6DE4/6CQ4*, 6DM4, 6DM4A, 6DM4A/6DQ4, 6DQ4, 6DT4
6AX4GTA	4CG	6AU4GT*, 6AU4GTA*, 6AX4GTB, 6CQ4*, 6CQ4/6DE4*, 6DA4, 6DA4A, 6DA4A/6DM4, 6DA4A/6DM4A, 6DE4*, 6DE4/6CQ4*, 6DM4, 6DM4A, 6DM4A/6DQ4, 6DQ4, 6DT4
6AX4GTB	4CG	6CQ4*, 6CQ4/6DE4*, 6DA4A*!, 6DA4A/6DM4*!, 6DA4A/6DM4A*!, 6DE4*, 6DE4/6CQ4*, 6DM4*!, 6DM4A*!, 6DM4A/6DQ4*!, 6DQ4*!, 6DT4*!
6AX5	6S	6AX5GT
6AX5GT	6S	6AX5
6AX6	7Q	6AX6G
6AX6G	7Q	6AX6
6AX7	9A	6AU7#, 6GU7*, 7AU7#
6AX8	9AE	6CQ8#, 6EA8, 6GH8, 6GH8A, 6GJ8#, 6LM8, 6KD8, 6U8, 6U8A, 6U8A/6AX8/6KD8
6AY3	9HP	6AY3A, 6AY3B, 6BA3, 6BH3*, 6BH3A*, 6BS3, 6BS3A, 6CK3, 6CK3/6CL3, 6CL3, 6CL3/6CK3, 6DW4, 6DW4A, 6DW4B
6AY3A	9HP	6AY3, 6AY3B, 6BA3, 6BH3*, 6BH3A*, 6BS3, cont'd

Tube Type	Basing	Replacement
6AY3A, cont'd		6BS3A, 6CK3, 6CK3/6CL3, 6CL3, 6CL3/6CK3, 6DW4, 6DW4A, 6DW4B
6AY3B	9HP	6AY3, 6AY3A, 6BA3, 6BH3*, 6BH3A*, 6BS3, 6BS3A, 6CK3, 6CK3/6CL3, 6CL3, 6CL3/6CK3, 6DW4, 6DW4A, 6DW4B
6AY11	12DA	-
6AZ5	8DF	-
6AZ6	-	-
6AZ8	9ED	-
6B3	9BD	6V3*, 6V3A*
6B4	5S	6B4G
6B4G	5S	6B4
6B5	6AS	-
6B6	7V	6BG6, 6Q7, 6Q7G, 6Q7GT, 6Q7MG, 6T7G*
6B6G	7V	6B6, 6Q7, 6Q7G, 6Q7GT, 6Q7MG, 6T7G*
6B7	7D	6B7S
6B7S	7D	6B7
6B8	8E	6B8G, 6B8GT, EBF32*
6B8G	8E	6B8, 6B8GT
6B8GT	8E	6B8, 6B8G
6B10	12BF	-
6B32	-	6AL5, 6AL5/6EB5, 6AL5/5726/6663, 6AL5/EAA91, 6AL5W, 6EB5, 5726, 5726/6AL5W, 5726/6AL5W/6097, 6097, 6663, 6663/6AL5, EAA91, EAA91/6AL5
6BA3	9HP	6AY3, 6AY3A, 6AY3B, 6BH3*, 6BH3A*, 6BS3, 6BS3A, 6CK3, 6CK3/6CL3, 6CL3, 6CL3/6CK3, 6DW4, 6DW4A, 6DW4B
6BA5	-	5638
6BA6	7BK	6AU6, 6AU6/6136, 6AU6A, 6AU6A/EF94, 6BA6/5749/6660, 6BA6/EF93, 6BA6W, 6BA6WA, 6BD6, 6BZ6#, 6BZ6/6JH6#, 6CG6, 6F31, 6JH6#, 5749, 5749/6BA6W, 6136, 6660, 6660/6BA6, 7496, 7543, 8425A, 8425A/6AU6A, EF93, EF94, M8101, PM04, W727
6BA6/5749/6660	7BK	6AU6, 6AU6/6136, 6AU6A, 6AU6A/EF94, 6BA6, 6BA6/EF93, 6BA6W, 6BA6WA, 6BD6, 6BZ6#, 6BZ6/6JH6#, 6CG6, 6F31, 6JH6#, 5749, 5749/6BA6W, 6136, 6660, 6660/6BA6, 7496, 7543, 8425A, 8425A/6AU6A, EF93, EF94
6BA6/EF93	7BK	6AU6, 6AU6/6136, 6AU6A, 6AU6A/EF94, 6BA6, 6BA6/5749/6660, 6BA6W, 6BA6WA, 6BD6, 6BZ6#, 6BZ6/6JH6#, 6CG6, 6JH6#, 6F31, 5749, 5749/6BA6W, 6136, 6660, 6660/6BA6, 7496, 7543, 8425A, 8425A/6AU6A, EF93, EF94, M8101, PM04, W727
6BA6W	7BK	6BA6/5749/6660, 6BA6/EF93#, 6BA6WA, 5749, 5749/6BA6W, 6660, 6660/6BA6#, EF93#
6BA6WA	7BK	6BA6/5749/6660, 6BA6/EF93#, 6BA6W, 5749, 5749/6BA6W, 6660, 6660/6BA6#, EF93#
6BA7	8CT	-
6BA8	9DX	6AU8, 6AU8A, 6AU8A/6BH8, 6AW8, 6AW8A, 6BA8A, 6BH8, 2081, 2081/6AW8A
6BA8A	9DX	6AU8, 6AU8A, 8AU8A/6BH8, 6AW8, 6AW8A, 6BA8, 6BH8, 2081, 2081/6AW8A
6BA11	12ER	-
6BC4	9DR	-
6BC5	7BD	6AG5, 6AG5/EF96, 6AK5*, 6AK5/5654*, 6AK5/EF95*, 6AU6#, 6AU6/6136#, 6AU6A#, 6AU6A/EF94#, 6AW6#, 6BC5/6CE5, 6CB6#, 6CB6A#, 6CB6A/6CF6#, 6CE5, 6CE5/6BC5, 6CF6#, 6CY5*#, 6DC6#, 6DE6#, 6DK6#, 6EA5*#, 6EV5*#, 5654#, 6136#, 6676, 6676/6CB6A, 8425A#, 8425A/6AU6A#, EF94#, EF95*, EF96
6BC5/6CE5	7BD	6AG5, 6AG5/EF96, 6AK5*, 6AK5/5654*, 6AK5/EF95*, 6AU6#, 6AU6/6136#, 6AU6A#, 6AU6A/EF94#, 6AW6#, 6BC5, 6CB6#, 6CB6A#, 6CB6A/6CF6#, 6CE5, 6CE5/6BC5, 6CF6#, 6CY5*#, 6DC6#, 6DE6#, 6DK6#, 6EA5*#, 6EV5*#, 5654#, 6136#, 6676, 6676/6CB6A, 8425A#, 8425A/6AU6A#, EF94#, EF95*, EF96
6BC7	9AX	-
6BC8	9AJ	6BC8/6BQ7A, 6BC8/6BZ8, 6BK7*, 6BK7A*, 6BK7B*, 6BQ7, 6BQ7A, 6BQ7A/6BZ7, 6BQ7A/6BZ7/6BS8, 6BS8, 6BS8/6BK7B, 6BX8, 6BZ7, 6BZ7/6BQ7A, 6BZ8, 6HK8, X155
6BC8/6BQ7A	9AJ	6BC8, 6BC8/6BZ8, 6BK7*, 6BK7A*, 6BK7B*, 6BQ7, 6BQ7A, 6BQ7A/6BZ7, 6BQ7A/6BZ7/6BS8, 6BS8, 6BS8/6BK7B, 6BX8, 6BZ7, 6BZ7/6BQ7A, 6BZ8, 6HK8, X155
6BC8/6BZ8	9A	6BC8, 6BC8/6BQ7A, 6BK7*, 6BK7A*, 6BK7B*, 6BQ7, 6BQ7A, 6BQ7A/6BZ7, 6BQ7A/6BZ7/6BS8, 6BS8, 6BS8/6BK7B, 6BX8, 6BZ7, 6BZ7/6BQ7A, 6BZ8, 6HK8, X155
6BC32	7BT	6AV6, 6AV6/EBC91, cont'd

Tube Type	Basing	Replacement
6BC32, cont'd		6AT6, 6AT6/6BK6, 6BK6, EBC91
6BD4	8FU	6BD4A, 6BK4*#, 6BK4A*#, 6BK4B*#, 6BK4C, 6BK4C/6EL4A
6BD4A	8FU	6BK4*#, 6BK4A*#, 6BK4B*#, 6BK4C, 6BK4C/6EL4A
6BD5	6CK	6AU5GT*, 6AV5GA*, 6AV5GT*, 6BD5, 6FW5*
6BD5GT	6CK	6AU5GT*, 6AV5GA*, 6AV5GT*, 6FW5*
6BD6	7BK	6BA6, 6BA6/5749/6660, 6BA6/EF93, 6CG6, 5749, 6660, 6660/6BA6, EF93
6BD7	9Z	6LD13*, EBC80, EBC81
6BD7A	9Z	6BD7A/EBC81, EBC81
6BD7A/EBC81	9Z	6BD7A, EBC81
6BD11	12DP	6AF11
6BE3	12GA	6BE3A, 6BE3/6BZ3, 6BE3A, 6BW3*, 6BZ3, 6CD3*, 6CD3/6CG3*, 6CE3*, 6CE3/6CD3*, 6CE3/6CD3/6DT3*, 6CG3*, 6CG3/6BW3*, 6CG3/6BW3/6DQ3*, 6CG3/6CD3*, 6CG3/6CE3/6CD3/6BW3*, 6DT3*
6BE3/6BZ3	12GA	6BE3, 6BE3A, 6BW3*, 6BZ3, 6CD3*, 6CD3/6CG3*, 6CE3*, 6CE3/6CD3*, 6CE3/6CD3/6DT3*, 6CG3*, 6CG3/6BW3*, 6CG3/6BW3/6DQ3*, 6CG3/6CD3*, 6CG3/6CE3/6CD3/6BW3*, 6DT3*
6BE3A	12GA	6BE3, 6BE3/6BZ3, 6BE3A, 6BW3*, 6BZ3, 6CD3*, 6CD3/6CG3*, 6CE3*, 6CE3/6CD3*, 6CE3/6CD3/6DT3*, 6CG3, 6CG3/6BW3*, 6CG3/6BW3/6DQ3*, 6CG3/6CD3*, 6CG3/6CE3/6CD3/6BW3*, 6DT3*
6BE6	7CH	6BE6/5750, 6BE6/EK90, 6BE6W, 6BY6, 6CS6, 6CS6/EH90, 6H31, 5750, 5750/6BE6W, 7502, EH90, EK90, HM04, X77, X727, 6H31, EH90, EK90, HM04
6BE6/5750	7CH	6BE6, 6BE6/5750, 6BE6/EK90, 6BE6W, 6CS6, 6CS6/EH90, 5750, 5750/6BE6W, EH90, EK90
6BE6/EK90	7CH	6BE6, 6BE6/5750, 6BE6W, 6CS6, 6CS6/EH90, 5750, 5750/6BE6W, EH90, EK90
6BE6W	7CH	6BE6, 6BE6/5750, 6BE6/EK90, 6CS6, 6CS6/EH90, 5750, 5750/6BE6W, EH90, EK90
6BE7	9AA	6BE7/EQ80, EQ80
6BE7/EQ80	9AA	6BE7, EQ80
6BE8	9EG	6BE8A, 6BR8#, 6BR8A#, 6BR8A/6FV8#, 6BR8A/6FV8A, 6CL8#, 6CL8A#, 6FG7#, 6FV8#, cont'd
6BE8, cont'd		6FV8A#, 6FV8A/6BR8A
6BE8A	9EG	6BE8*!, 6BR8A#, 6BR8A/6FV8#, 6BR8A/6FV8A, 6CL8#, 6CL8A#, 6FG7#, 6FV8#, 6FV8A#, 6FV8A/6BR8A#
6BF5	7BZ	-
6BF6	7BT	6BF6/6BU6, 6BU6
6BF6/6BU6	7BT	6BF6, 6BU6
6BF7	8DG	6BF7A, 6BF7W
6BF7A	8DG	6BF7, 6BF7W
6BF7W	8DG	6BF7#, 6BF7A#
6BF8	9NX	-
6BF11	12EZ	-
6BG6	5BT	6BG6G, 6BG6GA
6BG6G	5BT	6BG6GA
6BG6GA	5BT	6BG6G
6BG7	8DG	-
6BH3	9HP	6BH3A, 6BH3B, 6CH3*, 6CH3/6CJ3*, 6CJ3*, 6CJ3/6CH3*, 6CJ3/6DW4B*, 6CK3*, 6CK3/6CL3*, 6CL3*, 6CL3/6CK3*, 6DW4A*, 6DW4B*
6BH3A	9HP	6BH3, 6BH3B, 6CH3*, 6CH3/6CJ3*, 6CJ3*, 6CJ3/6CH3*, 6CJ3/6DW4B*, 6CK3*, 6CK3/6CL3*, 6CL3*, 6CL3/6CK3*, 6DW4A*
6BH3B	9HP	6BH3, 6BH3A, 6CH3*, 6CH3/6CJ3*, 6CJ3*, 6CJ3/6CH3*, 6CJ3/6DW4B*, 6CK3*, 6CK3/6CL3*, 6CL3*, 6CL3/6CK3*, 6DW4A*, 6DW4B*
6BH5	9AZ	EF81
6BH6	7CM	6AW6*, 6BH6/6265/6661, 6CB6*, 6CB6A*, 6CB6A/6CF6*, 6CF6*, 6DC6*, 6265*, 6661, 6661/6BH6, 6676, 6676/6CB6A, 7693, E90F
6BH6/6265/6661	7CM	6AW6*, 6BH6, 6CB6*, 6CB6A*, 6CB6A/6CF6*, 6CF6*, 6DC6*, 6265*, 6661, 6661/6BH6, 6676*, 6676/6CB6A*, 7693, E90F
6BH8	9DX	6AU8, 6AU8A, 6AU8A/6BH8, 6AW8, 6AW8A, 6BA8, 6BA8A, 2081, 2081/6AW8A
6BH11	12FP	-
6BJ3	12BL	6AX3
6BJ5	6CH	-
6BJ6	7CM	6BJ6/6662, 6BJ6A, 6662, 6662/6BJ6, 7694, E99F, N78
6BJ6/6662	7CM	6BJ6, 6BJ6A, 6662, 6662/6BJ6, 7694, E99F, N78
6BJ6A	7CM	6BJ6, 6BJ6/6662, 6662, 6662/6BJ6, 7694, E99F, N78
6BJ7	9AX	-
6BJ8	9ER	6BN8#
6BK4	8GC	6BK4C, 6BK4C/6EL4A, 6EL4, 6EL4A, 6EN4
6BK4A	8GC	6BK4C, 6BK4C/6EL4A, cont'd

Tube Type	Basing	Replacement
6BK4A, cont'd		6EL4, 6EL4A, 6EN4
6BK4B	8GC	6BK4C, 6BK4C/6EL4A, 6EL4, 6EL4A, 6EN4
6BK4C	8GC	6EL4, 6BK4C/6EL4A, 6EL4A, 6EN4
6BK4C/6EL4A	8GC	6BK4C, 6EL4A, 6EN4
6BK5	9BQ	-
6BK6	7BT	6AT6, 6AT6/6BK6, 6AV6, 6AV6/EBC91, 6AQ6*, 6BC32, 6BT6, DH77, EBC90, EBC91
6BK7	9AJ	6BC8*, 6BC8/6BQ7A*, 6BC8/6BZ8*, 6BK7A, 6BK7B, 6BQ7*, 6BQ7A*, 6BQ7A/6BZ7, 6BQ7A/6BZ7/6BS8*, 6BS8*, 6BS8/6BK7B, 6BX8*, 6BZ7*, 6BZ7/6BQ7A*, 6BZ8*, 6BZ8/6BQ7A*, 6HK8*, X155*
6BK7A	9AJ	6BC8*, 6BC8/6BQ7A*, 6BC8/6BZ8*, 6BK7, 6BK7B, 6BQ7*, 6BQ7A*, 6BQ7A/6BZ7*, 6BQ7A/6BZ7/6BS8*, 6BS8*, 6BS8/6BK7B, 6BX8*, 6BZ7*, 6BZ7/6BQ7A*, 6BZ8*, 6BZ8/6BQ7A*, 6HK8*, X155*
6BK7B	9AJ	6BC8*, 6BC8/6BQ7A*, 6BC8/6BZ8*, 6BK7*, 6BK7A*, 6BQ7*, 6BQ7A*, 6BQ7A/6BZ7*, 6BQ7A/6BZ7/6BS8*, 6BS8*, 6BS8/6BK7B, 6BX8*, 6BZ7*, 6BZ7/6BQ7A*, 6BZ8*, 6BZ8/6BQ7A*, 6HK8*, X155*
6BK8	9BJ	6BK8/EF83, EF83
6BK8/EF83	9BJ	6BK8, EF83
6BK11	12BY	6AC10, 6K11#, 6K11/6Q11#, 6Q11#
6BL4	8GB	6AU4GTA*, 6CQ4*, 6CQ4/6DE4*, 6DA4A*, 6DA4A/6DM4*, 6DA4A/6DM4A*, 6DE4*, 6DE4/6CQ4*, 6DM4A*, 6DM4A/6DQ4, 6DQ4, 6DT4*
6BL7	8BD	6BL7GT, 6BL7GTA, 6BX7GT, 6DN7
6BL7GT	8BD	6BL7GTA, 6BX7GT, 6DN7
6BL7GTA	8BD	6BL7GT, 6BX7GT, 6DN7
6BL8	9DC	6BL8/ECF80, 6BL8/6LN8, 6C16, 6LN8, 7643*, ECF80, ECF80/6BL8, E80CF*
6BL8/ECF80	9DC	6BL8, 6BL8/6LN8, 6C16, 6LN8, 7643*, ECF80, ECF80/6BL8, E80CF*
6BL8/6LN8	9DC	6BL8, 6BL8/ECF80, 6C16, 6LN8, 7643*, ECF80, ECF80/6BL8, E80CF*
6BM5	7BZ	6AQ5, 6AQ5/6005/6669, 6AQ5A, 6AQ5A/6HG5, 6DL5*, 6DS5*, 6HG5, 6L31, 6P9, 6005, 6669, 6669/6AQ5A, EL90, N727
6BM8	9EX	6BM8/ECL82, 6PL12, ECL82, ECL82/6BM8
6BM8/ECL82	9EX	6BM8, 6PL12, ECL82, ECL82/6BM8
6BN4	7EG	6BN4A
6BN4A	7EG	6BN4
6BN5	9CR	6BN5/EL85, EL85, N155
6BN5/EL85	9CR	6BN5, EL85, N155
6BN6	7DF	6BN6/6KS6, 6KS6
6BN6/6KS6	7DF	6BN6, 6KS6
6BN7	9AJ	-
6BN8	9ER	6BJ8#
6BN11	12GF	-
6BQ5	9CV	6BQ5/EL84, 6P15, 6267, 7189, 7189A, 7320, 7320/E84L, EF86, EL84, EL84/6BQ5, EL84L, N709, Z729
6BQ5/EL84	9CV	6BQ5, 6P15, 6267, 7189, 7189A, 7320, 7320/E84L, EF86, EL84, EL84/6BQ5, EL84L, N709, Z729
6BQ5P	9CV	6BQ5, 6BQ5/EL84, 6P15, 6267, 7189, 7189A, 7320, 7320/E84L, EF86, EL84, EL84/6BQ5, EL84L, N709, Z729
6BQ5WA	9CV	6BQ5#, 6BQ5/EL84#, E84#, EL84/6BQ5#, EL84M, EL84M/6BQ5WA
6BQ6	6AM	6BQ6G, 6BQ6GA, 6BQ6GA/6CU6, 6BQ6GT, 6BQ6GTA, 6BQ6GTB, 6BQ6GTB/6CU6, 6CU6, 6DQ6, 6DQ6A, 6DQ6B, 6DQ6B/6GW6, 6FH6, 6GW6
6BQ6G	6AM	6BQ6GA, 6BQ6GA/6CU6, 6BQ6GT, 6BQ6GTA, 6BQ6GTB, 6BQ6GTB/6CU6, 6CU6, 6DQ6, 6DQ6A, 6DQ6B, 6DQ6B/6GW6, 6FH6, 6GW6
6BQ6GA	6AM	6BQ6G, 6BQ6GA/6CU6, 6BQ6GT, 6BQ6GTA, 6BQ6GTB, 6BQ6GTB/6CU6, 6CU6, 6DQ6, 6DQ6A, 6DQ6B, 6DQ6B/6GW6, 6FH6, 6GW6
6BQ6GA/6CU6	6AM	6BQ6G, 6BQ6GA, 6BQ6GT, 6BQ6GTA, 6BQ6GTB, 6BQ6GTB/6CU6, 6CU6, 6DQ6, 6DQ6A, 6DQ6B, 6DQ6B/6GW6, 6FH6, 6GW6
6BQ6GT	6AM	6BQ6G, 6BQ6GA, 6BQ6GA/6CU6, 6BQ6GTA, 6BQ6GTB, 6BQ6GTB/6CU6, 6CU6, 6DQ6, 6DQ6A, 6DQ6B, 6DQ6B/6GW6, 6FH6, 6G-B6, 6GW6
6BQ6GTA	6AM	6BQ6G, 6BQ6GA, 6BQ6GA/6CU6, 6BQ6GT, 6BQ6GTB, 6BQ6GTB/6CU6, 6CU6, 6DQ6, 6DQ6A, 6DQ6B, 6DQ6B/6GW6, 6FH6, 6GW6
6BQ6GTB	6AM	6BQ6GA, 6BQ6GA/6CU6, 6BQ6GTB/6CU6, 6CU6, 6DQ6, 6DQ6A, cont'd

Tube Type	Basing	Replacement
6BQ6GTB, cont'd		6DQ6B, 6DQ6B/6GW6, 6FH6, 6GW6
6BQ6GTB/6CU6	6AM	6BQ6GA, 6BQ6GA/6CU6,6BQ6GTA, 6BQ6GTB, 6CU6, 6DQ6, 6DQ6A, 6DQ6B, 6DQ6B/6GW6, 6FH6, 6GW6
6BQ7	9AJ	6BC8, 6BC8/6BQ7A, 6BC8/6BZ8, 6BK7*, 6BK7A*, 6BK7B*, 6BQ7A, 6BQ7A/6BZ7, 6BQ7A/6BZ7/6BS8, 6BS8, 6BS8/6BK7B, 6BX8, 6BZ7, 6BZ7/6BQ7A, 6BZ8, 6HK8, X155
6BQ7A	9AJ	6BC8, 6BC8/6BQ7A, 6BC8/6BZ8, 6BK7*, 6BK7A*, 6BK7B*, 6BQ7, 6BQ7A/6BZ7, 6BQ7A/6BZ7/6BS8, 6BS8, 6BS8/6BK7B, 6BX8, 6BZ7, 6BZ7/6BQ7A, 6BZ8, 6HK8, ECC180, X155
6BQ7A/6BZ7	9AJ	6BC8, 6BC8/6BQ7A, 6BC8/6BZ8, 6BK7*, 6BK7A*, 6BK7B*, 6BQ7, 6BQ7A, 6BQ7A/6BZ7/6BS8, 6BS8, 6BS8/6BK7B, 6BX8, 6BZ7, 6BZ7/6BQ7A, 6BZ8, 6HK8, ECC180, X155
6BQ7A/6BZ7/6BS8	9AJ	6BC8, 6BC8/6BQ7A, 6BC8/6BZ8, 6BK7*, 6BK7A*, 6BK7B*, 6BQ7, 6BQ7A, 6BQ7A/6BZ7, 6BS8, 6BS8/6BK7B, 6BX8, 6BZ7, 6BZ7/6BQ7A, 6BZ8, 6HK8, ECC180, X155
6BR3	9CB	6AL3*, 6AL3/EY88*, 6BR3/6RK19, 6RK19, EY88*
6BR3/6RK19	9CB	6AL3*, 6AL3/EY88*, 6BR3, 6RK19, EY88*
6BR5	9DB	6BR5/EM80, 6DA5, 6DA5/EM81, 65ME, EM80, EM80/6BR5, EM81, EM81/6DA5
6BR5/EM80	9DB	6BR5, 6DA5, 6DA5/EM81, 65ME, EM80, EM80/6BR5, EM81, EM81/6DA5
6BR7	9BC	8D5, 6059
6BR8	9FA	6BE8#, 6BE8A#, 6BR8A, 6BR8A/6FV8, 6BR8A/6FV8A, 6CL8#, 6CL8A#, 6FV8, 6FV8A, 6FV8A/6BR8A, 6JN8
6BR8A	9FA	6BE8A#, 6BR8*!, 6BR8A, 6BR8A/6FV8, 6BR8A/6FV8A, 6CL8#, 6CL8A#, 6FV8, 6FV8A, 6FV8A/6BR8A, 6JN8
6BR8A/6FV8	9FA	6BR8A, 6BR8A/6FV8A, 6FV8, 6FV8A, 6FV8A/6BR8A
6BR8A/6FV8A	9FA	6BR8A, 6BR8A/6FV8, 6FV8, 6FV8A, 6FV8A/6BR8A
6BS3	9HP	6BS3A, 6CK3, 6CK3/6CL3, 6CL3, 6CL3/6CK3, 6DW4, 6DW4A, 6DW4B
6BS3A	9HP	6BS3, 6CK3, 6CK3/6CL3, 6CL3, 6CL3/6CK3, 6DW4, 6DW4A, 6DW4B
6BS4	-	EC93
6BS5	9BK	-
6BS7	9BB	8D7
6BS8	9AJ	6BC8, 6BC8/6BQ7A, 6BC8/6BZ8, 6BK7*, 6BK7A*, 6BK7B*, 6BQ7, 6BQ7A, 6BQ7A/6BZ7, 6BQ7A/6BZ7/6BS8, 6BS8/6BK7B, 6BX8, 6BZ7, 6BZ7/6BQ7A, 6BZ8, 6HK8, 6RHH2, X155
6BS8/6BK7B	9AJ	6BC8, 6BC8/6BQ7A, 6BC8/6BZ8, 6BK7*, 6BK7A*, 6BK7B*, 6BQ7, 6BQ7A, 6BQ7A/6BZ7, 6BQ7A/6BZ7/6BS8, 6BS8, 6BX8, 6BZ7, 6BZ7/6BQ7A, 6BZ8, 6HK8, 6RHH2, X155
6BT4	8HA	66KU, EZ40, U150, U718*, UU9*, V61
6BT6	7BT	6AQ6*, 6AV6, 6AV6/EBC91, 6AT6, 6AT6/6BK6, 6BC32, 6BK6, DH77, EBC90, EBC91
6BT8	9FE	-
6BU4	8GC	-
6BU5	8FP	-
6BU6	7BT	6BF6, 6BF6/6BU6
6BU8	9FG	6BU8A, 6GS8, 6HS8, 6KF8
6BU8A	9FG	6BU8, 6GS8, 6HS8, 6KF8
6BV7	9BU	-
6BV8	9FJ	-
6BV11	12HB	-
6BW3	12FX	6AX3, 6BE3, 6BE3/6BZ3, 6BE3A, 6BZ3, 6CD3, 6CD3/6CG3, 6CE3, 6CE3/6CD3/6DT3, 6CG3, 6CG3/6BW3, 6GC3/6BW3/6DQ3, 6CG3/6CD3, 6GC3/6CE3/6CD3/6BW3, 6DQ3, 6DT3
6BW4	9DJ	-
6BW6	9AM	6061
6BW7	9AQ	6BX6, 6BX6/EF80, 6EJ7*, 6EL7, 6F23, 6HM6#, 6JC6A*, 8D6, EF80
6BW8	9HK	-
6BW11	12HD	-
6BX4	5BS	6AV4*, EZ91*
6BX6	9AQ	6BW7, 6BX6/EF80, 6BY7, 6BY7/EF85, 6EL7, 6EC7*, 8D6, 64SPT, EF80, EF85, EF85/6BY7, Z152, Z719
6BX6/EF80	9AQ	6BX6, 6BW7, 6BY7, 6BY7/EF85, 6EL7, 8D6, 64SPT, EF80, EF85, EF85/6BY7, Z152, Z719
6BX7	8BD	6BL7GT, 6BL7GTA, 6BX7GT, 6DN7
6BX7GT	8BD	6BL7GT, 6BL7GTA, 6DN7
6BX8	9AJ	6BC8, 6BC8/6BQ7A, 6BC8/6BZ8, 6BK7*, 6BK7A*, 6BK7B*, 6BQ7, 6BQ7A, 6BQ7A/6BZ7, 6BQ7A/6BZ7/6BS8, 6BS8, 6BS8/6BK7B, 6BZ7, 6BZ7/6BQ7A, 6BZ8, 6HK8, 6RHH2, ECC180, X155
6BY4	-	7077
6BY5	6CN	6BY5G, 6BY5GA
6BY5G	6CN	6BY5GA

Tube Type	Basing	Replacement
6BY5GA	6CN	6BY5G
6BY6	7CH	6BE6, 6BE6/5750, 6BE6/EK90, 6BE6W, 6CS6, 6CS6/EH90, 6CS6/EH90, 6H31, 5750, 5750/6BE6W, 5915, 7036, EH90, EK90
6BY7	9AQ	6BX6, 6BX6/EF80, 6BY7/EF85, 6BY8, 6EC7*, 6F19, 6F26, EF80, EF85, EF85/6BY7, W719
6BY7/EF85	9AQ	6BX6, 6BX6/EF80, 6BY7, 6BY8, 6F19, 6F26, EF80, EF85, EF85/6BY7, W719
6BY8	9FN	-
6BY11	12EZ	-
6BZ3	12FX	6BE3, 6BE3/6BZ3, 6BE3A, 6BW3*, 6CD3*, 6CD3/6CG3*, 6CE3*, 6CE3/6CD3/6DT3*, 6CE3/6CD3*, 6CG3*, 6CG3/6BW3*, 6CG3/6BW3/6DQ3*, 6CG3/6CD3*, 6CG3/6CE3/6CD3/6BW3*, 6DT3*
6BZ6	7CM	6BA6#, 6BA6/5749/6660#, 6BA6/EF93#, 6BZ6/6JH6, 6DC6, 6GM6*, 6HQ6, 6JH6, 5749#, 6660#, 6660/6BA6#, EF93#
6BZ6/6JH6	7CM	6BA6#, 6BA6/5749/6660#, 6BA6/EF93#, 6BZ6, 6DC6, 6GM6*, 6HQ6, 6JH6, 5749#, 6660#, 6660/6BA6#, EF93#
6BZ7	9AJ	6BC8, 6BC8/6BQ7A, 6BC8/6BZ8, 6BK7*, 6BK7A*, 6BK7B*, 6BQ7, 6BQ7A, 6BQ7A/6BZ7, 6BQ7A/6BZ7/6BS8, 6BS8, 6BS8/6BK7B, 6BX8, 6BZ7/6BQ7A, 6BZ8, 6HK8, X155
6BZ7/6BQ7A	9AJ	6BC8, 6BC8/6BQ7A, 6BC8/6BZ8, 6BK7*, 6BK7A*, 6BK7B*, 6BQ7, 6BQ7A, 6BQ7A/6BZ7, 6BQ7A/6BZ7/6BS8, 6BS8, 6BS8/6BK7B, 6BX8, 6BZ7, 6BZ8, 6HK8, X155
6BZ8	9AJ	6BC8, 6BC8/6BQ7A, 6BC8/6BZ8, 6BK7*, 6BK7A*, 6BK7B*, 6BQ7, 6BQ7A, 6BQ7A/6BZ7, 6BQ7A/6BZ7/6BS8, 6BS8, 6BS8/6BK7B, 6BX8, 6BZ7, 6BZ7/6BQ7A, 6HK8, X155
6C4	6BG	6C4/6135, 6C4/EC90, 6C4W, 6C4WA, 6100, 6100/6CWA, 6135*, EC90, L77, M8080, QA2401, QL77
6C4/6135	6BG	6C4, 6C4/EC90, 6C4W, 6C4WA, 6100, 6100/6CWA, 6135, EC90, L77, M8080, QA2401, QL77
6C4/EC90	6BG	6C4, 6C4/6135, 6C4W, 6C4WA, 6100, 6100/6CWA, 6135, EC90, L77, M8080, QA2401, QL77
6C4A	6BG	6C4, 6C4/6135, 6C4/EC90, 6C4WA, 6100, 6100/6CWA, 6135*, EC90
6C4W	6BG	6C4, 6C4/6135, 6C4/EC90, 6C4WA, 6100, 6100/6CWA, 6135*, EC90
6C4WA	6BG	6C4#, 6C4/6135#, 6C4/EC90#, 6100, 6100/6C4WA, 6135*, EC90
6C5	6Q	6C5G, 6C5GT, 6C5GT/G, 6C5MG, 6J5, 6J5G, 6J5GT, 6J5GT/G, 6J5GTX, 6J5GX, 6J5MG, 6L5G, L63, OSW3112, WT390
6C5G	6Q	6C5, 6C5GT, 6C5GT/G, 6C5MG, 6J5, 6J5G, 6J5GT, 6J5GT/G, 6J5GTX, 6J5GX, 6J5MG, 6L5G
6C5GT	6Q	6C5, 6C5G, 6C5GT/G, 6C5MG, 6J5, 6J5G, 6J5GT, 6J5GT/G, 6J5GTX, 6J5GX, 6J5MG, 6L5G
6C5GT/G	6Q	6C5, 6C5G, 6C5GT, 6C5MG, 6J5, 6J5G, 6J5GT, 6J5GT/G, 6J5GTX, 6J5GX, 6J5MG, 6L5G
6C5MG	6Q	6C5, 6C5G, 6C5GT, 6C5GT/G, 6J5, 6J5G, 6J5GT, 6J5GT/G, 6J5GTX, 6J5GX, 6J5MG, 6L5G
6C6	6F	77, 57A*, 57AS*, 1221, 1223, 7700, WTT131
6C7	7G	-
6C8	8G	6C8G, 6F8G*#
6C8G	8G	6C8#, 6F8G*#
6C9	10F	6CU7*, 6CU7/ECH42*, ECH42*
6C10	12BQ	6CU7*, 6CU7/ECH42*, 6K11#, 6Q11!#
6C12	9CA	6AJ8, 6AJ8/ECH81, ECH81, ECH81/6AJ8
6C15	8GW	6CJ5, 6CJ5/EF41, EF41
6C16	9DC	6BL8, 6BL8/6LN8, 6BL8/ECF80, 6LN8, ECF80, ECF80/6BL8
6C18	9KN	6GV7
6C25	-	6973
6C31	8K	6K8
6CA4	9M	6CA4/EZ81, EZ81, EZ81/6CA4, U709, UU12
6CA4/EZ81	9M	6CA4, EZ81, EZ81/6CA4, U709, UU12
6CA5	7CV	6AS5*, 6CA5/6EH5, 6EH5
6CA5/6EH5	7CV	6AS5*, 6CA5, 6EH5
6CA7	8EP	6CA7/EL34, 7D11, 12E13, EL34, EL34/6CA7, KT77, KT88
6CA7/EL34	8EP	6CA7, 7D11, 12E13, EL34, EL34/6CA7, KT77, KT88
6CA11	12HN	-
6CB5	8GD	6CB5A, 6CL5
6CB5A	8GD	6CL5
6CB6	7CM	6AG5#, 6AG5/EF96#, 6AU6#, 6AU6/6136#, 6AU6A#, 6AU6A/EF94#, 6AW6, 6BC5#, 6BC5/6CE5#, 6BH6#, 6BH6/6265/6661#, 6CB6A, cont'd

Tube Type	Basing	Replacement
6CB6, cont'd		6CB6A/6CF6, 6CE5#, 6CE5/6BC5#, 6CF6, 6DC6, 6DE6, 6DK6, 6HQ6, 6HS6*#, 6136#, 6265#, 6661#, 6661/6BH6#, 6676, 6676/6CB6A, 7732, 8425A#, 8425A/6AU6A#, EF94#, EF96, EF190
6CB6A	7CM	6AU6A#, 6AU6A/EF94#, 6BC5#, 6BC5/6CE5#, 6CB6*!, 6CB6A/6CF6, 6CE5#, 6CE5/6BC5#, 6CF6*!, 6DC6*!, 6DE6*!, 6DK6*!, 6HQ6*!, 6HS6*#, 6136#, 6676, 6676/6CB6A, 8425A#, 8425A/6AU6A#, EF94#, EF190
6CB6A/6CF6	7CM	6AU6A#, 6AU6A/EF94#, 6BC5#, 6BC5/6CE5#, 6CB6*!, 6CB6A, 6CE5#, 6CE5/6BC5#, 6CF6*!, 6DC6*!, 6DE6*!, 6DK6*!, 6HQ6*!, 6HS6*#, 6136#, 6676, 6676/6CB6A, 8425A#, 8425A/6AU6A#, EF94#, EF190
6CB11	-	-
6CC10	-	5692
6CC31	7BF	6J6, 6J6A
6CC42	-	5670
6CC43	9DE	6AQ8, 6AQ8/ECC85, ECC85, ECC85/6AQ8
6CD3	12FX	6BW3, 6CD3/6CG3, 6CE3, 6CE3/6CD3, 6CE3/6CD3/6DT3, 6CG3, 6CG3/6BW3, 6CG3/6BW3/6DQ3, 6CG3/6CD3, 6CG3/6CE3/6CD3/6BW3, 6DT3
6CD3/6CG3	12FX	6BW3, 6CD3, 6CE3, 6CE3/6CD3, 6CD3, 6CE3/6CD3/6DT3, 6CG3, 6CG3/6BW3, 6CG3/6BW3/6DQ3, 6CG3/6CD3, 6GC3/6CE3/6CD3/6BW3, 6DT3
6CD6	5BT	6CD6G, 6CD6G/6DN6, 6CD6GA, 6CD6GA/6EX6, 6DN6, 6EX6
6CD6G	5BT	6CD6G/6DN6, 6CD6GA, 6CD6GA/6EX6, 6DN6, 6EX6
6CD6G/6DN6	5BT	6CD6G, 6CD6GA, 6CD6GA/6EX6, 6DN6, 6EX6
6CD6GA	5BT	6CD6G, 6CD6G/6DN6, 6CD6GA/6EX6, 6DN6, 6EX6
6CD6GA/6EX6	5BT	6CD6G, 6CD6G/6DN6, 6CD6GA, 6CD6GA, 6DN6, 6EX6
6CD7	8EV	6AF7G*, 6CD7/EM34, 6CD7/EM34-35, 64ME, EM34, EM34/6CD7, EM34-35, EM34-35/6CD7
6CD7/EM34	8EV	6CD7, 6CD7/EM34-35, EM34, EM34/6CD7, EM34-35, EM34-35/6CD7, 64ME
6CD7/EM34-35	8EV	6CD7, 6CD7/EM34, EM34, EM34/6CD7, cont'd

Tube Type	Basing	Replacement
6CD7/EM34-35, cont'd		EM34-35, EM34-35/6CD7, 64ME
6CE3	12GK	6BW3, 6CD3, 6CD3/6CG3, 6CE3/6CD3, 6CE3/6CD3/6DT3, 6CG3, 6CG3/6BW3, 6CG3/6BW3/6DQ3, 6CG3/6CD3, 6GC3/6CE3/6CD3/6BW3, 6DT3, 6DU3
6CE3/6CD3	12GK	6BW3, 6CD3, 6CD3/6CG3, 6CE3, 6CE3/6CD3/6DT3, 6CG3, 6CG3/6BW3, 6CG3/6BW3/6DQ3, 6CG3/6CD3, 6GC3/6CE3/6CD3/6BW3, 6DT3, 6DU3
6CE3/6CD3/6DT3	12GK	6BW3, 6CD3, 6CD3/6CG3, 6CE3, 6CE3/6DC3, 6CE3/6DT3, 6CG3, 6CG3/6BW3, 6CG3/6BW3/6DQ3, 6CG3/6CD3, 6GC3/6CE3/6CD3/6BW3, 6DT3, 6DU3
6CE5	7BD	6AG5*!, 6AG5/EF96*!, 6AK5*, 6AK5/5654*, 6AK5/EF95*, 6AU6A#, 6AU6A/EF94#, 6BC5*!, 6BC5/6CE5, 6CB6A#, 6CB6A/6CF6#, 6CE5/6BC5, 6CF6#, 6CY5*#, 6EA5*#, 6EV5*#, 6HS6*#, 5654#, 6676#, 6676/6CB6A#, 8425A#, 8425A/6AU6A#, EF94#, EF95*, EF96
6CE5/6BC5	7BD	6AG5*, 6AG5/EF96*!, 6AK5*, 6AK5/5654*, 6AK5/EF95*, 6AU6A#, 6AU6A/EF94#, 6BC5*!, 6BC5/6CE5, 6CB6A#, 6CB6A/6CF6#, 6CE5, 6CF6#, 6CY5*#, 6EA5*#, 6EV5*#, 6HS6*#, 5654#, 6676#, 6676/6CB6A#, 8425A#, 8425A/6AU6A#, EF95*, EF96
6CF6	7CM	6AG5#, 6AG5/EF96#, 6AU6#, 6AU6/6136#, 6AU6A#, 6AU6A/EF94#, 6AW6, 6BC5#, 6BC5/6CE5#, 6BH6*, 6BH6/6265/6661*, 6CB6, 6CB6A, 6CB6A/6CF6, 6CE5#, 6CE5/6BC5#, 6DC6, 6DE6, 6DK6, 6HS6*#, 6136#, 6265*, 6661*, 6661/6BH6*, 6676, 6676/6CB6A, 8425A#, 8425A/6AU6A#, EF94#, EF96#, EF190
6CF8	-	6267
6CG3	12HF	6BW3, 6CD3, 6CD3/6CG3, 6CE3, 6CE3/6CD3/6DT3, 6CG3/6BW3, 6CG3/6BW3/6DQ3, 6CG3/6CD3, 6CG3/6CE3/6BW3, 6CG3/6CE3/6CD3/6BW3, 6DQ3, 6DT3
6CG3/6BW3	12HF	6BW3, 6CD3, 6CD3/6CG3, 6CE3, 6CE3/6CD3/6DT3, 6CG3, 6CG3/6BW3/6DQ3, 6CG3/6CD3, cont'd

Tube Type	Basing	Replacement
6CG3/6BW3,cont'd		6CG3/6CE3/6BW3,6CG3/6CE3/6CD3/6BW3,6DQ3, 6DT3
6CG3/6BW3/6DQ3	12HF	6BW3, 6CD3, 6CD3/6CG3, 6CE3, 6CE3/6CD3/6DT3, 6CG3, 6CG3/6BW3, 6CG3/6CD3, 6CG3/6CE3/6BW3, 6CG3/6CE3/6CD3/6BW3, 6DQ3, 6DT3
6CG3/6CD3	12HF	6BW3, 6CD3, 6CD3/6CG3, 6CE3, 6CE3/6CD3/6DT3, 6CG3, 6CG3/6BW3, 6CG3/6BW3/6DQ3, 6CG3/6CE3/6BW3, 6CG3/6CE3/6CD3/6BW3, 6DQ3, 6DT3
6CG3/6CE3/6CD3/6BW3	12HF	6BW3, 6CD3, 6CD3/6CG3, 6CE3, 6CE3/6CD3/6DT3, 6CE3/6CD3, 6CG3, 6CG3/6BW3, 6CG3/6BW3/6DQ3, 6CG3/6CD3, 6CG3/6CE3/6BW3, 6DQ3, 6DT3
6CG6	7BK	6BA6, 6BA6/5749/6660, 6BA6/EF93, 6BD6, 5749, 6660, 6660/6BA6, EF93
6CG7	9AJ	6CG7/6FQ7, 6FQ7, 6FQ7/6CG7
6CG7/6FQ7	9AJ	6CG7, 6FQ7, 6FQ7/6CG7
6CG8	9GF	6AT8#, 6AT8A#, 6BE8#, 6BE8A#, 6BR8#, 6BR8A#, 6BR8A/6FV8#, 6BR8A/6FV8A#, 6CG8A, 6FG7, 6FV8#, 6FV8A#, 6FV8A/6BR8A#
6CG8A	9GF	6AT8A#, 6BE8A#, 6BR8A#, 6BR8A/6FV8#, 6BR8A/6FV8A#, 6CG8*!, 6FG7, 6FV8#, 6FV8A#, 6FV8A/6BR8A#
6CH3	9SD	6CH3/6CJ3, 6CJ3, 6CJ3/6CH3, 6CJ3/6DW4B, 6DW4B
6CH3/6CJ3	9SD	6CH3, 6CJ3, 6CJ3/6CH3, 6CJ3/6DW4B, 6DW4B
6CH6	9BA	6CH6/EL821, 7D10, 6132, 7499, EF82, EL821, EL822
6CH6/EL821	9BA	6CH6, 7D10, 6132, 7499, EF82, EL821, EL822
6CH7	9FC	6CX7
6CH8	9FT	6CU8#
6CH40	-	6AJ8, 6AJ8/ECH81, ECH81, ECH81/6AJ8
6CJ3	9SD	6CH3, 6CH3/6CJ3, 6CJ3/6CH3, 6CJ3/6DW4B, 6DW4B
6CJ3/6CH3	9HP	6CH3, 6CH3/6CJ3, 6CJ3, 6CJ3/6DW4B, 6DW4B
6CJ3/6DW4B	-	6CH3, 6CH3/6CJ3, 6CJ3, 6CJ3/6CH3, 6DW4B
6CJ5	8GW	6C15, 6CJ5/EF41, 6F16, 7F16, 62VP, EF41, HF61, W150
6CJ5/EF41	8GW	6C15, 6CJ5, 6F16, 7F16, 62VP, EF41, HF61, W150
6CJ6	9AS	6CJ6/EL81, 6DR6, EL81, EL820
6CJ6/EL81	9AS	6CJ6, 6DR6, EL81, EL820
6CJ8	-	-
6CK3	9HP	6CH3*, 6CH3/6CJ3*, 6CJ3*, 6CJ3/6CH3*, 6CJ3/6DW4B*, 6CK3/6CL3, 6CL3, 6CL3/6CK3, 6DW4, 6DW4A, 6DW4B
6CK3/6CL3	9HP	6CH3*, 6CH3/6CJ3*, 6CJ3*, 6CJ3/6CH*, 6CJ3/6DW4B3*, 6CK3, 6CL3, 6CL3/6CK3, 6DW4, 6DW4A, 6DW4B
6CK4	8JB	-
6CK5	8GW	6CK5/EL41, 67PT, BF61, EL41, N150
6CK5/EL41	8GW	6CK5, 67PT, BF61, EL41, N150
6CK6	9AR	6CK6/EL83, EL83, EL803#, EL820
6CK6/EL83	9AR	6CK6, EL83, EL803#, EL820
6CL3	9HP	6CH3*, 6CH3/6CJ3*, 6CJ3*, 6CJ3/6CH3*, 6CJ3/6DW4B*, 6CK3, 6CK3/6CL3, 6CL3/6CK3, 6DW4A, 6DW4B
6CL3/6CK3	9HP	6CH3*, 6CH3/6CJ3*, 6CJ3*, 6CJ3/6CH3*, 6CJ3/6DW4B*, 6CK3, 6CK3/6CL3, 6CL3, 6DW4A, 6DW4B
6CL5	8GD	-
6CL6	9BV	6CL6/6677, 6L43, 2014, 6197, 6297, 6677, 6677/6CL6
6CL6/6677	9BV	6CL6, 6L43, 2017, 6197, 6297, 6677
6CL8	9FX	6BE8A#, 6BR8A#, 6BR8A/6FV8#, 6BR8A/6FV8A#, 6CL8A, 6FV8#, 6FV8A#, 6FV8A/6BR8A#
6CL8A	9FX	6BE8A#, 6BR8A#, 6BR8A/6FV8#, 6BR8A/6FV8A#, 6CL8, 6FV8#, 6FV8A#, 6FV8A/6BR8A#
6CM3	9HP	6CM3/6DN3, 6DN3
6CM3/6DN3	9HP	6CM3, 6DN3
6CM4	9KG	6CM4/EC86, EC86, E86C*, EC806S*
6CM4/EC86	9KG	6CM4, EC86, E86C*, EC806S*
6CM5	8GT	6CM5/EL36, 6GB7, EL36, EL360
6CM5/EL36	8GT	6CM5, 6GB7, EL36, EL360
6CM6	9CK	-
6CM7	9ES	-
6CM8	9FZ	6CR8#, 6CS8#
6CN5	9CV	6CW5, 6CW5/EL86, EL86, EL86/6CW5
6CN6	8EW	5P29, EL38
6CN7	9EN	-
6CQ4	4CG	6CQ4/6DE4, 6DE4, 6DE4/6CQ4, 6DT4*
6CQ4/6DE4	4CG	6CQ4, 6DE4, 6DE4/6CQ4, 6DT4*
6CQ6	7DR	6CQ6/EF92, 6F21, 9D6, 6065, EF92, M8161, QA2400, QW77, V177, V884, VP6, W77
6CQ6/EF92	7DR	6CQ6, 6F21, 9D6, 6065, EF92, M8161, QA2400, QW77, V177, V884, VP6, W77

Tube Type	Basing	Replacement
6CQ8	9GE	6AX8#, 6EA8#, 6KD8#, 6U8A#, 6U8A/6AX8/6KD8#
6CR4	9BX	6AJ4*#
6CR5	9HC	-
6CR6	7EA	-
6CR8	9GJ	6CM8#, 6CS8#, 6KZ8#
6CS5	9CK	6DW5, 6CM6*
6CS6	7CH	6BE6, 6BE6/5750, 6BE6/EK90, 6BE6W, 6BY6, 6CS6/EH90, 6H31, 5750, 5750/6BE6W, EH90, EK90
6CS6/EH90	7CH	6BE6, 6BE6/5750, 6BE6/EK90, 6BE6W, 6BY6, 6CS6, 6H31, 5750, 5750/6BE6W, EH90, EK90
6CS7	9EF	6DA7*
6CS8	9FZ	6CM8#, 6CR8#, 6KZ8#
6CT3	9RX	-
6CT7	8GX	6CT7/EAF42, D61#, EAF42, WD150
6CT7/EAF42	8GX	6CT7, D61#, EAF42, WD150
6CU5	7CV	6AS5*
6CU6	6AM	6BQ6, 6BQ6GA, 6BQ6GA/6CU6, 6BQ6GTB, 6BQ6GTB/6CU6, 6DQ6, 6DQ6A, 6DQ6B, 6DQ6B/6GW6, 6FH6, 6G-B9, 6GW6
6CU7	8GY	6C9*, 6CU7/ECH42, 6C10*, 62TH, ECH42, ECH43, ECH113, X150*
6CU7/ECH42	8GY	6C9*, 6CU7, 6C10*, 62TH, ECH42, ECH43, ECH113, X150*
6CU8	9GM	-
6CV7	8GZ	6CV7/EBC41, 6LD3, 62DDT, DH150, DH718, EBC41
6CV7/EBC41	8GZ	6CV7, 6LD3, 62DDT, DH150, DH718, EBC41
6CW4	12AQ	6CW4/6DS4, 6DS4, 7895
6CW4/6DS4	12AQ	6CW4, 6DS4, 7895
6CW5	9CV	6CW5/EL86, EL86
6CW5/EL86	9CV	6CW5, EL86
6CW7	9DD	6CW7/ECC84, 6FC7, 6L16, ECC84
6CW7/ECC84	9DD	6CW7, 6CF7, 6L16, ECC84
6CX7	9FC	6CH7
6CX8	9DX	6AU8*, 6AU8A*, 6AU8A/6BH8*, 6AW8*, 6AW8A*, 6BH8*, 6CX8/6JA8, 6EB8, 6GN8, 6GN8/6EB8, 6HF8, 6JA8, 6JE8, 6JV8*, 2081*, 2081/6AW8A*
6CX8/6JA8	9DX	6AU8*, 6AU8A*, 6AU8A/6BH8*, 6AW8*, 6AW8A*, 6BH8*, 6CX8, 6EB8, 6GN8, 6GN8/6EB8, 6HF8, 6JA8, 6JE8, 6JV8*, 2081*, 2081/6AW8A*
6CY5	7EW	6AG5*#, 6AG5/EF96*#, 6AK5*#, 6AK5/5654*#, 6AK5/EF95*#, 6BC5*#, 6BC5/6CE5*#, 6CE5*#, 6CE5/6BC5*#, 6EA5, 6EV5, 6HS6*#, 5654*#, 7717, 8113, EF95*#, EF96*#
6CY7	9LG	-
6CZ5	9HN	6DW5*
6CZ7	8ET	-
6D1	-	2B35, 2B35/EA50, EA50
6D2	-	6AL5, 6AL5/6EB5, 6AL5/5726/6663, 6AL5/EAA91, 6AL5W, 6EB5, 5726#, 5726/6AL5W, 5726/6AL5W/6097, 6097, 6663, 6663/6AL5, EAA91, EAA91/6AL5
6D4	5AY	EN93
6D5	6Q	6D5G, 6D5MG
6D5G	6Q	6D5, 6D5MG
6D5MG	6Q	6D5, 6D5G
6D6	6F	78, 58AS*
6D7	7H	-
6D8	8A	6A8*, 6A8G*, 6A8GT*, 6A8GTX*, 6A8MG*, 6D8G, X73*, X73M*
6D8G	8A	6A8*, 6A8G*, 6A8GT*, 6A8GTX*, 6A8MG*, 6D8
6D10	12BQ	6AV11*#, 6C10#, 6K11*#, 6Q11*#
6DA4	4CG	6AU4GT*, 6AU4GTA*, 6AX4GTA, 6AX4GTB, 6CQ4*, 6CQ4/6DE4*, 6DA4A, 6DA4A/6DM4, 6DA4A/6DM4A, 6DE4*, 6DE4/6CQ4*, 6DM4, 6DM4A, 6DM4A/6DA4, 6DM4A/6DQ4, 6DQ4, 6DT4
6DA4A	4CG	6CQ4*, 6CQ4/6DE4*, 6DE4*, 6DE4/6CQ4*, 6DA4A/6DM4, 6DA4A/6DM4A, 6DM4A, 6DM4A/6DA4, 6DM4A/6DQ4, 6DQ4, 6DT4
6DA4A/6DM4	4CG	6DA4A, 6DA4A/6DM4A, 6DE4*, 6DE4/6CQ4*, 6DM4, 6DM4A, 6DM4A/6DA4, 6DM4A/6DQ4, 6DQ4, 6DT4
6DA4A/6DM4A	4CG	6DA4A, 6DA4A/6DM4, 6DM4, 6DM4A, 6DM4A/6DA4, 6DM4A/6DQ4, 6DQ4, 6DT4
6DA5	9DB	6BR5, 6BR5/EM80, 6DA5/EM81, 65ME, EM80, EM80/6BR5, EM81, EM81/6DA5
6DA5/EM81	9DB	6BR5, 6BR5/EM80, 6DA5, 65ME, EM80, EM80/6BR5, EM81, EM81/6DA5
6DA6	9AU	6DA6/EF89, EF89, EF89/6DA6
6DA6/EF89	9AU	6DA6, EF89, EF89/6DA6
6DA7	9EF	-
6DB5	9GR	-
6DB6	7CM	6AS6*, 6AS6W#, 5725#, 5725/6AS6W#, 6187*, 6187/6AS6*, 6954
6DC6	7CM	6AG5#, 6AG5/EF96#, 6AU6#, 6AU6/6136#, 6AU6A#, 6AU6A/EF94#, 6AW6, 6BC5#, 6BC5/6CE5#, 6BH6*, 6BH6/6265/6661*, 6BZ6, 6BZ6/6JH6, 6CB6, 6CB6A, cont'd

Tube Type	Basing	Replacement
6DC6, cont'd		6CB6A/6CF6, 6CE5#, 6CE5/6BC5#, 6CF6, 6DE6, 6DK6, 6JH6, 6136#, 6265*, 6661*, 6661/6BH6*, 6676, 6676/6CB6A, 8425A#, 8425A/6AU6A#, EF94#, EF96#, EF190
6DC8	9HE	6AD8, 6AD8/6DC8, 6FD12, 6N8, EBF85, EBF89
6DC8/EBF89	9HE	6AD8, 6AD8/6DC8, 6DC8, 6N8, 6FD12, EBF85, EBF89
6DE4	4CG	6CQ4, 6CQ4/6DE4, 6DA4A**, 6DA4A/6DM4*, 6DA4A/6DM4A*, 6DE4/6CQ4, 6DM4*, 6DM4A*, 6DM4A/6DQ4*, 6DQ4*, 6DT4*
6DE4/6CQ4	4CG	6CQ4, 6CQ4/6DE4, 6DA4A*, 6DA4A/6DM4*, 6DA4A/6DM4A*, 6DE4, 6DM4*, 6DM4A*, 6DM4A/6DQ4*, 6DQ4*, 6DT4*
6DE6	7CM	6AG5#, 6AG5/EF96#, 6AU6#, 6AU6/6136#, 6AU6A#, 6AU6A/EF94#, 6AW6, 6BC5#, 6BC5/6CE5#, 6BH6*, 6BH6/6265/6661*, 6CE5#, 6CE5/6BC5#, 6CB6, 6CB6A, 6CB6A/6CF6, 6CF6, 6DC6, 6DK6, 6HQ6, 6136#, 6265*, 6661*, 6661/6BH6*, 6676, 6676/6CB6A, 8425A#, 8425A/6AU6A#, EF94#, EF96#
6DE7	9HF	6EW7
6DF12	-	-
6DG6	7S	6DG6GT, 6EF6*, 6EY6*, 6EZ5*, 6W6, 6W6GT
6DG6GT	7S	6EF6*, 6EY6*, 6EZ5*, 6W6, 6W6GT
6DG6GT/6W6GT	7S	6DG6GT, 6W6GT
6DG7	9BA	6DG7/EM85, EF89F
6DG7/EM85	9BA	6DG7, EM85
6DHH12	-	-
6DHH13	-	6FX7
6DJ8	9DE	6DJ8/ECC88, 6ES8#, 6ES8/ECC189#, 6FW8*, 6KN8*, 6922*, ECC88, ECC189, E88CC*
6DJ8/ECC88	9AJ	6DJ8, 6ES8, ECC88, ECC189
6DK3	9SG	6DK3/6DL3, 6DL3
6DK3/6DL3	9SG	6DK3, 6DL3
6DK6	7CM	6BC5#, 6BC5/6CE5#, 6CB6, 6CB6A, 6CB6A/6CF6, 6CE5#, 6CE5/6BC5#, 6CF6, 6DC6, 6DE6, 6HQ6, 6HS6*#, 6676, 6676/6CB6A, 8136
6DL3	9GD	6DK3, 6DK3/6DL3
6DL4	9NY	6DL4/EC88, EC88
6DL4/EC88	9NY	6DL4, 8255, EC88
6DL5	7DQ	6BM5*, 6DL5/EL95, EL95
6DL5/EL95	7DQ	6BM5*, 6DL5, EL95
6DL7	8EV	-
6DM4	4CG	6CQ4*, 6CQ4/6DE4*, 6DA4, 6DA4A, 6DA4A/6DM4, 6DA4A/6DM4A, cont'd
6DM4, cont'd		6DE4*, 6DE4/6CQ4*, 6DM4A, 6DM4A/6DA4, 6DM4A/6DQ4, 6DQ4, 6DT4
6DM4A	4CG	6CQ4*, 6CQ4/6DE4*, 6DA4, 6DA4A, 6DA4A/6DM4, 6DA4A/6DM4A, 6DM4A/6DA4, 6DM4A/6DQ4, 6DQ4, 6DT4
6DM4A/6DA4	4CG	6DA4, 6DA4A, 6DA4A/6DM4, 6DA4A/6DM4A, 6DM4A, 6DM4A/6DQ4, 6DQ4, 6DT4
6DM4A/6DA4A	4CG	6DA4A, 6DA4A/6DM4, 6DA4A/6DM4A, 6DM4A, 6DM4A/6DA4, 6DM4A/6DQ4, 6DQ4, 6DT4
6DM4A/6DQ4	4CG	6DA4, 6DA4A, 6DA4A/6DM4, 6DA4A/6DM4A, 6DM4A, 6DM4A/6DA4, 6DQ4, 6DT4
6DM8	-	-
6DN3	9HP	6CM3, 6CM3/6DN3
6DN6	5BT	6CD6G, 6CD6G/6DN6, 6CD6GA, 6CD6GA/6EX6, 6EX6
6DN7	8BD	-
6DQ3	12HF	6BW3, 6CD3, 6CD3/6CG3, 6CE3, 6CE3/6CD3/6DT3, 6CG3, 6CG3/6BW3, 6CG3/6BW3/6DQ3, 6CG3/6CD3, 6CG3/6CE3/6CD3/6BW3, 6DQ3A, 6DT3
6DQ3/6DT3	12GF	6BW3, 6CD3, 6CD3/6CG3, 6CE3, 6CE3/6CD3/6DT3, 6CG3, 6CG3/6BW3, 6CG3/6BW3/6DQ3, 6CG3/6CD3, 6CG3/6CE3/6CD3/6BW3, 6CG3/6CD3, 6DQ3, 6DQ3A, 6DT3
6DQ3A	12HF	6DQ3A/6DU3, 6DU3
6DQ3A/6DU3	12HF	6DQ3A, 6DU3
6DQ4	4CG	6AX4GTB, 6CQ4*, 6CQ4/6DE4*, 6DA4A, 6DA4A/6DM4, 6DA4A/6DM4A, 6DE4*, 6DE4/6CQ4, 6DM4A/6DQ4, 6DT4
6DQ5	8JC	-
6DQ6	6AM	6DQ6A, 6DQ6B, 6DQ6B/6GW6, 6FH6, 6GW6, 6GW6/6DQ6B
6DQ6A	6AM	6DQ6B, 6DQ6B/6GW6, 6FH6, 6GB9, 6GW6, 6GW6/6DQ6B
6DQ6B	6AM	6DQ6B/6GW6, 6GW6*!, 6GW6/6DQ6B
6DQ6B/6GW6	6AM	6DQ6B, 6GW6*!, 6GW6/6DQ6B
6DR4	6BG	-
6DR6	9AS	6JC6, EL81, EL820
6DR7	9HF	6FD7, 6FR7
6DR8	9HE	6DR8/EBF83, EBF83
6DR8/EBF83	9HE	6DR8, EBF83
6DS4	12AQ	6CW4, 6CW4/6DS4, 7895
6DS5	7BZ	6AQ5*, 6AQ5/6005/6669*, 6AQ5A*, 6BM5*, 6HG5*, 6005*, 6669*, 6669/6AQ5A*
6DS8	9CA	6DS8/ECH83, ECH83

Tube Type	Basing	Replacement
6DS8/ECH83	9CA	6DS8, ECH83
6DT3	12HF	6BW3, 6CD3, 6CD3/6CG3, 6CE3, 6CE3/6CD3, 6CE3/6CD3/6DT3, 6CG3, 6CG3/6BW3, 6CG3/6BW3/6DQ3, 6CG3/6CD3, 6GC3/6CE3/6CD3/6BW3, 6DQ3, 6DQ3/6DT3, 6DT3
6DT4	4CG	6AU4GTA, 6DA4A, 6CQ4, 6CQ4/6DE4, 6DA4A/6DM4, 6DA4A/6DM4A, 6DE4, 6DE4/6CQ4, 6DM4A, 6DM4A/6DQ4, 6DQ4
6DT5	9HN	6EM5*
6DT6	7EN	6DT6A
6DT6A	7EN	6DT6
6DT8	9AJ	6AT7N
6DU3	12JK	6DQ3A, 6DQ3A/6DU3
6DV4	12EA	-
6DW4	9HP	6CH3, 6CH3/6CJ3, 6CJ3, 6CJ3/6CH3, 6CJ3/6DW4B, 6CK3, 6CK3/6CL3, 6CL3, 6CL3/6CK3, 6DW4, 6DW4A, 6DW4B
6DW4A	9HP	6CH3, 6CH3/6CJ3, 6CJ3, 6CJ3/6CH3, 6CJ3/6DW4B, 6CK3, 6CK3/6CL3, 6CL3, 6CL3/6CK3, 6DW4, 6DW4A, 6DW4B
6DW4B	9HP	6CH3, 6CH3/6CJ3, 6CJ3, 6CJ3/6CH3, 6CJ3/6DW4B, 6CK3, 6CK3/6CL3, 6CL3, 6CL3/6CK3, 6DW4, 6DW4A, 6DW4B
6DW5	9CK	6CM6
6DX4	7DK	6DY4*#, 6DY4A*#
6DX8	9HX	6DX8/ECL84, ECL84
6DX8/ECL84	9HX	6DX8, ECL84
6DY4	7DK	6DY4A, 6DX4*#
6DY4A	7DK	6DY4, 6DX4*#
6DY5	9CV	6BQ5, 6BQ5/EL84, 6CW5*, 6CW5/EL86*, EL82, EL84, EL84/6BQ5, EL86*
6DY7	8JP	6DZ7#
6DZ4	7DK	6AF4#, 6AF4A#, 6AF4A/6DZ4, 6DZ4/6AF4A, 6T4#
6DZ4/6AF4A	7DK	6AF4#, 6AF4A, 6AF4A/6DZ4, 6DZ4, 6T4#
6DZ7	8JP	6DY7#
6DZ8	9JE	6FY8*
6E5	6R	6G5#, 6S5G, OSW3110
6E6	7B	-
6E7	7H	-
6E8	8O	6E8G, ECH35*, TH62*, X61M, X147
6E8G	8O	6E8, ECH35*, TH62*, X61M, X147
6EA4	12FA	6EH4, 6EH4A
6EA4/6EH4A	12FA	6EA4, 6EH4A
6EA5	7EW	6AG5*#, 6AG5/EF96*#, 6AK5*#, 6AK5/5654*#, 6AK5/EF95#, 6BC5*#, 6BC5/6CE5*#, 6CE5*#, 6CE5/6BC5*#, 6CY5, 6EV5, 5654*#, EF95*#, EF96*#
6EA5/6EV5	7EW	6EA5, 6EV5
6EA7	8BD	6EM7, 6EM7/6EA7, 6GL7
6EA7/6EM7	8BD	6EA7, 6EM7
6EA8	9AE	6AX8*, 6CQ8#, 6GH8#, 6KD8#, 6MQ8*, 6U8A*, 6U8A/6AX8/6KD8*
6EA8/6U8A	9AE	6AX8#, 6EA8, 6KD8#, 6U8A#, 6U8A/6AX8/6KD8#
6EB5	6BT	6AL5, 6AL5/6EB5, 6AL5/5726/6663, 6AL5/EAA91, 6AL5W, 6B32, 6D2, 5726, 5726/6AL5W, 5726/6AL5W/6097, 6097, 6663, 6663/6AL5, D2M9, D77, D152, D717, DD6, EAA91, EAA91/6AL5, EB91, EAA901, EAA901S
6EB8	9DX	6AU8*, 6AU8A*, 6AU8A/6BH8*, 6BH8*, 6CX8, 6CX8/6JA8, 6EB8/6GN8, 6GN8, 6GN8/6EB8, 6HF8, 6HZ8*, 6JA8, 6JE8, 6JV8*
6EB8/6GN8	9DX	6AU8*, 6AU8A*, 6AU8A/6BH8*, 6BH8*, 6CX8, 6CX8/6JA8, 6EB8, 6GN8, 6GN8/6EB8, 6HF8, 6HZ8*, 6JA8, 6JE8, 6JV8*
6EC4	6EC4	6EC4/EY500, 6EC4/EY500A, 6EC4A, 6EC4A/EY500, EY500, EY500A
6EC4/EY500	6EC4	6EC4, 6EC4/EY500A, 6EC4A, EY500, EY500A
6EC4/EY500A	6EC4	6EC4, 6EC4/EY500, 6EC4A, EY500, EY500A
6EC4A	6EC4	6EC4/EY500, 6EC4/EY500A, 6EC4A/EY500, EY500, EY500A
6EC4A/EY500	6EC4	6EC4/EY500, 6EC4/EY500A, 6EC4A, EY500, EY500A
6EC7	9AQ	6BX6*, 6BX6/EF80*, 6BY7*, 6BY7/EF85*, 6F18, EF80*, EF85, EF85/6BY7, W739
6ED4	-	ED500
6EF4	12HC	6EF4/6EJ4A, 6EJ4, 6EJ4A
6EF4/6EJ4A	12HC	6EF4, 6EJ4A
6EF6	7S	6DG6GT#, 6EY6*, 6EZ5*, 6W6GT#
6EH4	12FA	6EH4A
6EH4A	12FA	6EA4, 6EH4A
6EH5	7CV	6CA5, 6CA5/6EH5
6EH7	9AQ	6EH7/EF183, 6EJ7, 6F25, 6F29, 6HM6#, 6HT6#, 6JC6#, 6JC6A#, 6JD6#, EF183, EF811
6EH7/EF183	9AQ	6EH7, 6EJ7, 6F25, 6F29, EF811, EF183
6EH8	9JG	6AU8*#, 6AU8A*#, 6AU8A/6BH8*#, 6AW8*#, 6AW8A*#, 6BH8*#, 6X8A#, 2081*#, 2081/6AW8A*#
6EJ4	12HC	6EJ4A
6EJ4A	12HC	6EF4, 6EF4/6EJ4A
6EJ7	9AQ	6EH7, 6EJ7/EF184, 6F24, 6F30, 6HM6#, 6HT6#, 6JC6A#, 6JD6#, EF184, EF814

Tube Type	Basing	Replacement
6EJ7/EF184	9AQ	6EH7, 6EJ7, 6F24, 6F30, EF184, EF814
6EL4	8MW	6BK4C, 6BK4C/6EL4A, 6EL4A, 6EN4
6EL4A	8MW	6BK4C, 6BK4C/6EL4A, 6EN4
6EL7	9AQ	6BW7, 6BX6, 6BX6/EF80, 6EJ7#, 6F23, 6HM6#, 6JC6A#, EF80, EF812, Z749
6EM5	9HN	-
6EM7	8BD	6EA7, 6EA7/6EM7, 6EM7/6EA7, 6GL7
6EM7/6EA7	8BD	6EA7, 6EA7/6EM7, 6EM7, 6GL7
6EN4	8NJ	-
6EQ7	9LQ	6KL8#
6ER5	7FN	6ER5/EC95, 6FY5*, EC95, EC97*
6ER5/EC95	7FN	6ER5, EC95
6ES5	7FN	-
6ES6	7EN	6ES6/6FD6, 6ET6, 6ET6/EF98, 6FD6, EF97, EF98
6ES6/6FD6	7EN	6ES6, 6ET6, 6ET6/EF98, 6FD6, EF97, EF98
6ES8	9DE	6DJ8, 6ES8/ECC189, 6FW8*, 6KN8*, ECC88, ECC189
6ES8/ECC189	9AT	6ES8, 6DJ8, ECC189, ECC88
6ET6	7EN	6DT6A#, 6ES6, 6ET6/EF98, 6FD6, 6GY6/6GX6#, 6HZ6#, EF97, EF98
6ET6/EF98	7EN	6DT6A#, 6ES6, 6ET6, 6FD6, 6GY6/6GX6#, 6HZ6#, EF97, EF98
6ET7	9LT	6KU8#
6EU7	9LS	-
6EU8	9JF	-
6EV5	7EW	6AG5*#, 6AG5/EF96*#, 6AK5*#, 6AK5/5654*#, 6AK5/EF95*#, 6BC5*#, 6BC5/6CE5*#, 6CE5*#, 6CE5/6BC5*#, 6EA5, 6EA5/6EV5, 5654*#, EF95*#, EF96*#
6EV7	9LP	-
6EV8	-	-
6EW6	7CM	6GM6, 6HS6*#
6EW7	9HF	-
6EX6	5BT	6CD6GA, 6CD6GA/6EX6
6EY5	-	6EY5/6EZ5, 6EZ5
6EY5/6EZ5	-	6EY5, 6EZ5
6EY6	7AC	6EZ5*
6EZ5	7AC	6EY5, 6EY5/6EZ5
6EZ8	9KA	-
6F4	7BR	6L4
6F5	5M	6F5G, 6F5GT, 6F5MG, H63
6F5G	5M	6F5, 6F5GT, 6F5MG
6F5GT	5M	6F5, 6F5G, 6F5MG, H63
6F5MG	5M	6F5, 6F5G, 6F5GT
6F6	7S	6F6G, 6F6GT, 6F6GT/G, 6F6MG, 1611, 1621
6F6G	7S	6F6, 6F6GT, 6F6GT/G, 6F6MG, KT63
6F6GT	7S	6F6, 6F6G, 6F6GT/G, 6F6MG
6F6GT/G	7S	6F6, 6F6G, 6F6GT, 6F6MG
6F6MG	7S	6F6, 6F6G, 6F6GT, 6F6GT/G
6F7	7E	6F7S
6F7S	7E	6F7
6F8	8G	6F8G
6F8G	8G	6C8G*#
6F10	-	6AC7, 6AC7/1852, 6AC7/1852/6134, 6AC7/EL34, 1852, 6134, EL34
6F11	-	6AM6, 6AM6/EF91, EF91
6F12	-	6AM6, 6AM6/EF91, EF91
6F15	-	6CJ5, 6CJ5/EF41, EF41
6F16	-	6CJ5, 6CJ5/EF41, EF41
6F17	-	-
6F18	-	6EC7
6F19	-	6BY7, 6BY7/EF85, EF85, EF85/6BY7
6F21	-	6CQ6, 6CQ6/EF92, EF92
6F22	-	6267
6F23	-	6EL7
6F24	-	6EJ7, 6EJ7/EF184
6F25	-	6EH7
6F26	-	6BY7, 6BY7/EF85, EF85, EF85/6BY7
6F29	-	6EH7, 6EH7/EF183
6F30	-	6EJ7, 6EJ7/EF184
6F31	-	6AU6A, 6AU6A/EF94, 6BA6, 6BA6/5749/6660, 6BA6/EF93, 5749, 6660, 6660/6BA6, 8425A, 8425A/6AU6A, EF93, EF94
6F32	-	6AK5, 6AK5/5654, 6AK5/EF95, 5654, EF95
6F33	-	6AS6, 6AS6W, 5725, 5725/6AS6W, 6187, 6187/6AS6
6F35	-	6AJ5
6F36	-	6AH6
6FA7	9MR	-
6FB8	-	-
6FC7	9DD	6CW7, 6CW7/ECC84, ECC84, ECC89
6FD5	-	-
6FD6	7BK	6ES6, 6ES6/6FD6, 6ET6, 6ET6/EF98, EF97, EF98
6FD7	9HF	6FR7
6FD12	-	6DC8
6FE5	8KB	-
6FE8	9AJ	-
6FG5	7GA	6FS5, 6HS6*#
6FG6	9GA	6FG6/EM84, EM84, EM840
6FG6/EM84	9GA	6FG6, EM84, EM840
6FG6G	9GA	6FG6/EM84
6FG7	9GF	6BE8A#, 6BR8A#, 6BR8A/6FV8#, 6BR8A/6FV8A#, 6FV8A#, 6FV8A/6BR8A#
6FH5	7FP	6ES5, 6FQ5*, 6FQ5A*, 6GK5*
6FH6	6AM	6DQ6B, 6DQ6B/6GW6, 6GW6, 6GW6/6DQ6B
6FH8	9KP	-
6FJ7	12BM	-
6FM7	12EJ	-
6FM8	9KR	-
6FN5	8GD	-
6FQ5	7FP	6ES5*, 6FQ5A, 6FQ5A/6GK5, 6GK5, 6GK5/6FQ5A
6FQ5A	7FP	6ES5*, 6FQ5, 6FQ5A/6GK5, 6GK5, 6GK5/6FQ5A
6FQ5A/6GK5	7FP	6ES5*, 6FQ5, 6FQ5A, 6GK5, 6GK5/6FQ5A
6FQ7	9LP	6CG7, 6CG7/6FQ7, 6FQ7/6CG7
6FQ7/6CG7	9LP	6CG7, 6CG7/6FQ7, 6FQ7

Tube Type	Basing	Replacement
6FR7	9HF	6FD7
6FS5	7GA	6FG5, 6GU5*
6FV6	7FQ	-
6FV8	9FA	6BE8A#, 6BR8*!, 6BR8A, 6BR8A/6FV8, 6BR8A/6FV8A, 6CL8#, 6CL8A#, 6FV8A, 6FV8A/6BR8A
6FV8A	9FA	6BE8A, 6BR8*!, 6BR8A, 6BR8A/6FV8, 6BR8A/6FV8A, 6CL8#, 6CL8A#, 6FV8, 6FV8A/6BR8A
6FV8A/6BR8A	9FA	6BE8A, 6BR8*!, 6BR8A, 6BR8A/6FV8, 6BR8A/6FV8A, 6CL8#, 6CL8A#, 6FV8, 6FV8A
6FW5	6CK	-
6FW8	9AJ	6DJ8*, 6DJ8/ECC88, 6ES8*, 6ES8/ECC189, 6KN8, 6KN8/6RHH8, 6RHH8, ECC88*, ECC189*
6FX4	-	6AV4*
6FX7	-	6DHH13
6FY5	7FP	6ER5*, 6FY5/EC97, EC95*, EC97
6FY5/EC97	7FP	6ER5*, 6FY5, 6FY5/EC97, EC95*, EC97
6FY7	12EO	-
6FY8	9EX	-
6G3	-	1V/6Z3, 1V/KR1, 6G3AD, 6Z3AD, KR1
6G3AD	-	1V, 1V/6Z3, 1V/KR1, 6Z3, KR1
6G5	6R	6G5/6H5, 6H5, 6T5, 6U5/6G5, 6U5
6G5/6H5	6R	6G5, 6H5, 6U5, 6U5/6G5
6G5/6U5	6R	6G5, 6U5
6G6	7S	6CG6, 6G6GT, WTT130
6G6G	7S	6G6GT, WTT130
6G6GT	7S	6G6G, WTT130
6G11	12BU	-
6GA7	12EB	-
6GA8	9AJ	6CG7*, 6CG7/6FQ7*, 6FQ7#, 6FQ7/6CG7#, 6GU7#, 6/30L2, B729, ECC804
6GB3A	6AM	6BQ6GA, 6BQ6GA/6CU6, 6BQ6GT, 6BQ6GTB/6CU6, 6CU6
6GB5	9NH	6GB5/EL500, 6GB5/EL504, EL500, EL504
6GB5/EL500	9NH	6GB5, 6GB5/EL504, EL500, EL504
6GB5A	9NH	6GB5A/EL500
6GB5A/EL500	9NH	6GB5A, EL500
6GB6	6AM	6DQ6A, 6DQ6B, 6DQ6B/6GW6, 6GW6, 6GW6/6DQ6B
6GB7	6AM	6DQ6B, 6DQ6B/6GW6, 6GW6, 6GW6/6DQ6B
6GB9	-	6DQ6B, 6DQ6B/6GW6, 6GW6, 6GW6/6DQ6B
6GC5	9EU	-
6GC6	8JX	-
6GD7	9GF	6CG8A#, 6FG7#, 6LJ8#
6GE5	12BJ	-
6GE8	9LC	6GE8/7734, 7734
6GE8/7734	9LC	6GE8, 7734
6GF5	12BJ	6GE5
6GF7	9QD	6GF7A
6GF7A	9QD	6GF7
6GH8	9AE	6AX8*!, 6CQ8#, 6EA8, 6GH8A, 6GJ8*, 6HL8*!, 6KD8*!, 6KE8*, 6LM8*!, 6U8*!, 6U8A, 6U8A/6AX8/6KD8
6GH8A	9AE	6AX8*!, 6CQ8#, 6EA8, 6GH8, 6GJ8*, 6HL8*!, 6KD8*!, 6KE8*, 6LM8*!, 6U8*!, 6U8A, 6U8A/6AX8/6KD8
6GJ5	9QK	6GJ5A
6GJ5A	9QK	6GJ5
6GJ7	9QA	6GJ7/ECF801, ECF801
6GJ7/ECF801	9QA	6GJ7, ECF801
6GJ8	9AE	6AX8*, 6CQ8*#, 6EA8*, 6GH8*, 6GH8A*, 6HL8, 6KD8*, 6MU8, 6U8*, 6U8A*, 6U8A/6AX8/6KD8*
6GK5	7FP	6FQ5, 6FQ5A, 6FQ5A/6GK5, 6GK5/6FQ5A
6GK5/6FQ5A	7FP	6ES5*, 6FQ5A, 6FQ5A/6GK5, 6GK5
6GK6	9GK	-
6GK7	9AQ	-
6GK17	4CG	6AU4GT, 6AU4GTA
6GL7	8BD	6EA7, 6EM7
6GM5	9MQ	-
6GM6	7CM	6EW6, 6HQ6*, 6JH6*
6GM8	9DE	6GM8/ECC86, ECC86
6GM8/ECC86	9DE	6GM8, ECC86
6GN6	7FW	-
6GN8	9DX	6AU8*, 6AU8A*, 6AU8A/6BH8*, 6AW8*, 6AW8A*, 6BH8*, 6CX8, 6CX8/6JA8, 6EB8, 6EB8/6GN8, 6GN8/6EB8, 6HF8, 6HZ8*, 6JA8, 6JE8, 6JV8*, 6LF8*, 2081*, 2081/6AW8A*
6GN8/6EB8	9DX	6AU8*, 6AU8A*, 6AU8A/6BH8*, 6AW8*, 6AW8A*, 6BH8*, 6CX8, 6CX8/6JA8, 6EB8, 6EB8/6GN8, 6GN8, 6HF8, 6GZ8*, 6JA8, 6JE8, 6JV8*, 6LF8*, 2081*, 2081/6AW8A*
6GR7	-	-
6GQ7	9RB	6BC7#, 6BJ7#
6GS7	9GF	-
6GS8	9LW	6BU8, 6BU8A, 6HS8, 6KF8, 6MK8A
6GT5	9NZ	6GT5A
6GT5A	9NZ	6GT5
6GU5	7GA	6FS5*
6GU7	9LP	-
6GV5	12DR	-
6GV7	9KN	6C18, ECF805
6GV8	9LY	6GV8/ECL85, ECL85
6GV8/ECL85	9LY	6GV8, ECL85
6GW5	7GK	-
6GW6	6AM	6DQ6B, 6DQ6B/6GW6, 6GW6/6DQ6B
6GW6/6DQ6B	6AM	6DQ6B, 6DQ6B/6GW6, 6GW6
6GW8	9LZ	6GW8/ECL86, ECL86
6GW8/ECL86	9LZ	6GW8, ECL86
6GX6	7EN	6GY6, 6GY6/6GX6, 6HZ6
6GX6/6GY6	7EN	6GX6, 6GY6

Tube Type	Basing	Replacement
6GX7	9QA	-
6GX8	9HX	6GX8/ECL84, EAM86, ECL84
6GX8/ECL84	9HX	6GX8, EAM86, ECL84
6GY5	12DR	-
6GY6	7EN	6GY6/6GX6, 6GY6/6GX6/6HZ6, 6GX6, 6GX6/6GY6, 6HZ6
6GY6/6GX6	7EN	6GX6, 6GX6/6GY6, 6GY6, 6GY6/6GX6/6HZ6, 6HZ6
6GY6/6GX6/6HZ6	7EN	6GY6, 6GX6, 6GX6/6GY6, 6GY6/6GX6, 6HZ6
6GY8	9MB	-
6GZ5	7CV	-
6H4	5AF	6H4GT
6H4GT	5AF	6H4
6H5	6R	6G5, 6G5/6H5, 6H5G, 6M1, 6T5, 6T5, 6U5, 6U5/6G5
6H5G	6R	6G5, 6G5/6H5, 6M1, 6T5, 6T5, 6U5, 6U5/6G5
6H6	7Q	6H6G, 6H6GT, 6H6GT/G, 6H6MG, D63, EB34*, OSW3109, WTT103, WT261
6H6G	7Q	6H6, 6H6GT, 6H6GT/G, 6H6MG
6H6GT	7Q	6H6, 6H6G, 6H6GT/G, 6H6MG
6H6GT/G	7Q	6H6, 6H6G, 6H6GT, 6H6MG
6H6MG	7Q	6H6, 6H6G, 6H6GT, 6H6GT/G
6H8	8E	6H8G
6H8G	8E	6H8
6H31	-	6BE6, 6BE6/5750, 6BE6/EK90, 6BE6W, 5750, 5750/6BE6W, EK90
6HA5	7GM	6HA5/6HM5, 6HK5, 6HM5, 6HM5/6HA5, 6HQ5, EC900
6HA5/6HM5	7GM	6HA5, 6HK5, 6HM5, 6HM5/6HA5, 6HQ5, EC900
6HA5S	-	6HA5, 6HA5/6HM5, 6HK5, 6HM5, 6HM5/6HA5, 6HQ5, EC900
6HA6	9NW	6HB6, 6HB6/6HA6
6HA6/6HB6	9NW	6HA6, 6HB6
6HB5	12BJ	-
6HB6	9NW	6HA6, 6HA6/6HB6, 6HB6/6HA6
6HB6/6HA6	9NW	6HA6, 6HA6/6HB6, 6HB6, 6HB6/6HA6
6HB7	9QA	-
6HC8	9EX	6BM8*, 6BM8/ECL82, ECL82*, ECL82/6BM8*
6HD5	12ES	-
6HD7	9QA	6HD7/6HJ7, 6HJ7
6HD7/6HJ7	9QA	6HD7, 6HJ7
6HE5	12EY	6HE5/6JB5, 6JB5, 6JB5/6HE5, 6JC5, 6JC5/6JB5/6HE5
6HE5/6JB5	12EY	6HE5, 6JB5, 6JB5/6HE5, 6JC5, 6JC5/6JB5/6HE5
6HE7	12FS	-
6HF5	12FB	-
6HF8	9DX	6CX8, 6CX8/6JA8, 6EB8, 6GN8, 6GN8/6EB8, 6HZ8*, 6JA8, 6JE8, 6JV8*, 6LF8*
6HG5	7BZ	6AQ5*!, 6AQ5/6005/6669*!, 6AQ5A, 6AQ5A/6HG5, 6L31, 6P9, 6005*!, cont'd
6HG5, cont'd		6669*!, 6669/6AQ5A*!, EL90, N727
6HG8	9MP	6HG8/ECF86, ECF86
6HG8/ECF86	9MP	6HG8, ECF86
6HJ5	12FL	-
6HJ7	9QA	6HD7, 6HD7/6HJ7
6HJ8	9CY	6AM8*!, 6AM8A
6HK5	7GM	6HA5, 6HM5, 6HQ5
6HK8	9DE	6BC8, 6BC8/6BQ7A, 6BC8/6BZ8, 6BK7*, 6BK7A*, 6BK7B*, 6BQ7A, 6BQ7A/6BZ7, 6BQ7A/6BZ7/6BS8, 6BS8, 6BS8/6BK7B, 6BX8, 6BZ7, 6BZ7/6BQ7A, 6BZ8, X155
6HL5	9QW	-
6HL8	9AE	6C16, 6BL8*, 6BL8/6LN8*, 6BL8/ECF80*, 6GH8*!, 6GH8A*!, 6LN8*, ECF80*, ECF80/6BL8*
6HM5	7GM	6HA5, 6HA5/6HM5, 6HK5, 6HM5/6HA5, 6HQ5, EC900
6HM5/6HA5	7GM	6HA5, 6HA5/6HM5, 6HK5, 6HM5, 6HQ5, EC900
6HM6	9PM	6HM6/6HT6, 6HT6, 6JC6, 6JC6A, 6JD6
6HM6/6HT6	9PM	6HM6, 6HT6
6HQ5	7GM	6HA5, 6HK5, 6HM5, EC900
6HQ6	7CM	6BZ6, 6BZ6/6JH6, 6CB6, 5CB6A, 6DE6, 6DK6, 6GM6*, 6JH6, 6JK6, EF190
6HR5	7BZ	-
6HR6	7BK	-
6HS5	12GY	6HV5, 6HV5A
6HS6	7BK	6CB6*#, 6CB6A*#, 6CB6A/6CF6*#, 6CE5*#, 6CF6*#, 6CY5*#, 6DK6*#, 6EW6*#, 6FG5*#, 6676*#, 6676/6CB6A*#,
6HS8	9FG	6BU8, 6BU8A, 6GS8, 6KF8, 6MK8, 6MK8A
6HT5	-	N2ED
6HT6	9PM	6HM6, 6HM6/6HT6, 6JC6, 6JC6A, 6JD6
6HU6	9GA	6HU6/EM87, EM87
6HU6/EM87	9GA	6HU6, EM87
6HU8	9NJ	6HU8/ELL80, ELL80
6HU8/ELL80	9NJ	6HU8, ELL80
6HV5	12GY	6HS5, 6HV5A
6HV5A	12GY	-
6HW8	9NQ	-
6HZ5	12GY	6HZ5/6JD5
6HZ5/6JD5	12GY	6HZ5, 6JD5
6HZ6	7EN	6GX6, 6GY6
6HZ8	9DX	6AW8A, 6EB8*, 6GN8*, 6GN8/6EB8*, 6HF8*, 6JE8*, 6JV8*, 6LF8*, 2081, 2081/6AW8A
6J4	7BQ	6J4W, 6J4WA, 6MH1, 8532, M8232, TM12
6J4W	7BQ	6J4WA, 8532
6J4WA	7BQ	6J4#, 8532
6J5	6Q	6C5, 6C5G, 6C5GT, 6C5GT/G, 6C5MG, 6J5/L63, 6J5G, 6J5GT, 6J5GT/G, 6J5GTX, 6J5GX, 6J5MG, 6J5WGT, 6L5G, L63, L63B, cont'd

Tube Type	Basing	Replacement
6J5, cont'd		OSW3112, WTT129
6J5/L63	6Q	6C5, 6C5G, 6C5GT, 6C5GT/G, 6C5MG, 6J5, 6J5G, 6J5GT, 6J5GT/G, 6J5GTX, 6J5GX, 6J5MG, 6J5WGT, 6L5G, L63, L63B, OSW3112, WTT129
6J5G	6Q	6C5, 6C5G, 6C5GT, 6C5GT/G, 6C5MG, 6J5, 6J5/L63, 6J5GT, 6J5GT/G, 6J5GTX, 6J5GX, 6J5MG, 6L5G
6J5GT	6Q	6C5, 6C5G, 6C5GT, 6C5GT/G, 6C5MG, 6J5, 6J5/L63, 6J5G, 6J5GT/G, 6J5GTX, 6J5GX, 6J5MG, 6L5G
6J5GT/G	6Q	6C5, 6C5G, 6C5GT, 6C5GT/G, 6C5MG, 6J5, 6J5/L63, 6J5G, 6J5GT, 6J5GTX, 6J5GX, 6J5MG, 6L5G
6J5GTX	6Q	6C5, 6C5G, 6C5GT, 6C5GT/G, 6C5MG, 6J5, 6J5/L63, 6J5G, 6J5GT, 6J5GT/G, 6J5GX, 6J5MG, 6L5G
6J5GX	6Q	6C5, 6C5G, 6C5GT, 6C5GT/G, 6C5MG, 6J5, 6J5/L63, 6J5G, 6J5GT, 6J5GT/G, 6J5GTX, 6J5MG, 6L5G
6J5MG	6Q	6C5, 6C5G, 6C5GT, 6C5GT/G, 6C5MG, 6J5, 6J5/L63 ,6J5G, 6J5GT, 6J5GT/G, 6J5GTX, 6J5GX, 6L5G
6J5WGT	6Q	6C5GT#, 6C5GT/G#, 6C5MG#, 6J5GT#, 6J5GT/G#, 6J5GTX#, 6J5GX#, 6L5G
6J6	7BF	6CC31, 6J6A, 6J6W, 6J6WA, 6MHH3, 5964, 6030, 6045, 6099, 6099/6J6, 6101, 6927*, ECC91, M8081, T2M05
6J6A	7BF	6J6*!, 6J6A/ECC91, ECC91
6J6A/ECC91	7BF	6J6A, ECC91
6J6W	7BF	6J6A#, 6J6WA, 5964, 6099, 6101
6J6WA	7BF	6J6A#, 5964, 6099, 6101, 6101/6J6WA
6J7	7R	6J7G, 6J7GT, 6J7GTX, 6J7MG, 6W7G*, 1223, 1620, 7000, A863, EF37*, Z63
6J7G	7R	6J7, 6J7GT, 6J7GTX, 6J7MG, 6W7G*
6J7GT	7R	6J7, 6J7G, 6J7GTX, 6J7MG, 6W7G*, A863, KTW63, KTZ63, KTZ63M, Z63, EF37*
6J7GTX	7R	6J7, 6J7G, 6J7GT, 6J7MG, 6W7G*
6J7MG	7R	6J7, 6J7G, 6J7GT, 6J7GTX, 6W7G*
6J8	8H	6J8G
6J8G	8H	6J8
6J9	10G	-
6J10	12BT	6J10/6Z10, 6Z10, 6Z10/6J10
6J10/6Z10	12BT	6J10, 6Z10, 6Z10/6J10
6J11	12BW	-
6JA5	12EY	-
6JA8	9DX	6CX8, 6CX8/6JA8, cont'd
6JA8, cont'd		6EB8, 6GN8, 6GN8/6EB8, 6HF8, 6HZ8*, 6JE8, 6JT8, 6JV8*, 6LF8*
6JB5	12EY	6HE5, 6HE5/6JB5, 6JB5/6HE5, 6JB5/6JC5, 6JC5, 6JC5/6JB5/6HE5
6JB5/6HE5	12EY	6HE5, 6JB5, 6JB5/6JC5, 6JC5, 6JC5/6JB5/6HE5
6JB5/6JC5	12EY	6HE5, 6JB5, 6JB5/6HE5, 6JC5, 6JC5/6JB5/6HE5
6JB6	9QL	6JB6A
6JB6A	9QL	6JB6
6JB8	9AE	-
6JC5	12EY	6HE5, 6JB5, 6JB5/6HE5, 6JB5/6JC5, 6JC5/6JB5/6HE5
6JC5/6JB5/6HE5	12EY	6HE5, 6JB5, 6JB5/6HE5, 6JC5
6JC6	9PM	6HM6, 6HT6, 6JC6A, 6JD6
6JC6A	9PM	6HM6, 6HT6, 6JC6, 6JD6
6JC8	9PA	-
6JD5	12GY	6HV5A, 6HZ5/6JD5
6JD6	9PM	6HM6, 6HT6, 6JC6, 6JC6A
6JE6	9QL	6JE6A, 6JE6B, 6JE6C, 6LQ6, 6LQ6/6JE6C, 6MJ6, 6MJ6/6LQ6
6JE6A	9QL	6JE6B, 6JE6C, 6LQ6, 6LQ6/6JE6C, 6MJ6, 6MJ6/6LQ6
6JE6B	9QL	6JE6C, 6JE6C/6LQ6, 6LQ6, 6LQ6/6JE6C, 6ME6, 6MJ6, 6MJ6/6LQ6, 6LZ6
6JE6C	9QL	6JE6B, 6JE6C/6LQ6, 6LQ6, 6LQ6/6JE6C, 6MJ6, 6MJ6/6LQ6
6JE6C/6LQ6	9QL	6MJ6, 6MJ6/6LQ6, 6JE6C, 6LQ6, 6LQ6/6JE6C, 6MJ6, 6MJ6/6LQ6
6JE8	9DX	6EB8, 6GN8, 6GN8/6EB8, 6HF8, 6HZ8*, 6JA8, 6JV8*, 6LF8*
6JF5	-	-
6JF6	9QL	6JE6*, 6KM6, 6LQ6*, 6MJ6*, 6MJ6/6LQ6*
6JG5	9SF	-
6JG6	9QU	6JG6A, 6JR6, 6KV6A
6JG6A	9QU	6JG6, 6JR6, 6KV6A
6JH5	12JE	6JK5
6JH6	7CM	6BZ6, 6BZ6/6JH6, 6GM6*, 6HQ6
6JH8	9DP	-
6JK5	12JE	6JH5
6JK6	7CM	6EW6, 6JK6/6JL6, 6JL6
6JK6/6JL6	7CM	6EW6, 6JK6, 6JL6
6JK8	9AJ	-
6JL6	7CM	6EW6, 6GM6, 6JK6, 6JK6/6JL6
6JL8	9DX	-
6JM6	12FJ	6JM6A
6JM6A	12FJ	6JM6
6JN6	12FK	6JN6A
6JN6A	12FK	6JN6
6JN8	9FA	6BR8*!, 6BR8A, 6BR8A/6FV8, 6BR8A/6FV8A, 6FV8, 6FV8A, 6FV8A/6BR8A
6JQ6	9RA	-
6JR6	9QU	6JG6, 6JG6A
6JS6	12FY	6JS6A, 6JS6B, 6JS6C, 6KD6
6JS6A	12FY	6JS6, 6JS6B, 6JS6C, 6KD6
6JS6B	12FY	6JS6, 6JS6A, 6JS6C, 6LB6
6JS6C	12FY	6JS6B, 6LB6A

Tube Type	Basing	Replacement
6JT6	9QU	6JT6A, 6JG6*, 6JG6A*
6JT6A	9QU	6JT6, 6JG6*, 6JG6A*
6JT8	9DX	6JA8, 6KR8, 6KR8A, 6KV8, 6LB8, 6LQ8, 6LY8
6JU6	9QL	-
6JU8	9PQ	6JU8A
6JU8A	9PQ	6JU8
6JV8	9DX	6AW8, 6AW8A, 6CX8*, 6CX8/6JA8*, 6EB8*, 6GN8*, 6GN8/6EB8*, 6HF8*, 6HZ8*, 6JA8*, 6JE8*, 6KS8, 6LF8, 2081, 2081/6AW8A
6JW6	9PU	-
6JW8	9DC	6JW8/ECF802, 6LX8, ECF802
6JW8/ECF802	9AE	6JW8, 6LX8, ECF802
6JX8	-	ECH84
6JY8	-	-
6JZ6	12GD	-
6JZ8	12DZ	-
6K4	6K4	6K4A
6K4A	6K4	6K4
6K5	5U	6K5G, 6K5GT, 6K5WGT
6K5G	5U	6K5, 6K5GT, 6K5WGT
6K5GT	5U	6K5, 6K5G, 6K5WGT
6K5WGT	5U	6K5G#, 6K5GT#
6K6	7S	6K6G, 6K6GT, 6K6GT/G, 6K6MG
6K6G	7S	6K6, 6K6GT, 6K6GT/G, 6K6MG
6K6GT	7S	6K6, 6K6G, 6K6MG
6K6GT/G	7S	6K6, 6K6G, 6K6GT, 6K6MG
6K6MG	7S	6K6, 6K6G, 6K6GT, 6K6GT/G
6K7	7R	6K7G, 6K7GT, 6K7GTX, 6K7MG, 6U7G, 5732, PF9, W61, W63, EF39*, OM6*, W147*
6K7G	7R	6K7, 6K7GT, 6K7GTX, 6K7MG, 6U7G
6K7GT	7R	6K7, 6K7G, 6K7GTX, 6K7MG, 6U7G
6K7GTX	7R	6K7, 6K7G, 6K7GT, 6K7MG, 6U7G
6K7MG	7R	6K7, 6K7G, 6K7GT, 6K7GTX, 6U7G
6K8	8K	6K8G, 6K8GT, 6K8GTX, WTT128
6K8G	8K	6K8, 6K8GT, 6K8GTX
6K8GT	8K	6K8, 6K8G, 6K8GTX
6K8GTX	8K	6K8, 6K8G, 6K8GT
6K11	12BY	6K11/6Q11, 6Q11
6K11/6Q11	12BY	6K11, 6Q11
6KA8	9PV	-
6KD6	12GW	-
6KD8	9AE	6AX8, 6EA8, 6GH8, 6GH8A, 6KD8/6U8A, 6U8, 6U8A, 6U8A/6KD8, 6U8A/6AX8/6KD8
6KD8/6U8A	9AE	6AX8, 6EA8, 6GH8, 6GH8A, 6KD8, 6U8, 6U8A, 6U8A/6KD8, 6U8A/6AX8/6KD8
6KE6	12GM	-
6KE8	9DC	-
6KF8	9FG	6BU8, 6BU8A, 6GS8, 6HS8, 6MK8A
6KG6	9RJ	6KG6, 6KG6A/EL509, 6KG6A/EL519, EL509, EL519
6KG6A	9RJ	6KG6A/EL509, 6KG6A/EL519, EL509, EL519
6KG6A/EL509	9RJ	6KG6A, 6KG6A/EL519, EL509, EL519
6KG6A/EL519	9RJ	6KG6A, 6KG6A/EL509, EL509, EL519
6KH8	-	6KH8/ECLL800, ECLL800
6KH8/ECLL800	-	6KH8, ECLL800
6KL8	9LQ	6EQ7#
6KM6	9QL	6JF6, 6MJ6*, 6MJ6/6LQ6*, 6LQ6*
6KM8	9QG	-
6KN6	12GU	-
6KN8	9AJ	6KN8/6RHH8, 6RHH8, ECC88*, ECC189*
6KN8/6RHH8	9AJ	6KN8, 6RHH8, ECC88*, ECC189*
6KR8	9DX	6JT8, 6KR8A, 6KV8, 6LB8, 6LB8/6KR8, 6LQ8, 6LY8
6KR8A	9DX	6JT8, 6KR8, 6KV8, 6LB8, 6LQ8, 6LY8
6KS6	7DF	6BN6, 6BN6/6KS6
6KS8	9DX	6AU8, 6AU8A, 6AU8A/6BH8, 6AW8, 6AW8A, 6BH8, 6JV8, 6LF8, 6CX8*, 6CX8/6JA8*, 6EB8*, 6GN8*, 6GN8/6EB8*, 6HF8*, 6HZ8*, 6JA8*, 6JE8*, 6LF8, 2081, 2081/6AW8A
6KT6	9PM	-
6KT8	9QP	-
6KU8	9LT	-
6KV6	9QU	6KV6A
6KV6A	9QU	6KV6
6KV8	9DX	6JT8, 6KR8, 6KR8A, 6LB8, 6LQ8, 6LY8
6KW6	-	EL508
6KX6	-	-
6KX8	-	-
6KY6	9GK	-
6KY8	9QT	6KY8A
6KY8A	9QT	6KY8
6KZ8	9FZ	-
6L4	7BR	-
6L5	6Q	6C5, 6C5G, 6C5GT, 6C5GT/G, 6C5MG, 6J5, 6J5G, 6J5GT, 6J5GT/G, 6J5GTX, 6J5GX, 6J5MG, 6L5G
6L5G	6Q	6C5, 6C5G, 6C5GT, 6C5GT/G, 6C5MG, 6J5, 6J5G, 6J5GT, 6J5GT/G, 6J5GTX, 6J5GX, 6J5MG
6L6	7AC	6L6/5881, 6L6A, 6L6G, 6L6GA, 6L6GAY, 6L6WA, 6L6WGA, 6L6WGB, 6L6WGC, 6L6WGT, 6L6GB, 6L6GC, 6L6GT, 6L6GX, 6L6Y, 1622, 5881, 5881/6LCWGC, 5932, 7581, 7581A, WT6, EL37
6L6/5881	7AC	6L6, 6L6A, 6L6G, 6L6GA, 6L6GAY, 6L6WA, 6L6WGA, 6L6WGB, 6L6WGC, 6L6WGT, 6L6GB, 6L6GC, 6L6GT, 6L6GX, 6L6Y, 1622, 5881, 5881/6LCWGC, 5932, 7581, cont'd

Tube Type	Basing	Replacement
6L6/5881, cont'd		7581A, WT6, EL37
6L6A	7AC	6L6, 6L6/5881, 6L6G, 6L6GA, 6L6GAY, 6L6GB, 6L6GC, 6L6GT, 6L6GX, 6L6WGC, 6L6Y, 5881, 5881/6L6WGC, 7581, 7581A
6L6G	7AC	6L6, 6L6/5881, 6L6A, 6L6GA, 6L6GAY, 6L6GB, 6L6GC, 6L6GT, 6L6GX, 6L6WGC, 6L6Y, 5881, 5881/6L6WGC, 7581, 7581A
6L6GA	7AC	6L6, 6L6/5881, 6L6A, 6L6G, 6L6GAY, 6L6GB, 6L6GC, 6L6GT, 6L6GX, 6L6WGC, 6L6Y, 5881, 5881/6L6WGC, 7581, 7581A
6L6GAY	7AC	6L6, 6L6/5881, 6L6A, 6L6G, 6L6GA, 6L6GB, 6L6GC, 6L6GT, 6L6GX, 6L6WGC, 6L6Y, 5881, 5881/6L6WGC, 7581, 7581A
6L6GB	7AC	6L6, 6L6/5881, 6L6A, 6L6G, 6L6GA, 6L6GAY, 6L6GC, 6L6GT, 6L6GX, 6L6WGC, 6L6Y, 5881, 5881/6L6WGC, 7581, 7581A
6L6GC	7AC	6L6, 6L6/5881, 6L6WGC, 5881, 5881/6L6WGC, 7581, 7581A, KT66
6L6GCR	7AC	6L6, 6L6/5881, 6L6WGC, 5881, 5881/6L6WGC, 7581, 7581A, KT66
6L6GT	7AC	6L6, 6L6/5881, 6L6A, 6L6G, 6L6GA, 6L6GAY, 6L6GB, 6L6GC, 6L6GX, 6L6WGC, 6L6Y, 5881, 5881/6L6WGC, 7581, 7581A
6L6GX	7AC	6L6, 6L6/5881, 6L6A, 6L6G, 6L6GA, 6L6GAY, 6L6GB, 6L6GC, 6L6GT, 6L6WGC, 6L6Y, 5881, 5881/6L6WGC, 7581, 7581A
6L6W	7AC	6L6GC#, 6L6WA, 6L6WGA, 6L6WGB, 6L6WGC#, 5881#, 5881/6L6WGC#, 5932
6L6WA	7AC	6L6GC#, 6L6WGA, 6L6WGB, 6L6WGC#, 5881#, 5881/6L6WGC#, 5932
6L6WGA	7AC	6L6GC#, 6L6WGB, 6L6WGC, 5881#, 5881/6L6WGC, 5932
6L6WGB	7AC	6L6GC#, 6L6WGC, 5881#, 5881/6L6WGC
6L6WGC	7AC	5881, 5881/6L6WGC
6L6WGT	7AC	6L6GC#, 6L6WGC#, 5881#, 5881/6L6WGC#, 5932
6L6Y	7AC	6L6, 6L6A, 6L6G, 6L6GA, 6L6GAY, 6L6GB, 6L6GC, 6L6GT, 6L6GX, 6L6WGC, 5881, 5881/6L6WGC, 7581, 7581A
6L7	7T	6L7G, 1612, 1620, X64
6L7G	7T	6L7
6L8G	-	-
6L10	-	6AG7, 6AG7/6AK7, 6AG7W, 6AK7
6L12	-	6AQ8, 6AQ8/ECC85, ECC85, ECC85/6AQ8
6L13	-	12AX7, 12AX7A, 12AX7A/ECC83, 12AX7A/ECC83/7025, 12AX7WA, 12AX7WA/7025, 7025, ECC83
6L16	-	6CW7, 6CW7/ECC84, ECC84
6L31	-	6AQ5, 6AQ5/6005/6669, 6AQ5A, 6AQ5A/6HG5, 6HG5, 6005, 6669, 6669/6AQ5A
6L34	-	6AQ4, 6AQ4/EC91, EC91
6L40	-	-
6L43	-	6CL6, 6CL6/6677, 6677
6LB6	12JF	6JS6, 6JS6A, 6JS6B, 6JS6C
6LB6A	12JF	6JS6A, 6JS6B, 6JS6C
6LB8	9DX	6JT8, 6KR8, 6KR8A, 6KV8, 6LB8/6KR8, 6LQ8, 6LY8
6LB8/6KR8	9DX	6JT8, 6KR8, 6KR8A, 6KV8, 6LB8, 6LQ8, 6LY8
6LC6	8ML	6LH6, 6LH6A
6LC8	9QY	-
6LD3	8GZ	6LD3/EBC41, 6CV7, 6CV7/EBC41, EBC41
6LD3/EBC41	8GZ	6LD3, 6CV7, 6CV7/EBC41, EBC41
6LD6	-	EL802
6LD12	-	6AK8, 6AK8/EABC80, EABC80
6LD13	-	6BD7
6LE8	9QZ	-
6LF6	12GW	6LF6/6MH6, 6LF6/6LX6, 6LV6
6LF6/6LX6	12GW	6LF6, 6LF6/6MH6, 6LX6, 6MH6
6LF6/6MH6	12GW	6LF6, 6LF6/6LX6, 6MH6
6LF8	9DX	6AW8, 6AW8A, 6CX8*, 6CX8/6JA8*, 6EB8*, 6GN8*, 6GN8/6EB8*, 6HF8*, 6HZ8*, 6JA8*, 6JE8*, 6JV8, 6KS8, 2081, 2081/6AW8A
6LG6	12HL	6LG6C, WT210-0007
6LG6C	12HL	WT210-0007
6LH6	8ML	6LH6A, 6LH6A/6LJ6, 6LJ6A/6LH6A
6LH6A	8ML	6LH6A/6LJ6, 6LJ6A, 6LJ6A/6LH6A
6LH6A/6LJ6	8ML	6LH6A, 6LJ6, 6LJ6A/6LH6A
6LJ6	8MQ	6LH6A/6LJ6, 6LJ6A, 6LJ6A/6LH6A
6LJ6A	8MQ	6LH6A, 6LH6A/6LJ6, 6LJ6A/6LH6A
6LJ6A/6LH6A	8MQ	6LH6A, 6LH6A/6LJ6, 6LJ6A
6LJ8	9GF	6GD7#, 6MB8#
6LM8	9AE	6AX8, 6CQ8#, 6GH8, 6GH8A, 6KD8, 6LM8A, 6U8A, 6U8A/6AX8/6KD8
6LM8A	9AE	6AX8, 6CQ8#, 6GH8, 6GH8A, 6KD8, 6U8A, 6U8A/6AX8/6KD8
6LN8	9DC	6BL8*!, 6BL8/6LN8, 6LN8/ECF80, ECF80, ECF80/6BL8, LCF80
6LN8/LCF80	9DC	6BL8, 6BL8/6LN8, 6BL8/ECF80, 6LN8, ECF80, ECF80/6BL8, LCF80
6LP12	-	6BM8, 6BM8/ECL82, ECL82, ECL82/6BM8

Tube Type	Basing	Replacement
6LQ6	9QL	6JE6B, 6JE6C, 6JE6C/6LQ6, 6LQ6/6JE6B, 6LQ6/6JE6C, 6LZ6, 6ME6, 6MJ6, 6MJ6/6LQ6, 6MJ6/6LQ6/6JE6C
6LQ6/6JE6B	9QL	6JE6B, 6JE6C, 6JE6C/6LQ6, 6LQ6, 6LQ6/6JE6C, 6LZ6, 6ME6, 6MJ6, 6MJ6/6LQ6, 6MJ6/6LQ6/6JE6C
6LQ6/6JE6C	9QL	6JE6B, 6JE6C, 6JE6C/6LQ6, 6LQ6, 6LQ6/6JE6B, 6LZ6, 6ME6, 6MJ6, 6MJ6/6LQ6, 6MJ6/6LQ6/6JE6C
6LQ8	9DX	6JT8, 6KR8, 6KR8A, 6KV8, 6LB8, 6LY8
6LR6	12FY	6JS6C
6LR8	9QT	-
6LT8	9RL	-
6LU6	7CM	-
6LU8	12DZ	6MY8
6LV6	12GW	6LF6
6LW6	8NC	-
6LX6	12JA	6LF6/6LX6
6LX8	9DC	6JW8*!, 6LX8/LCF802, LCF802
6LX8/LCF802	9DC	6JW8*!, 6LX8, LCF802
6LY8	9DX	6JT8, 6KR8, 6KR8A, 6KV8, 6LB8, 6LQ8
6LZ6	9QL	6LQ6, 6ME6, 6MJ6, 6MJ6/6LQ6
6M1	-	6U5, 6U5G
6M2	-	6CD7, 6CD7/EM34, 6CD7/EM34-35, EM34, EM34/6CD7, EM34-35, EM34-35/6CD7
6M3	8GV	-
6M5	9N	6CK6#, 6CK6/EL83#, EL80, EL83
6M6	7S	6AG6G*, 6AG6G/EL33*, 6M6G, EL33*, KT61*, N147*, PP6BG#
6M6G	7S	6AG6G*, 6AG6G/EL33*, EL33*, KT61*, N147*, PP6BG#
6M7	7G	6K7#, 6K7G#, 6K7GT#, 6K7GTX#, 6K7MG#, 6M7G
6M7G	7R	6K7#, 6K7G#, 6K7GT#, 6K7GTX#, 6K7MG#
6M8	8AU	6M8G, 6M8GT
6M8G	8AU	6M8GT
6M8GT	8AU	6M8G
6M11	12CA	-
6MA6	8NP	-
6MB6	12FY	6JE6B, 6JE6C, 6LZ6
6MB8	9FA	6LJ8#
6MC6	9QL	6LQ6, 6MJ6, 6MJ6/6LQ6
6MD8	9RR	-
6ME6	9QL	6JE6B, 6JE6C, 6LQ6, 6MJ6, 6MJ6/6LQ6, 6LZ6
6ME8	9RU	-
6MF8	12DZ	-
6MG8	9DC	6AX8, 6CQ8#, 6EA8, 6KD8, 6U8, 6U8A, 6U8A/6AX8/6KD8
6MH1	-	6J4#
6MH6	12GW	6LF6/6MH6, 6LX6#
6MHH3	7BF	6J6
6MJ6	9QL	6JE6B, 6JE6C, 6LQ6, 6LZ6, 6ME6, 6MJ6/6LQ6, cont'd
6MJ6, cont'd		6MJ6/6LQ6/6JE6C
6MJ6/6LQ6	9QL	6JE6B, 6JE6C, 6LQ6, 6LZ6, 6ME6, 6MJ6, 6MJ6/6LQ6/6JE6C
6MJ6/6LQ6/6JE6C	9QL	6JE6B, 6JE6C, 6LQ6, 6LZ6, 6ME6, 6MJ6, 6MJ6/6LQ6
6MJ8	12HG	-
6MK8	9FG	6HS8, 6MK8A
6MK8A	9FG	6HS8, 6MK8
6ML8	9RQ	-
6MN8	12HU	-
6MP17	-	-
6MP18	-	-
6MP20	-	-
6MQ8	9AE	6AX8*, 6EA8*, 6KD8*, 6U8*, 6U8A*, 6U8A/6AX8/6KD8*
6MU8	9AE	-
6MV8	9DX	-
6MY8	12DZ	6LU8
6N3	9BM	6U3, EY80, EY82
6N3/EY82	9BM	6N3, 6U3, EY80, EY82
6N4	7CA	-
6N5	6R	6AB5, 6AB5/6N5, 6N5G
6N5G	6R	6AB5, 6AB5/6N5, 6N5
6N6	7AU	6AB6G*, 6N6G, 6N6
6N6G	7AU	6AB6G*, 6N6, 6N6MG
6N6MG	7AU	6AB6G*, 6N6, 6N6G
6N7	8B	6N7G, 6N7GT, 6N7GT/G, 6N7MG, 5694
6N7G	8B	6N7, 6N7GT, 6N7GT/G, 6N7MG
6N7GT	8B	6N7, 6N7G, 6N7GT/G, 6N7MG, WTT125
6N7GT/G	8B	6N7, 6N7G, 6N7GT, 6N7MG
6N7MG	8B	6N7, 6N7G, 6N7GT, 6N7GT/G
6N8	9T	6AD8, 6AD8/6DC8, 6DC8, EBF80, EBF81, WD709, ZD152
6N8/EBF80	9T	6AD8, 6AD8/6DC8, 6DC8, 6N8, EBF80, EBF81, WD709, ZD152
6NH10	-	-
6NK7	-	-
6NL7	-	-
6P5	6Q	6P5G, 6P5GT, 6P5GT/G
6P5G	6Q	6P5GT, 6P5GT/G
6P5GT	6Q	6P5G, 6P5GT/G
6P5GT/G	6Q	6P5G, 6P5GT
6P6	6AC	-
6P7	7U	6P7G
6P7G	7U	6P7
6P8	-	-
6P9	-	6AQ5A, 6AQ5A/6HG5, 6BM5, 6HG5, 6669, 6669/6AQ5A
6P15	-	6BQ5, 6BQ5/EL84, EL84, EL84/6BQ5
6P17	-	6AM5, 6AM5/EL91, EL91
6P22	-	-
6P25	-	6AG6
6PL12	-	6BM8, 6BM8/ECL82, ECL82, ECL82/6BM8
6Q4	9S	6Q4/EL80, EC80
6Q4/EL80	9S	6Q4, EL80
6Q5	6Q	6Q5G, 884, 884/6Q5, 884W
6Q5G	6Q	884, 884/6Q5, 884W, WT245
6Q6	6Y	6Q6G

Tube Type	Basing	Replacement
6Q6G	6Y	6Q6
6Q7	7V	6B6, 6B6G, 6Q7G, 6Q7GT, 6Q7MG, 6T7, 6T7G*, 6118, DH63
6Q7/DH63	7V	6B6, 6B6G, 6Q7, 6Q7G, 6Q7GT, 6Q7MG, 6T7G*, 6118, DH63
6Q7G	7V	6B6, 6B6G, 6Q7, 6Q7GT, 6Q7MG, 6T7G*
6Q7GT	7V	6B6, 6B6G, 6Q7, 6Q7G, 6Q7MG, 6T7G*
6Q7MG	7V	6B6, 6B6G, 6Q7, 6Q7G, 6Q7GT, 6T7G*
6Q8	-	6A8
6Q11	12BY	6K11/6Q11
6QL6	-	-
6R3	9CB	6AF3*, 6AL3*, 6AL3/EY88*, 6AL3/EY88, 6BR3*, 6BR3/6RK19*, 6RK19*, 63VP*, EY81, EY83*, EY88*
6R3/EY81	9CB	6R3, EY81
6R4	9R	EC81
6R6	6AW	6R6G
6R6G	6AW	6R6
6R7	7V	6R7G, 6R7GT, 6R7GT/G, 6R7MG
6R7G	7V	6R7, 6R7GT, 6R7GT/G, 6R7MG
6R7GT	7V	6R7, 6R7G, 6R7GT/G, 6R7MG
6R7GT/G	7V	6R7, 6R7G, 6R7GT, 6R7MG
6R7MG	7V	6R7, 6R7G, 6R7GT, 6R7GT/G
6R8	9E	6LD12, 6T8, 6T8A, DH719, EABC80
6RA6	-	-
6RA9	-	-
6RHH2	7BD	6BC8, 6BC8/6BQ7A, 6BC8/6BZ8, 6BK7B, 6BQ7A, 6BQ7A/6BZ7, 6BQ7A/6BZ7/6BZ8, 6BS8, 6BS8/6BK7B, 6BZ7, 6BZ7/6BQ7A, 6BZ8
6RHH8	9DE	6KN8, 6KN8/6RHH8
6RK19	9CB	6AV3, 6BR3, 6BR3/6RK19
6RP15	-	-
6RP22	-	-
6R	-	5847, 5847/404A
6RR8C	-	5847, 5847/404A
6S2	9DT	6S2A, EY86, EY87
6S2A	9DT	6S2, EY87
6S2A/EY87	9DT	6S2A, EY87
6S4	9AC	6S4A
6S4A	9AC	6S4
6S5G	-	6E5
6S6	5AK	6S6GT
6S6GT	5AK	6S6
6S7	7R	6K7*, 6K7G*, 6K7GT*, 6K7GTX*, 6K7MG*, 6S7G, OF1, OM5*
6S7G	7R	6K7*, 6K7G*, 6K7GT*, 6K7GTX*, 6K7MG* 6S7
6S8	8CB	6S8GT
6S8GT	8CB	6S8
6SA7	8R	6SA7G, 6SA7GT, 6SA7GT/G, 6SA7GTX, 6SA7GTY, 6SA7WGT, 6SA7Y, 6SB7, 6SB7Y, 6SB7GTY, OSW3104
6SA7G	8R	6SA7, 6SA7GT, 6SA7GT/G, 6SA7GTX, 6SA7GTY, 6SA7WGT, 6SA7Y, 6SB7, 6SB7Y, 6SB7GTY
6SA7GT	8R	6SA7, 6SA7G, 6SA7GT/G, 6SA7GTX, 6SA7GTY, 6SA7WGT, 6SA7Y, 6SB7, 6SB7Y, 6SB7GTY
6SA7GT/G	8R	6SA7, 6SA7G, 6SA7GT, 6SA7GTX, 6SA7GTY, 6SA7WGT, 6SA7Y, 6SB7, 6SB7Y, 6SB7GTY
6SA7GTX	8R	6SA7, 6SA7G, 6SA7GT, 6SA7GT/G, 6SA7GTY, 6SA7WGT, 6SA7Y, 6SB7, 6SB7Y, 6SB7GTY
6SA7GTY	8R	6SA7, 6SA7G, 6SA7GT, 6SA7GT/G, 6SA7GTX, 6SA7WGT, 6SA7Y, 6SB7, 6SB7Y, 6SB7GTY
6SA7WGT	8R	6SA7GT#, 6SA7GT/G#, 6SA7GTX#, 6SA7WGT, 6SB7GTY
6SA7Y	8R	6SA7#, 6SA7G, 6SA7GT, 6SA7GT/G, 6SA7GTX, 6SA7GTY, 6SA7WGT, 6SB7, 6SB7Y, 6SB7GTY
6SB7	8R	6SB7GTY, 6SB7Y
6SB7GTY	8R	6SB7, 6SB7Y
6SB7Y	8R	6SB7, 6SB7GTY
6SC7	8S	6SC7GT, 6SC7GTY, 1655
6SC7GT	8S	6SC7, 6SC7GTY
6SC7GTY	8S	6SC7, 6SC7GT
6SD7	8N	6SD7GT, 6SE7GT
6SD7GT	8N	6SD7, 6SE7GT
6SE7	8N	6SD6GT, 6SE7GT
6SE7GT	8N	6SD7GT, 6SE7
6SF5	6AB	6SF5GT
6SF5GT	6AB	6SF5
6SF7	7AZ	6SF7GT
6SF7GT	7AZ	6SF7
6SG7	8BK	6AB7*#, 6AB7/1853*#, 6AB7Y*#, 6AC7*#, 6AC7/1852*#, 6AC7/1852/6134, 6AC7/EL34*#, 6AC7A*#, 6AC7Y*#, 6AJ7*#, 6SG7Y, 6SH7, 6SH7GT, 1852*#, 1853*#, 6006, 6134*#, EL34#
6SG7GT	8BK	6AB7*#, 6AB7/1853*#, 6AB7Y*#, 6AC7*#, 6AC7/1852*#, 6AC7/1852/6134, 6AC7/EL34*#, 6AC7A*#, 6AC7Y*#, 6AJ7*#, 6SG7, 6SG7Y, 6SH7, 6SH7GT, 1852*#, 1853*#, 6134*#, EL34#
6SG7Y	8BK	6AB7*#, 6AB7/1853*#, 6AB7Y*#, 6AC7*#, 6AC7/1852*#, 6AC7/1852/6134, 6AC7/EL34*#, 6AC7A*#, 6AC7Y*#, 6AJ7*#, 6SG7#, 6SG7GT, 6SH7, 6SH7GT1852*#, 1853*#, 6134*#, EL34#
6SH7	8BK	6AB7*#, 6ABY/1853*#, 6AB7Y*#, cont'd

Tube Type	Basing	Replacement
6SH7, cont'd		6AC7*#, 6AC7/1852*#, 6AC7/1852/6134, 6AC7/EL34*#, 6AC7A*#, 6AC7Y*#, 6AJ7*#, 6SH7GT, 6SH7L, 6SG7, 6SG7GT, 6SG7Y, 1852*#, 1853*#, 6134*#, EL34#
6SH7GT	8BK	6AB7*#, 6AB7/1853*#, 6AB7Y*#, 6AC7*#, 6AC7/1852*#, 6AC7/1852/6134, 6AC7/EL34*#, 6AC7A*#, 6AC7Y*#, 6AJ7*#, 6SG7, 6SG7GT, 6SG7Y, 6SH7, 6SH7L, 1852*#, 1853*#, 6134*#, EL34#
6SH7L	8BK	6AB7*#, 6AB7/1853*#, 6AB7Y*#, 6AC7*#, 6AC7/1852*#, 6AC7/1852/6134, 6AC7/EL34*#, 6AC7A*#, 6AC7Y*#, 6AJ7*#, 6SG7, 6SG7GT, 6SG7Y, 6SH7, 6SH7GT, 1852*#, 1853*#, 6134*#, EL34#
6SH8	-	-
6SJ7	8N	6SJ7GT, 6SJ7GTX, 6SJ7GTY, 6SJ7W, 6SJ7WGT, 6SJ7WGTY, 6SJ7Y, 5693, WTT122
6SJ7/5693	8N	6SJ7, 6SJ7GT, 6SJ7GTX, 6SJ7GTY, 6SJ7W, 6SJ7WGT, 6SJ7WGTY, 6SJ7Y, 5693, WTT122
6SJ7GT	8N	6SJ7, 6SJ7GTX, 6SJ7GTY, 6SJ7Y
6SJ7GTX	8N	6SJ7, 6SJ7GT, 6SJ7GTY, 6SJ7Y
6SJ7GTY	8N	6SJ7, 6SJ7GT, 6SJ7GTX, 6SJ7Y
6SJ7W	8N	6SJ7#, 6SJ7WGT, 6SJ7WGTY, 5963
6SJ7WGT	8N	6SJ7#, 6SJ7W, 6SJ7WGTY, 5963
6SJ7WGTY	8N	6SJ7#, 6SJ7W, 6SJ7WGT, 5963
6SJ7Y	8N	6SJ7#, 6SJ7GT, 6SJ7GTX, 6SJ7GTY, 5963
6SK7	8N	6SG7, 6SK7/6137, 6SK7GT, 6SK7GT/G, 6SK7GTX, 6SK7GTY, 6SK7W, 6SK7WA, 6SK7WGT, 6SK7Y, 6SS7*, 6SS7GT*, 6137, OSW3111
6SK7/6137	8N	6SG7, 6SK7, 6SK7GT, 6SK7GT/G, 6SK7GTX, 6SK7GTY, 6SK7W, 6SK7WA, 6SK7WGT, 6SK7Y, 6SS7*, 6SS7GT*, 6137, OSW3111
6SK7G	8N	6SG7, 6SH7, 6SK7, 6SK7GT, 6SK7GT/G, 6SK7GTX, 6SK7GTY, 6SK7Y, 6SS7*, 6SS7GT*
6SK7GT	8N	6SG7, 6SH7, 6SK7, 6SK7G, 6SK7GT/G, 6SK7GTX, 6SK7GTY, 6SK7Y, 6SS7*, 6SS7GT*
6SK7GT/G	8N	6SK7, 6SK7G, 6SK7GT, 6SK7GTX, 6SK7GTY, 6SK7Y, 6SS7*, 6SS7GT*
6SK7GTX	8N	6SK7, 6SK7G, 6SK7GT, 6SK7GT/G, 6SK7GTY, 6SK7Y, 6SS7*, 6SS7GT*
6SK7GTY	8N	6SK7, 6SK7G, 6SK7GT, 6SK7GT/G, 6SK7GTX, 6SK7Y, 6SS7*, 6SS7GT*
6SK7W	8N	6SK7, 6SK7GT, 6SK7WA, 6SK7WGT
6SK7WA	8N	6SK7, 6SK7GT, 6SK7W, 6SK7WGT, 6137, 6137/6SK7WA
6SK7WGT	8N	6SK7, 6SK7GT, 6SK7W, 6SK7WA
6SK7Y	8N	6SK7, 6SK7G, 6SK7GT, 6SK7GT/G, 6SK7GTX, 6SK7GTY, 6SS7*, 6SS7GT*
6SL7	8BD	6SL7/5691, 6SL7W, 6SL7WGT, 5691*, 6113, ECC35*
6SL7/5691	8BD	6SL7, 6SL7W, 6SL7WGT, 5691*, 6113, ECC35*
6SL7A	8BD	6SL7, 6SL7/5691, 6SL7GT, 6SL7L
6SL7GT	8BD	6SL7A, 6SL7GTY, 6SL7L
6SL7GTY	8BD	6SL7A, 6SL7GT, 6SL7L
6SL7L	8BD	6SL7A, 6SL7GT, 6SL7GTY
6SL7W	8BD	6SL7GT#, 6SL7WGT, 5691, 6113
6SL7WGT	8BD	6SL7GT#, 6SL7W, 5691, 6113
6SN7	8BD	6SN7/5692, 6SN7W, 6SN7WGT, 6SN7WGTA, 65W7, 5692, B65, ECC33
6SN7/5692	8BD	6SN7, 6SN7W, 6SN7WGT, 6SN7WGTA, 65W7, 5692, B65, ECC33
6SN7A	8BD	6SN7GT, 6SN7GTA, 6SN7GTB, 6SN7GTY, 6SN7L
6SN7GT	8BD	6SN7A, 6SN7GTA, 6SN7GTB, 6SN7GTY, 6SN7L, 13D2, B65, ECC32*, QA2408, QB65
6SN7GTA	8BD	6SN7GTB
6SN7GTB	8BD	6SN7GTA*!
6SN7GTY	8BD	6SN7A, 6SN7GT, 6SN7GTA, 6SN7GTB, 6SN7L, 5692
6SN7L	8BD	6SN7A, 6SN7GT, 6SN7GTA, 6SN7GTB, 6SN7GTY
6SN7W	8BD	6SN7GTB#, 6SN7WGT, 6SN7WGTA, 5692
6SN7WGT	8BD	6SN7GTB#, 6SN7W, 6SN7WGTA, 5692
6SN7WGTA	8BD	6SN7GTB#, 6SN7W, 6SN7WGT, 5692
6SQ7	8Q	6SQ7G, 6SQ7GT, 6SQ7GT/G, 6SQ7W, 6SZ7*, OSW3105
6SQ7G	8Q	6SQ7, 6SQ7GT, 6SQ7GT/G, 6SZ7*
6SQ7GT	8Q	6SQ7, 6SQ7G, 6SQ7GT/G, 6SZ7*
6SQ7GT/G	8Q	6SQ7, 6SQ7G, 6SQ7GT, 6SZ7*
6SQ7W	8Q	6SQ7#, 6SQ7G, 6SQ7GT
6SR7	8Q	6SR7G, 6SR7GT, 6ST7*
6SR7G	8Q	6SR7, 6SR7GT, 6ST7*
6SR7GT	8Q	6SR7, 6SR7G, 6ST7*
6SS7	8N	6SS7GT, 6SK7*, 6SK7G*, 6SK7GT*, 6SK7GT/G*, cont'd

Tube Type	Basing	Replacement
6SS7, cont'd		6SK7GTX*, 6SK7GTY*, 6SK7Y*
6SS7GT	8N	6SS7, 6SK7*, 6SK7G*, 6SK7GT*, 6SK7GT/G*, 6SK7GTX*, 6SK7GTY*, 6SK7Y*
6ST7	8Q	6SR7, 6SR7G*, 6SR7GT*
6SU7	8BD	6SU7GTY/6188, 6SU7WGT, 6188
6SU7GT	8BD	6SU7GTX, 6SU7GTY, 6SU7GTY/6188, 6188
6SU7GTX	8BD	6SU7GT, 6SU7GTY, 6SU7GTY/6188, 6188
6SU7GTY	8BD	6SU7GT, 6SU7GTX, 6SU7GTY/6188, 6188
6SU7GTY/6188	8BD	6SU7GT, 6SU7GTY, 6188
6SU7WGT	8BD	6SU7GTY/6188, 6188
6SV7	7AZ	6SV7GT
6SV7GT	7AZ	6SV7
6SZ7	8Q	6SQ7*, 6SQ7G, 6SQ7GT*, 6SQ7GT/G*
6T1	-	6AF4, 6AF4, 6AF4A, 6AF4A/6DZ4, 6DZ4, 6DZ4/6AF4A
6T4	7DK	6AF4#, 6AF4A#, 6AF4A/6DZ4#, 6AN4#, 6DZ4#, 6DZ4/6AF4A#
6T5	6R	6G5, 6G5/6H5, 6H5, 6U5, 6U5/6G5
6T6	6Z	6T6GM
6T6GM	6Z	6T6
6T7	7V	6Q7, 6T7/676, 676, DH63
6T7/676	7V	6Q7, 6T7, 676, DH63
6T7G	7V	6B6*, 6B6G*, 6Q7*, 6Q7G*, 6Q7GT*, 6Q7MG*, 6T7, 6T7/676, 676, DH63
6T8	9E	6AK8, 6AK8/EABC80, 6LD12, 6T8A, DH719, EABC80
6T8/EABC80	9E	6AK8, 6AK8/EABC80, 6LD12, 6T8, 6T8A, DH719, EABC80
6T8A	9E	6AK8*!, 6AK8/EABC80*!, 6T8*!, EABC80*!
6T9	12FM	-
6T10	12EZ	6AD10#
6TE8	-	-
6TE9	-	-
6TH8	-	-
6U3	9BM	EY80
6U4	4CG	6AS4GT, 6AU4GT*, 6AU4GTA*, 6AX4GT, 6AX4GTA, 6AX4GTB, 6CQ4*, 6CQ4/6DE4*, 6DA4, 6DA4A, 6DA4A/6DM4, 6DA4A/6DM4A, 6DE4*, 6DE4/6CQ4*, 6DM4, 6DM4A, 6DM4A/6DQ4, 6DQ4, 6DT4, 6U4GT, 6W4GT, 6W4GTA
6U4GT	4CG	6AS4GT, 6AU4GT*, 6AU4GTA*, 6AX4GT, 6AX4GTA, 6AX4GTB, 6CQ4*, 6CQ4/6DE4*, 6DA4, 6DA4A, 6DA4A/6DM4, 6DA4A/6DM4A, 6DE4*, 6DE4/6CQ4*, 6DM4, 6DM4A, 6DM4A/6DQ4, 6DQ4, 6DT4, 6W4GT, 6W4GTA
6U5	6R	6G5, 6G5/6H5, 6G5/6U5, 6H5, 6T5, 6U5, 6U5/6G5
6U5/6G5	6R	6G5, 6G5/6H5, 6G5/6U5, 6H5, 6T5, 6U5
6U5G	6R	6G5, 6G5/6H5, 6G5/6U5, 6H5, 6T5, 6U5, 6U5/6G5
6U6	7AC	6U6GT, 6U6WGT, 6Y6G*, 6Y6GA*, 6Y6GA/6Y6G*, 6Y6GT*
6U6GT	7AC	6U6, 6U6WGT, 6Y6G*, 6Y6GA*, 6Y6GA/6Y6G*, 6Y6GT*
6U6WGT	7AC	6U6#, 6U6GT#, 6Y6G*, 6Y6GA*, 6Y6GA/6Y6G*, 6Y6GT*
6U7	7R	6J7, 6K7, 6K7G, 6K7GT, 6K7GTX, 6K7MG, 6S7*, 6S7G*, 6U7G
6U7G	7R	6J7, 6K7, 6K7G, 6K7GT, 6K7GTX, 6K7MG, 6S7*, 6S7G*
6U8	9AE	6AX8, 6CQ8#, 6EA8, 6KD8*, 6KD8/6U8A, 6LN8, 6MQ8*, 6U8/6678, 6U8A, 6U8A/6KD8, 6U8A/6AX8/6KD8, 1252, 6678, 7731, ECF80, ECF82
6U8/6678	9AE	6AX8, 6CQ8#, 6EA8, 6KD8*, 6KD8/6U8A, 6LN8, 6MQ8*, 6U8, 6U8A, 6U8A/6KD8, 6U8A/6AX8/6KD8, 1252, 6678, 7731, ECF80, ECF82
6U8A	9AE	6AX8, 6CQ8#, 6EA8, 6LN8, 6KD8#, 6KD8/6U8A, 6MQ8#, 6U8*!, 6U8A/6KD8, 6U8A/6AX8/6KD8
6U8A/6KD8	9AE	6AX8, 6EA8, 6GH8, 6GH8A, 6KD8, 6KD8/6U8A, 6U8A, 6U8A/6AX8/6KD8
6U8A/6AX8/6KD8	9AE	6AX8, 6EA8, 6GH8, 6GH8A, 6KD8, 6KD8/6U8A, 6U8A, 6U8A/6KD8
6U9	10K	6U9/ECF201, ECF201
6U9/ECF201	10K	6U9, ECF201
6U10	12FE	6AC10#
6V3	9BD	6V3A, EY81F
6V3A	9BD	6V3
6V3P	-	6R3*
6V4	9M	6CA4*, 6CA4/EZ81*, 6V4/EZ80, EZ80, EZ81*, EZ81/6CA4*
6V4/EZ80	9M	6CA4*, 6CA4/EZ81*, 6V4, EZ80, EZ81*, EZ81/6CA4*
6V5	6AO	6V5G, 6V5GT
6V5G	6AO	6V5GT
6V5GT	6AO	6V5G
6V6	7AC	6V6G, 6V6GT, 6V6GTA, 6V6GT/G, 6V6GTX, 6V6GTY, 6V6GX, 6V6Y, 5871, 7184, 7408, WTT123
6V6G	7AC	6V6, 6V6GT, 6V6GTA, 6V6GT/G, 6V6GTX, 6V6GTY, 6V6GX, 6V6Y, 7408
6V6GT	7AC	6V6, 6V6G, 6V6GTA, 6V6GT/G, 6V6GTX, 6V6GTY, 6V6GX, cont'd

Tube Type	Basing	Replacement
6V6GT, cont'd		6V6Y, 7408, OSW3106*
6V6GT/G	7AC	6V6, 6V6G, 6V6GT, 6V6GTA, 6V6GTX, 6V6GTY, 6V6GX, 6V6Y, 7408
6V6GTA	7AC	6V6*!, 6V6G*!, 6V6GT*!, 6V6GT/G*!, 6V6GTX*!, 6V6GTY*!, 6V6GX*!, 6V6Y*!, 7408*!
6V6GTX	7AC	6V6, 6V6G, 6V6GT, 6V6GTA, 6V6GT/G, 6V6GTY, 6V6GX, 6V6Y, 7408
6V6GTY	7AC	6V6, 6V6G, 6V6GT, 6V6GTA, 6V6GT/G, 6V6GTX, 6V6GX, 6V6Y, 7408
6V6GX	7AC	6V6, 6V6G, 6V6GT, 6V6GTA, 6V6GT/G, 6V6GTY, 6V6GTX, 6V6Y, 7408
6V6Y	7AC	6V6, 6V6G, 6V6GT, 6V6GTA, 6V6GT/G, 6V6GTX, 6V6GTY, 6V6GX, 7408
6V7	7V	6V7G
6V7G	7V	6V7
6V8	9AH	-
6V9	-	ECH200
6W4	4CG	6AS4GT, 6AU4GT*, 6AU4GTA*, 6AX4GT, 6AX4GTA, 6AX4GTB, 6CQ4*, 6CQ4/6DE4*, 6DA4, 6DA4A, 6DA4A/6DM4, 6DA4A/6DM4A, 6DE4*, 6DE4/6CQ4*, 6DM4, 6DM4A, 6DM4A/6DQ4, 6DQ4, 6DT4, 6U4GT, 6W4GT, 6W4GTA
6W4GT	4CG	6AS4GT, 6AU4GT*, 6AU4GTA*, 6AX4GT, 6AX4GTA, 6AX4GTB, 6CQ4*, 6CQ4/6DE4*, 6DA4, 6DA4A, 6DA4A/6DM4, 6DA4A/6DM4A, 6DE4*, 6DE4/6CQ4*, 6DM4, 6DM4A, 6DM4A/6DQ4, 6DQ4, 6DT4, 6U4GT, 6W4GTA
6W4GTA	4CG	6AU4GT*, 6AU4GTA*, 6AX4GTA, 6AX4GTB, 6CQ4*, 6CQ4/6DE4*, 6CQ4/6DE4*, 6DA4, 6DA4A, 6DA4A/6DM4, 6DA4A/6DM4A, 6DE4*, 6DE4/6CQ4*, 6DM4, 6DM4A, 6DM4A/6DQ4, 6DQ4, 6DT4, 6U4GT
6W5	6S	6AX5GT*, 6W5G, 6W5GT
6W5G	6S	6AX5GT*, 6W5, 6W5GT, 6X5GT
6W5GT	6S	6AX5GT*, 6W5, 6W5G, 6X5GT
6W6	7AC	6DG6GT, 6DG6GT/6W6GT, 6EF6*, 6EY6*, 6EZ5*, 6W6GT
6W6GT	7AC	6DG6GT, 6DG6GT/6W6GT, 6EF6*, 6EY6*, 6EZ5*, 6W6
6W7	7R	6J7*, 6J7G*, 6J7GT*, 6J7GTX*, 6J7MG*, 6W7G, 8D4*
6W7G	7R	6J7*, 6J7G*, 6J7GT*, cont'd
6W7G, cont'd		6J7GTX*, 6J7MG*, 8D4*
6W9	-	-
6X2	-	6X2/EY51, EY51, R12, SU61, U43, U45, U151
6X2/EY51	-	6X2, EY51, R12, SU61, U43, U45, U151
6X4	5BS	6AV4*, 6BX4, 6X4/6202, 6X4/EZ90, 6X4W, 6X4WA, 6Z31, 6063, 6202, 6202/6X4WA, E90Z, EZ90, EZ900, QA2407*, U78*, U707, V2M70, WTT100
6X4/6202	5BS	6AV4*, 6BX4, 6X4, 6X4/EZ90, 6X4W, 6X4WA, 6Z31, 6063, 6202, 6202/6X4WA, E90Z, EZ90, EZ900, QA2407*, U78*, U707, V2M70, WTT100
6X4/EZ90	5BS	6AV4*, 6BX4, 6X4, 6X4/6202, 6X4W, 6X4WA, 6Z31, 6063, 6202, 6202/6X4WA, E90Z, EZ90, EZ900, QA2407*, U78*, U707, V2M70, WTT100
6X4W	5BS	6X4#, 6X4WA, 6202, 6202/6X4WA
6X4WA	5BS	6202, 6202/6X4WA
6X5	6S	6AX5GT*, 6W5G*, 6W5GT*, 6X5G, 6X5GT, 6X5GT/G, 6X5L, 6X5MG, 6X5W, 6X5WGT, 5852*, WT308
6X5G	6S	6W5G*, 6W5GT*, 6X5, 6X5GT, 6X5GT/G, 6X5L, 6X5MG, 6X5GT*, EZ35, U147, U70*
6X5GT	6S	6AX5GT*, 6W5G*, 6W5GT*, 6X5, 6X5G, 6X5GT/EZ35, 6X5GT/G, 6X5L, 6X5MG, EZ35
6X5GT/EZ35	6S	6X5GT, EZ35
6X5GT/G	6S	6X5, 6X5G, 6X5GT, 6X5L, 6X5MG, 6X5GT*, 6W5G*, 6W5GT*
6X5L	6S	6AX5GT*, 6W5G*, 6W5GT*, 6X5, 6X5G, 6X5GT, 6X5GT/G, 6X5MG
6X5MG	6S	6AX5GT*, 6W5G*, 6W5GT*, 6X5, 6X5G, 6X5GT, 6X5L, 6X5GT/G
6X5W	6S	6X5GT, 6X5WGT
6X5WGT	6S	6X5GT, 6X5W
6X6	-	-
6X8	9AK	6AU8*#, 6AU8A*#, 6AU8A/6BH8*#, 6AW8*#, 6AW8A*#, 6BH8*#, 6EH8#, 6X8A, 2081*#, 2081/6AW8A*#
6X8A	9AK	6AU8*#, 6AU8A*#, 6AU8A/6BH8*#, 6AW8*#, 6AW8A*#, 6BH8*#, 6EH8#, 6X8*!, 2081*#, 2081/6AW8A*#
6X9	10K	6X9/ECF200, ECF200
6X9/ECF200	10K	6X9, ECF200
6Y3	4AC	6Y3G
6Y3G	4AC	6Y3
6Y5	6J	6Y5G, 6Y5GT, 6Y5V

Tube Type	Basing	Replacement
6Y5G	6J	6Y5, 6Y5GT, 6Y5V
6Y5GT	6J	6Y5, 6Y5GT, 6Y5V
6Y5V	6J	6Y5, 6Y5G, 6Y5GT
6Y6	7AC	6U6GT*, 6Y6G, 6Y6GA, 6Y6GA/6Y6G, 6Y6GT
6Y6G	7AC	6U6GT*, 6Y6GA, 6Y6GA/6Y6G, 6Y6GT
6Y6GA	7AC	6U6GT*, 6Y6G, 6Y6GA/6Y6G, 6Y6GT
6Y6GA/6Y6G	7AC	6Y6GA, 6Y6G, 6Y6GT
6Y6GT	7AC	6U6GT*, 6Y6G, 6Y6GA, 6Y6GA/6Y6G
6Y7	8B	6Y6G, 1635
6Y7G	8B	1635
6Y9	10L	6Y9/EFL200, 6Y9/EFL201, EFL200, EF201
6Y9/EFL200	10L	6Y9, 6Y9/EFL201, EFL200
6Y9/EFL201	10L	6Y9, 6Y9/EFL200, EFL201
6Y10	12EZ	6AD10
6Z3	4G	1V, 1V/6Z3, 1V/KR1, AD
6Z4	5D	6BX4, 6Z4/84, 84, 84/6Z4, 98, WT263
6Z4/84	5D	6BX4, 6Z4, 6Z4/84, 84, 84/6Z4, 98, WT263
6Z5	6K	12Z5, 6Z5/12Z5
6Z5/12Z5	6K	6Z5, 12Z5
6Z6	7Q	6Z6G, 6Z6GT
6Z6G	7Q	6Z6GT, 6Z6MG
6Z6GT	7Q	6Z6G, 6Z6MG
6Z6MG	7Q	6Z6G, 6Z6GT
6Z7	8B	6Z7G
6Z7G	8B	6Z7
6Z10	12BT	6J10, 6J10/6Z10, 6Z10/6J10
6Z10/6J10	12BT	6J10, 6J10/6Z10, 6Z10
6Z31	-	6X4, 6202#
6ZY5	6S	6X5*, 6X5G*, 6X5GT*, 6X5GT/G*, 6X5L*, 6X5MG*, 6ZY5G
6ZY5G	6S	6X5*, 6X5G*, 6X5GT*, 6X5GT/G*, 6X5L*, 6X5MG*
7A4	5AC	XXL
7A4/XXL	5AC	7A4, XXL
7A5	6AA	-
7A6	7AJ	5679
7A7	8V	7A7LM, 7B7*, 7H7, EF22*, W81, W148, W143*
7A7LM	8V	7A7, 7B7*, 7H7
7A8	8U	7B8*, 7B8LM*
7AB7	8BO	7AB7/1204, 1204
7AB7/1204	8B	7AB7, 1204
7AD7	8V	-
7AF7	8AC	-
7AG7	8V	7AG7/7AH7, 7AH7, 7G7*, 7T7*
7AG7/7AH7	8V	7AG7, 7AH7, 7G7*, 7T7*
7AH7	8V	7AG7, 7AG7/7AH7, 7B7, 7H7*
7AJ7	8V	7C7*, 1273
7AK7	8V	-
7AN7	9DD	7AN7/PCC84, 7EK7, 30L1, 30L15, B319, PCC84
7AN7/PCC84	9DD	7AN7, 7EK7, 30L1, 30L15, B319, PCC84
7AU7	9A	6AU7, 6AX7#, PCC186, XCC82
7B4	5AC	7A4#, XXL#
7B5	6AE	7B5LT, 7C5*, cont'd

Tube Type	Basing	Replacement
7B5, cont'd		7C5LT*, EL22, KT81, N148
7B5LT	6AE	7B5, 7C5*, 7C5LT*
7B6	8W	7B6LM, 7C6*, DH81, DL82
7B6LM	8W	7B6, 7C6*
7B7	8V	7A7*, 7A7LM*, 7AH7, 7H7*, W149
7B8	8X	7A8*, 7B8LM
7B8LM	8X	7A8*, 7B8
7C4	4AH	7C4/1203A, 1203, 1203A
7C4/1203A	4AH	7C4, 1203A
7C5	6AA	7B5*, 7B5LT*, 7C5LT, EL22, KT81, N148
7C5LT	6AA	7B5*, 7B5LT*, 7C5
7C6	8W	7B6*, 7B6LM*, DH149
7C7	8V	7AJ7*, 7G7*, 7L7*
7C23	-	5680, 5680/7C23
7C24	-	5762, 5762/7C24
7D4	-	-
7D7	8AR	-
7D9	-	6AM5, 6AM5/EL91, EL91
7D10	-	6CH6, 6CH6/EL821, EL821
7D11	-	6AC7, 6AC7/1852, 6AC7/1852/6134, 6AC7/EL34, 1852, 6134, 6550#, EL34
7DJ8	9DE	7DJ8/PCC88, 7ES8, PCC88, PCC189
7DJ8/PCC88	9DE	7DJ8, 7ES8, PCC88, PCC189
7E5	8BN	7E5/1201, 1201, 1201A
7E5/1201	8BN	7E5, 1201, 1201A
7E6	8W	-
7E7	8AE	7R7
7ED7	9AQ	30F5, Z329
7EK7	9DD	7AN7, 7EK7/7FC7, 30L15, B349, PCC805
7EK7/7FC7	9DD	7AN7, 7EK7, 30L15, B349, PCC805
7ES8	9DE	7DJ8, PCC88, PCC189
7EY6	7AC	-
7F7	8AC	-
7F8	8BW	7F8W
7F8W	8BW	7F8#
7F16	-	6CJ5, 6CJ5/EF41, EF41
7FC7	9DD	7EK7, 7EK7/7FC7, PCC84, PCC89
7G7	8V	7AG7*, 7C7*, 7G7/1232, 7L7*, 7V7, 1232
7G7/1232	8V	7AG7*, 7C7*, 7G7, 7L7*, 7V7, 1232
7G8	8BV	1206
7GS7	9GF	-
7GV7	9KN	30C18, PCF805
7H7	8V	7A7, 7A7LM, 7AH7*, 7B7*, EF22*, W143*, W148
7HG8	9MP	7HG8/8HG8, 7HG8/PCF86, 8HG8, PCF86
7HG8/8HG8	9MP	7HG8, 7HG8/PCF86, 8HG8, PCF86
7HG8/PCF86	9MP	7HG8, 7HG8/8HG8, 8HG8, PCF86
7J7	8BL	7J7/7S7, 7S7
7J7/7S7	8BL	7J7, 7S7
7K7	8BF	WTT124
7KY6	9GK	7KY6/9KX6, 9KX6
7KY6/9KX6	9GK	7KY6, 9KX6
7KZ6	9GK	-
7L7	8V	7C7*, 7G7*, 7V7*
7MP18	-	-

Tube Type	Basing	Replacement
7N7	8AC	7AF7*
7Q7	8AL	-
7R7	8AE	7E7
7S7	8BL	7J7, 7J7/7S7, XB1, X148
7T7	8V	7AG7*, 7V7*
7V7	8V	7G7, 7L7*, 7T7*
7W7	8BJ	-
7X6	7DX	-
7X7	8BZ	7X7/XXFM, XXFM
7X7/XXFM	8BZ	7X7, XXFM
7Y4	5AB	7Z4*, 1274*, U82*, U149
7Z4	5AB	-
8A8	9DC	8A8/9A8/9U8, 9A8, 9A8/PCF80, 9EA8, 9GH8A, 9U8, 9U8A, 30C1, LZ319, LZ329, PCF80, PCF82
8A8/9A8/9U8	9DC	8A8, 9A8, 9A8/PCF80, 9EA8, 9GH8A, 9U8, 9U8A, 30C1, LZ319, LZ329, PCF80, PCF82
8AC9	12GN	-
8AC10	12FE	8AC10A
8AC10A	12FE	-
8AG10	-	-
8AL9	12HE	-
8AR11	12DM	8BQ11#
8AU8	9DX	8AU8A, 8AU8A/8BA8A, 8AW8A, 8BA8A, 8BH8
8AU8A	9DX	8AU8, 8AU8A/8BA8A, 8AW8A, 8BA8A, 8BH8
8AU8A/8BA8A	9DX	8AU8A, 8BA8A
8AV11	-	-
8AW8	9DX	-
8AW8A	9DX	8AU8, 8AU8A, 8BA8A, 8BH8, 8JV8, 8KS8
8B8	9EX	8RHP1, XCL82
8B10	12BF	-
8BA8	9DX	8AU8, 8AU8A, 8AU8A/8BA8A, 8AW8A, 8BA8A, 8BH8
8BA8A	9DX	8AU8, 8AU8A, 8AU8A/8BA8A, 8AW8A, 8BH8
8BA11	12ER	-
8BH8	9DX	8AU8, 8AU8A, 8AW8A, 8BA8A, 8JV8
8BM11	12FU	-
8BN8	9ER	-
8BN11	12GF	-
8BQ5	9CV	XL84
8BQ7	9AJ	-
8BQ7A	9AJ	-
8BQ11	12DM	8AR11#
8BU11	12FP	-
8BV7	-	-
8CB11	12DM	-
8CG7	9AJ	8CG7/8FQ7, 8FQ7#, 8FQ7/8CG7
8CG7/8FQ7	9AJ	8CG7, 8FQ7, 8FQ7/8CG7
8CM7	9ES	-
8CN7	9EN	-
8CS7	9EF	-
8CW5	9CV	8CW5/XL86, 8CW5A, XL86
8CW5/XL86	9CV	8CW5, 8CW5A, XL86
8CW5A	9CV	8CW5!, 8CW5/XL86
8CX8	9DX	8EB8, 8GN8, 8JE8
8CY7	9EF	-
8D3	-	6AM6, 6AM6/EF91, EF91
8D4	-	6W7G*
8D5	-	6BR7
8D6	-	6BW7
8D7	-	6BS7
8D8	-	6269
8DX8	-	-
8EB8	9DX	8CX8, 8EB8/8GN8, 8GN8, 8GN8/8EB8, 8JE8
8EB8/8GN8	9DX	8CX8, 8EB8, 8GN8, 8GN8/8EB8, 8JE8
8EM5	9HN	-
8ET7	9LT	-
8FQ7	9LP	8CG7, 8CG7/8FQ7, 8FQ7/8CG7
8FQ7/8CG7	9LP	8CG7, 8CG7/8FQ7, 8FQ7
8GJ7	9QA	8GJ7/PCF801, 8GX7, PCF801, PCF806
8GJ7/PCF801	9QA	8GJ7, 8GX7, PCF801, PCF806
8GK6	9GK	-
8GM6	-	-
8GM8	-	-
8GN8	9DX	8CX8, 8EB8, 8EB8/8GN8, 8GN8/8EB8, 8JE8
8GN8/8EB8	9DX	8CX8, 8GN8, 8EB8, 8EB8/8GN8, 8JE8
8GU7	9LP	-
8GX7	9QA	8GJ7, 8GJ7/PCF801#
8HA6	9NW	-
8HG8	9MP	7HG8, 8HG8/PCF86, PCF86
8HG8/PCF86	9MP	7HG8, 8HG8, PCF86
8JE8	9DX	8CX8, 8EB8, 8GN8, 8GN8/8EB8
8JK8	9AJ	-
8JL8	9DX	-
8JT8	9DX	-
8JU8	9PQ	-
8JU8A	9PQ	-
8JV8	9DX	8AW8A, 8KS8
8KA8	9PV	-
8KR8	9DX	-
8KS8	9DX	8AU8, 8AU8A, 8AW8A, 8JV8
8LC8	9QY	-
8LE8	9QZ	-
8LN8	-	-
8LS6	9GK	-
8LT8	9RL	-
8MU8	9AE	-
8RHP1	-	8B8
8RLP1	-	-
8SN7	8BD	-
8SN7GTB	8BD	-
8TL8	-	-
8U9	-	PCF201
8X9	-	PCF200
9A8	9DC	8A8, 8A8/9A8/9U8, 9A8/8A8/PCF80, 9A8/PCF80, 9EA8, 9U8, 9U8A, 30C1, LZ319, LZ329, PCF80
9A8/8A8/PCF80	9DC	8A8, 8A8/9A8/9U8, 9A8, 9EA8, 9U8, 9U8A, 30C1, LZ319, LZ329, PCF80
9A8/9U8	9DC	8A8, 8A8/9A8/9U8, 9A8, 9A8/PCF80, 9EA8, 9U8, 9U8A, C0C1, LZ319, LZ329, PCF80
9A8/PCF80	9DC	8A8, 8A8/9A8/9U8, 9A8, 9A8/8A8/PCF80, 9EA8, 9U8, 9U8A, 30C1, cont'd

Tube Type	Basing	Replacement
9A8/PCF80, cont'd		LZ319, LZ329, PCF80
9AB4	-	UC92
9AH9	12HJ	-
9AK8	9E	9AK8/PABC80, PABC80
9AK8/PABC80	9E	9AK8, PABC80
9AK10	12FE	-
9AM10	12FE	-
9AQ8	9DE	9AQ8/PCC85, PCC85
9AQ8/PCC85	9AJ	9AQ8, PCC85
9AU7	9A	-
9BJ11	12FU	-
9BM5	7DQ	9P9
9BR7	9CF	-
9BR8	9FA	9CL8#
9BW6	9AM	-
9C25	-	-
9CG8	9GF	-
9CG8A	9GF	9BR8!#
9CL8	9FX	9BR8#!
9D6	-	6CQ6, 6CQ6/EF92, EF92
9D7	-	-
9DZ8	9JE	-
9EA8	9AE	8A8, 9A8!, 9GH8A, 9U8!, 9U8A, 30C1, LZ319, LZ329, PCF80, PCF82
9ED4	-	-
9EF6	7S	-
9EN7	9LM	30C15, LZ339, PCF800
9GB8	9DA	30FL1, PCE800
9GH8	9AE	-
9GH8A	9AE	9EA8
9GV8	9LY	9GV8/XCL85, XCL85
9GV8/XCL85	9LY	9GV8, XCL85
9JC8	-	-
9JW8	9DC	9JW8/PCF802, PCF802
9JW8/PCF802	9DC	9JW8, PCF802
9KC6	9RF	-
9KX6	9GK	7KY6, 7KY6/9KX6
9KZ6	-	-
9KZ8	9FZ	-
9LA6	9GK	-
9MHH3	-	-
9ML8	9RQ	-
9MN8	12HU	-
9MP12	-	-
9P9	-	9BM5
9RA1L	9HF	10DE7
9RA6	-	-
9RAL1	9HF	10DE7, 10EW7
9RHH2	-	9GH8, 9GH8A
9RHR2	-	-
9U8	9AE	8A8, 8A8/9A8/9U8, 9A8, 9A8/9U8, 9EA8, 9U8A, 30C1, LZ319, LZ329, PCF80, PCF82
9U8A	9AE	8A8, 8A8/9A8/9U8, 9A8!, 9A8/9U8, 9EA8, 9U8!, 30C1, LZ319, LZ329, PCF80, PCF82
9V9	-	PCH200
9X8	9AK	-
10	4D	10X, 10Y
10AJ5	-	-
10AL11	12BU	-
10BQ5	9CV	-
10C8	9DA	-
10C14	-	19D8
10CW5	9CV	10CW5/LL86, LL86
10CW5/LL86	9CV	10CW5, LL86
10D2	-	12AL5
10DA7	9EF	-
10DE7	9HF	9RAL1, 10EW7
10DR7	9HF	10DR7/10FD7, 10FD7, 10FR7
10DR7/10FD7	9HF	10DR7, 10FD7, 10FR7
10DX8	9HX	10DX8/LCL84, LCL84
10DX8/LCL84	9HX	10DX8, LCL84
10EB8	9DX	10EB8/10GN8, 10GN8, 10HF8, 10JA8, 11JE8
10EB8/10GN8	9DX	10EB8, 10GN8, 10HF8, 10JA8, 11JE8
10EG7	8BD	10EM7
10EM7	8BD	10EG7
10EW7	9HF	-
10F9	-	12AC5
10F18	-	13EC7
10FD7	9HF	10DR7, 10DR7/10FD7
10FD12	-	-
10FH6	-	-
10FR7	9HF	10FD7
10GF7	9QD	10GF7A
10GF7A	9QD	10GF7
10GK6	9GK	-
10GN8	9DX	10EB8, 10EB8/10GN8, 10HF8, 10JA8, 10JY8, 10LZ8, 11JE8
10GV8	9LY	10GV8/LCL85, LCL85
10GV8/LCL85	9LY	10GV8, LCL85
10HA6	9NW	-
10HE8	9DX	-
10HF8	9DX	10EB8, 10GN8, 10LZ8, 10JA8, 10JY8, 11JE8
10J10	12BT	-
10JA5	12EY	-
10JA8	9DX	10EB8, 10GN8, 10HF8, 10JA8/10LZ8, 10JY8, 11JE8
10JA8/10LZ8	9DX	10EB8, 10GN8, 10HF8, 10JA8, 10JY8, 10LZ8, 11JE8
10JT8	9DX	10JT8/10LY8, 10KR8, 10LB8*, 10LW8, 10LY8, 11KV8, 11LQ8
10JT8/10LY8	9DX	10JT8, 10KR8, 10LB8*, 10LW8, 10LY8, 11KV8, 11LQ8
10JY8	9DX	10EB8, 10GN8, 10HF8, 10JY8/10LZ8, 10LZ8, 10JA8, 11JE8
10JY8/10LZ8	9DX	10EB8, 10GN8, 10HF8, 10JY8, 10LZ8, 10JA8, 11JE8
10KR8	9DX	10JT8, 10LB8, 10LB8/10KR8, 10LW8, 11KV8, 11LQ8
10KU8	9LT	-
10L14	-	-
10LB8	9DX	10JT8, 10KR8, 10LB8/10KR8, 10LW8, 11KV8, 11LQ8
10LB8/10KR8	9DX	10JT8, 10KR8, 10LB8, 10LW8, 11KV8, 11LQ8
10LD3	-	14L7
10LD6	-	-
10LD12	-	-
10LD13	-	14G6
10LE8	9QZ	-
10LT8	9QZ	10LT8/10LW8, 10LW8
10LT8/10LW8	9QZ	10LT8, 10LW8
10LW8	9DX	10JT8, 10KR8, 10LB8, 10LT8, 10LT8/10LW8, 11KV8, 11LQ8
10LY8	9DX	10JT8, 10JT8/10LY8

Tube Type	Basing	Replacement
10LZ8	9DX	10EB8, 10GN8, 10HF8, 10JA8/10LZ8, 10JY8, 10JY8/10LZ8, 11JE8
10P14	-	45A5, 45A5/UL41, 451PT, BF451, N142, UL41, U446
10P18	-	45B5
10PL12	-	50BM8, 50BM8/UCL82
10T10	12EZ	-
10X	4D	10, 10Y
10Y	4D	10, 10X
10Z10	12BT	-
11	4F	-
11AF9	10L	-
11AR11	12DM	11BQ11#
11BM8	9EX	LCL82
11BQ11	12DM	11AR11#
11BT11	12GS	-
11C5	7CV	12DM5
11CA11	12HN	-
11CF11	12HW	-
11CH11	12GS	-
11CY7	9LG	-
11D12	-	-
11DS5	7BZ	-
11E13	-	-
11FY3	-	-
11FY7	12EO	-
11HM7	9BF	-
11JE8	9DX	10EB8, 10GN8, 10HF8, 10JA8, 10JY8, 10LZ8
11KV8	9DX	10JT8, 10KR8, 10LB8, 11LQ8, 10LW8
11LQ8	9DX	10LB8, 10LW8, 10KR8, 10JT8, 11KV8
11LT8	9RL	-
11LY6	9GK	12BV7!, 12BV7/12BY7A!, 12BY7A!, 12BY7A/12BV7/ 12DQ7!, 12DQ7!
11MS8	9LY	-
11R3	-	LY81
11Y9	10L	LFL200
11Y9/LFL200	10L	11Y9, LFL200
12	4D	-
12A	-	-
12A4	9AG	-
12A5	7F	-
12A6	7AC	12A6G, 12A6GT, 12A6GTY, 12A6Y, 12V6GT*#, 5659
12A6G	7AC	12A6, 12A6GT, 12A6GTY, 12A6Y, 12V6GT*#, 5659
12A6GT	7AC	12A6, 12A6G, 12A6GT, 12A6Y, 12V6GT*#, 5659
12A6GTY	7AC	12A6, 12A6G, 12A6GT, 12A6Y, 12V6GT*#, 5659
12A6Y	7AC	12A6, 12A6G, 12A6GT, 12A6GTY, 12V6GT*#, 5659
12A7	7K	-
12A8	8A	12A8G, 12A8GT, OH4
12A8G	8A	12A8, 12A8GT
12A8GT	8A	12A8, 12A8G
12A	4D	12A/112A, 112A
12A/112A	4D	12A, 112A
12AB5	9EU	-
12AC5	8GW	10F9, 12AC5/UF41, 121VP, HF121, UF41, W118, W142, W145
12AC5/UF41	8GW	12AC5, 121VP, HF121, cont'd

Tube Type	Basing	Replacement
12AC5/UF41, cont'd		UF41, W118, W142, W145
12AC6	7BK	12AF6#, 12BL6
12AC10	12FE	12AC10A
12AC10A	12FE	-
12AD5	9AZ	-
12AD6	7CH	12AD6/12AG6, 12AG6
12AD6/12AG6	7CH	12AD6, 12AG6
12AD7	9A	6L13*, 12AX7*, 12AX7A*, 12AX7A/ECC83*, 12AX7A/ ECC83/7025*, 12AX7WA, 12AX7WA/7025, 12BZ7*, 12DF7*, 12DM7*, 12DT7*, 7025*, 7025A*, ECC83*
12AE6	7BT	12AE6A, 12FT6
12AE6A	7BT	12AE6, 12FT6
12AE7	9A	-
12AE10	12EZ	13V10
12AF3	9CB	12AF3/12BR3/12RK19, 12BR3, 12RK19
12AF3/12BR3/ 12RK19	9CB	12AF3, 12BR3, 12RK19
12AF6	7BK	12AC6, 12BL6, 12EK6#
12AG6	7CH	12AD6, 12AD6/12AG6
12AG7	-	-
12AH6	7BK	-
12AH7	8BE	-
12AH7GT	8BE	-
12AH8	9BP	20D3
12AJ6	7BT	-
12AJ7	9CA	HCH81
12AJ8	-	-
12AK7	-	-
12AL5	6BT	10D2, HAA91, UB91
12AL8	9GS	-
12AL11	12BU	-
12AQ5	7BZ	12BM5*
12AQ6	-	-
12AQ7	-	-
12AS5	7CV	12CA5*, 12R5*
12AT6	7BT	12AT6/HBC90, 12AT6A, 12AV6, 12AV6A, 12BK6, 12BT6, 12BC32, HBC90, HBC91
12AT6/HBC90	7BT	12AT6, 12AT6A, 12AV6, 12AV6A, 12BK6, 12BT6, 12BC32, HBC90, HBC91
12AT6A	7BT	12AT6*!, 12AV6A, 12BK6*!, 12BT6*!
12AT7	9A	12AT7/6201/6679, 12AT7/ ECC81, 12AT7WA, 12AT7WB, 12AZ7*, 12AZ7A*, 6060, 6201, 6671, 6679, 7492, 7728, A2900, B152, B309, B739, CV4024, E81CC, ECC81, ECC801, ECC801S, M8162, QA2406, QB309
12AT7/6201/6679	9A	12AT7, 12AT7/ECC81, 12AT7WA, 12AT7WB, 12AZ7*, 12AZ7A*, 6060, 6201, 6671, 6679, 7492, 7728, A2900, B152, B309, B739, CV4024, E81CC, ECC81, ECC801, ECC801S, M8162, QA2406, QB309
12AT7/ECC81	9A	12AT7, 12AT7WA, cont'd

Tube Type	Basing	Replacement
12AT7/ECC81, cont'd		12AT7WB, 6060, 6201, 6671, 6679, 7492, 7728, A2900, B152, B309, B739, E81CC, ECC81, ECC801, ECC801S, M8162, QA2406, QB309
12AT7WA	9A	12AT7WB, 6201#, 6201/12AT7WA, 6679/12AT7#
12AT7WB	9A	12AT7WA#, 6201#, 6201/12AT7WA, 6679/12AT7#
12AU6	7BK	12AU6A, 12BA6, 12BA6A, 12AW6#, 12BA6A, 12F31, HF93, HF94
12AU6A	7BK	12AU6*!, 12BA6A, 12AW6#, 12BA6A, 12F31, HF93, HF94
12AU7	9A	12AU7/5814/6680, 12AU7A, 12AU7A/ECC82, 12AU7W, 12AU7WA, 12AX7#, 12AX7A#, 12AX7A/ECC83#, 12AX7A/ECC83/7025#, 12AX7WA#, 12AX7WA/7025#, 5814*, 5814A*, 5814WA*, 5963, 6067, 6189, 6670, 6670/12AU7A, 6680, 7025#, 7316, 7489, 7730, B329, B749, CV4003, E82CC, ECC82, ECC82/12AU7A, ECC83#, ECC186, ECC802, ECC802S, M8136
12AU7/5814/6680	9A	12AU7, 12AU7A, 12AU7A/ECC82, 12AU7W, 12AU7WA, 12AX7#, 12AX7A#, 12AX7A/ECC83#, 12AX7A/ECC83/7025#, 12AX7WA#, 12AX7WA/7025#, 5814*, 5814A, 5814WA*, 5963, 6067, 6189, 6670, 6670/12AU7A, 6680, 7025#, 7316, 7489, 7730, B329, B749, CV4003, E82CC, ECC82, ECC82/12AU7A, ECC83#, ECC186, ECC802, ECC802S, M8136
12AU7A	9A	12AU7, 12AU7/5814/6680, 12AU7A/ECC82, 12AU7W, 12AU7WA, 12AX7#, 12AX7A#, 12AX7A/ECC83#, 12AX7A/ECC83/7025#, 12AX7WA#, 12AX7WA/7025#, 6670, 6670/12AU7A, 7025#, ECC82, ECCC82/12AU7A, ECC83#
12AU7A/ECC82	9A	12AU7A, 12AX7A#, 12AX7A/ECC83#, 12AX7A/ECC83/7025#, 12AU7W, 12AU7WA, 12AX7WA#, 12AX7WA/7025#, 6670, 6670/12AU7A, 7025#, ECC82, ECC83#
12AU7W	9A	12AT7WB, 12AU7A/ECC82#, 12AU7WA, 12AX7WA#, 12AX7WA/7025, 5814A#, 6189, 6189W, cont'd
12AU7W, cont'd		6201#, 6670#, 6670/12AU7A#, 7025#
12AU7WA	9A	12AT7WB, 12AU7A/ECC82#, 12AX7WA, 12AX7WA/7025#, 5814A#, 6189, 6189W, 6201#, 6670#, 6670/12AU7A#, 7025#
12AU8	9DX	-
12AV5	6CK	-
12AV5GA	6CK	12AV5GT*!
12AV5GT	6CK	12AV5GA
12AV6	7BT	12AT6, 12AT6A, 12AV6A, 12BC32, 12BK6, 12BT6, HBC90, HBC91
12AV6A	7BT	12AT6A, 12AV6*!, 12BK6*!, 12BT6*!
12AV7	9A	12RLL3#, 5965, 6829
12AW6	7CM	12AU6#, 12AU6A#
12AX3	12BL	-
12AX4	4CG	-
12AX4GT	4CG	12AX4GTA, 12AX4GTB, 12D4, 12D4A, 12DM4, 12DM4A, 12DQ4
12AX4GTA	4CG	12AX4GTB, 12D4, 12D4A, 12DM4, 12DM4A, 12DQ4, 12AX4GT*!
12AX4GTB	4CG	12D4A, 12DM4, 12DM4A, 12DQ4
12AX7	9A	6L13, 12AD7*, 12AU7#, 12AU7A#, 12AX7/5751/6681, 12AX7/ECC83, 12AX7A, 12AX7A/ECC83, 12AX7A/ECC83/7025, 12AX7WA, 12AX7WA/7025, 12BZ7*, 12DM7*, 12DF7, 12DT7, 5751*, 5751WA*, 6057, 6681, 7025, 7025A, 7494, 7729, B339, B759, CV4004, E83CC, ECC83, ECC803, M8137
12AX7/5751/6681	9A	6L13, 12AD7*, 12AU7#, 12AU7A#, 12AX7, 12AX7/ECC83, 12AX7A, 12AX7A/ECC83, 12AX7A/ECC83/7025, 12AX7WA, 12AX7WA/7025, 12BZ7*, 12DM7*, 12DF7, 12DT7, 5751*, 5751WA*, 6057, 6681, 7025, 7025A, 7494, 7729, B339, B759, CV4004, E83CC, ECC83, ECC803, M8137
12AX7/ECC83	9A	6L13, 12AX7, 12AX7A, 12AX7A/ECC83, 12AX7A/ECC83/7025, 12AX7WA, 12AX7WA/7025, 12DF7, 12DT7, 6057, 6681, 7025, 7025A, 7494, 7729, B339, B759, E83CC, ECC83, ECC803, M8137
12AX7A	9A	12AD7*, 12AU7#, 12AU7A#, 12AX7, 12AX7/ECC83, 12AX7A/ECC83, 12AX7A/ECC83/7025, 12BZ7*, 12DF7, 12DM7*, 12DT7, 7025, 7025A, ECC83

Tube Type	Basing	Replacement
12AX7A/ECC83	9A	12AX7, 12AX7/ECC83, 12AX7A, 12AX7A/ECC83/7025, 12AX7WA, 12AX7WA/7025 5761, 6057, 6681, 6681/12AX7A, 7025, 7494, 7729, ECC83
12AX7A/ECC83/7025	9A	12AX7, 12AX7/ECC83, 12AX7A, 12AX7A/ECC83, 12AX7WA, 12AX7A/7025, 5761, 6057, 6681, 6681/12AX7A, 7025, 7494, 7729, ECC83
12AX7WA	9A	12AX7A/ECC83#, 12AX7A/ECC83/7025#, 12AX7WA/7025, 5721, 6057, 6681, 6681/12AX7A, 7025#, 7494, 7729, ECC83#
12AX7WA/7025	9A	12AX7A/ECC83#, 12AX7A/ECC83/7025#, 12AX7WA, 5721, 6057, 6681, 6681/12AX7A, 7025, 7494, 7729, ECC83#
12AY3	9HP	12AY3A, 12AY3A/12BS3A, 12BS3, 12BS3A, 12CK3, 12CL3, 12DW4A
12AY3A	9HP	12AY3, 12AY3A/12BS3A, 12BS3, 12BS3A, 12CK3, 12CL3, 12DW4A
12AY3A/12BS3A	9HP	12AY3, 12AY3A, 12BS3, 12BS3A, 12CK3, 12CL3, 12DW4A
12AY7	9A	2082, 2082/12AY7, 6072*
12AZ7	9A	12AT7*, 12AZ7A
12AZ7A	9A	12AT7*, 12AZ7*!
12B3	9BD	-
12B4	9AG	12B4A
12B4A	9AG	12B4*!
12B6	6Y	-
12B6M	6Y	-
12B7	8V	12B7/14A7, 12B7ML, 14A7, 14A7/12B7, 14A7ML, 14A7ML/12B7ML, 14H7
12B7/14A7	8V	12B7, 12B7ML, 14A7, 14A7/12B7, 14A7ML, 14A7ML/12B7ML, 14H7
12B7ML	8V	12B7, 14A7, 14H7, 14A7/12B7, 14A7ML, 14A7ML/12B7ML
12B8	8T	12B8GT
12B8GT	8T	12B8
12BA6	7BK	12AU6, 12AU6A, 12BA6/HF93, 12BA6A, 12BZ6#, 12F31, HF93, HF94
12BA6/HF93	7BK	12AU6, 12AU6A, 12BA6, 12BA6A, 12BZ6#, 12F31, HF93, HF94
12BA6A	7BK	12AU6A, 12BA6*!
12BA7	8CT	-
12BB14	-	13GB5, 13GB5/XL500, XL500
12BC22	-	12AV6
12BC32	-	12AV6, 6550
12BD5	-	-
12BD6	7BK	12BA6, 12BA6A
12BE3	12GA	12BE3A
12BE3A	12GA	12BE3
12BE6	7CH	12BE6/HK90, 12BE6A, cont'd
12BE6, cont'd		12CS6, 12H31, HK90
12BE6/HK90	7CH	12BE6, 12BE6A, 12CS6, 12H31, HK90
12BE6A	7CH	12BE6*!, 12CS6*!
12BE7	-	UQ80
12BF6	7BT	12BF6/12BU6, 12BU6
12BF6/12BU6	7BT	12BF6, 12BU6
12BF11	12EZ	-
12BH7	9A	12BH7A, 6913
12BH7A	9A	12BH7*!
12BJ3	-	-
12BK5	9BQ	-
12BK6	7BT	12AT6, 12AT6A, 12AV6, 12AV6A, 12BC32, 12BT6, HBC90, HBC91
12BL6	7BK	12AF6, 12EK6#
12BM5	7DQ	12AQ5*
12BN6	7DF	12BN6A
12BN6A	7DF	12BN6*!
12BQ6	6AM	12BQ6GA, 12BQ6GA/12CU6, 12BQ6GTA, 12BQ6GTB, 12BQ6GTB/12CU6, 12CU6, 12DQ6, 12DQ6A, 12DQ6B, 12GW6
12BQ6GA	6AM	12BQ6GA/12CU6, 12BQ6GTA, 12BQ6GTB, 12BQ6GTB/12CU6, 12CU6, 12DQ6, 12DQ6A, 12DQ6B, 12GW6
12BQ6GA/12CU6	6AM	12BQ6GA, 12BQ6GA/12CU6, 12BQ6GTA, 12BQ6GTB, 12BQ6GTB/12CU6, 12CU6, 12DQ6, 12DQ6A, 12DQ6B, 12GW6
12BQ6GT	6AM	12BQ6GA, 12BQ6GA/12CU6, 12BQ6GTA, 12BQ6GTB, 12BQ6GTB/12CU6, 12CU6, 12DQ6, 12DQ6A, 12DQ6B, 12GW6, 12G-B6
12BQ6GTA	6AM	12BQ6GA, 12BQ6GTB, 12BQ6GTB/12CU6, 12CU6, 12DQ6, 12DQ6A, 12DQ6B, 12GW6
12BQ6GTB	6AM	12BQ6GTB/12CU6, 12CU6, 12DQ6A, 12DQ6B, 12GW6
12BQ6GTB/12CU6	6AM	12BQ6GTB, 12CU6, 12DQ6, 12DQ6A, 12DQ6B, 12GW6
12BR3	9CB	12AF3/12BR3/12RK19, 12BR3/12RK19, 12RK19
12BR3/12RK19	9CB	12AF3/12BR3/12RK19, 12BR3, 12RK19
12BR7	9CF	12BR7A, 8447
12BR7A	9CF	12BR7*!
12BS3	9HP	12AY3A, 12AY3A/12BS3A, 12BS3A, 12BS3A/12DW4A, 12CK3, 12CL3, 12DW4
12BS3A	9HP	12AY3A, 12AY3A/12BS3A, 12BS3A, 12BS3A/12DW4A, 12CK3, 12CL3, 12DW4
12BS3A/12DW4A	9HP	12AY3A, 12AY3A/12BS3A, 12BS3A, 12CK3, 12CL3, 12DW4A
12BT3	12BL	-
12BT6	7BT	12AT6, 12AT6A, 12AV6, 12AV6A, 12BC32, 12BK6, HBC90, HBC91
12BU6	7BT	12BF6, 12BF6/12BU6

Tube Type	Basing	Replacement
12BV7	9BF	12BV7/12BY7A, 12BY7, 12BY7A, 12BY7A/12BV7/12DQ7, 12DQ7
12BV7/12BY7A	9BF	12BV7, 12BY7, 12BY7A, 12BY7A/12BV7/12DQ7, 12DQ7
12BV11	12HB	-
12BW4	9DJ	-
12BX6	9AQ	10F18*, 13EC7*, W119*
12BY3	9CB	-
12BY7	9BF	12BY7A, 12BY7A/12BV7/12DQ7, 12BV7, 12BV7/12BY7A, 12DQ7, 7733, 8448, EL180
12BY7A	9BF	12BY7A/12BV7/12DQ7, 12DQ7, 12BV7*!, 12BV7/12BY7A, 12BY7*!
12BY7A/12BV7/12DQ7	9BF	12BV7#, 12BV7/12BY7A, 12BY7A, 12DQ7
12BZ6	7CM	12BA6#, 12BA6A#
12BZ7	9A	-
12C5	7CV	12AS5*, 12C5/12CU5, 12CU5, 12CU5/12C5, 12DM5*, 12R5
12C5/12CU5	7CV	12AS5*, 12C5, 12CU5, 12CU5/12C5, 12DM5*, 12R5
12C8	8E	12C8Y, 1664, 5660, VP12D
12C8Y	8E	12C8
12CA5	7CV	12AS5*, 12CA5/12EH5, 12ED5*, 12EH5, 12FX5*
12CA5/12EH5	7CV	12AS5*, 12CA5, 12ED5*, 12EH5, 12FX5*
12CD6	-	-
12CK3	9HP	12AY3A, 12AY3A/12BS3A, 12BS3A, 12BS3A/12DW4A, 12CK3/12CL3, 12CL3, 12DW4A
12CK3/12CL3	9HP	12AY3A, 12AY3A/12BS3A, 12BS3A, 12BS3A/12DW4A, 12CK3, 12CL3, 12DW4A
12CL3	9HP	12AY3A, 12AY3A/12BS3A, 12BS3A, 12BS3A/12DW4A, 12CK3, 12CK3/12CL3, 12DW4A
12CM6	9CK	-
12CN5	7CV	-
12CR5	9HC	-
12CR6	7EA	-
12CS5	9CK	12CM6*, 12DW5
12CS6	7CH	12BE6, 12BE6A, 12H31, HK90
12CT3	9RX	-
12CT8	9DA	-
12CU5	7CV	12AS5*, 12C5, 12C5/12CU5, 12CU5/12C5, 12DM5*, 12R5
12CU5/12C5	7CV	12AS5*, 12C5, 12CU5, 12C5/12CU5, 12DM5*, 12R5
12CU6	6AM	12BQ6, 12BQ6GTB, 12BQ6GTB/12CU6, 12DQ6, 12DQ6A, 12DQ6B, 12GW6
12CX6	7BK	12AF6, 12BL6, 12EZ6
12CY6	7BK	-
12D4	4CG	12AX4GTB, 12D4/12DM4, 12D4A, 12DM4, cont'd
12D4, cont'd		12DM4A, 12DQ4
12D4/12DM4	4CG	12AX4GTB, 12D4, 12D4A, 12DM4, 12DM4A, 12DQ4
12D4A	4CG	12DM4, 12DM4A, 12DQ4
12DA6/UF89	-	12DA6, UF89
12DA6	-	UF89
12DB5	9GR	-
12DE8	9HG	-
12DF5	9BS	-
12DF7	9A	6L13, 12AD7*, 12AX7, 12AX7A, 12AX7A/ECC83, 12AX7A/ECC83/7025, 12AX7WA, 12AX7WA/7025, 12BZ7*, 12DF7/12DT7, 12DM7*, 12DT7, 7025, 7025A, B339, B759, ECC83, ECC803
12DF7/12DT7	9A	6L13, 12AD7*, 12AX7, 12AX7A, 12AX7A/ECC83, 12AX7A/ECC83/7025, 12AX7WA, 12AX7WA/7025, 12BZ7*, 12DF7, 12DM7*, 12DT7, 7025, 7025A, B339, B759, ECC83, ECC803
12DJ8	9DE	-
12DK5	9GT	-
12DK6	7CM	-
12DK7	9HZ	-
12DL8	9HR	12DS7#, 12DS7A#
12DM4	4CG	12D4, 12D4/12DM4, 12D4A, 12DM4A, 12DQ4
12DM4A	4CG	12D4, 12D4/12DM4, 12D4A, 12DM4, 12DQ4
12DM5	7CV	11C5, 12AS5*, 12C5*, 12C5/12CU5*, 12CU5*, 12FX5, 12R5*
12DM7	9A	12AD7*, 12AX7*, 12AX7A*, 12AX7A/ECC83#, 12AX7A/ECC83/7025#, 12AX7WA#, 12AX7WA/7025#, 12BZ7*, 12DF7*, 12DT7*, 7025*, 7025A*, ECC83#
12DM8	-	-
12DN6	-	-
12DQ4	4CG	-
12DQ6	6AM	12DQ6A, 12DQ6B, 12DQ6B/12GW6, 12GW6, 12GW6/12DQ6B
12DQ6A	6AM	12DQ6B, 12DQ6B/12GW6, 12GW6, 12GW6/12DQ6B
12DQ6B	6AM	12DQ6B/12GW6, 12GW6, 12GW6/12DQ6B
12DQ6B/12GW6	6AM	12DQ6B, 12GW6, 12GW6/12DQ6B
12DQ7	9BF	12BV7*!, 12BV7/12BY7A, 12BY7A, 12BY7A/12BV7/12DQ7
12DS7	9JU	12DS7A
12DS7A	9JU	12DS7
12DT5	9HN	-
12DT6	7EN	-
12DT7	9A	6L13, 12AD7*, 12AX7, 12AX7A, 12AX7A/ECC83, 12AX7A/ECC83/7025, 12AX7WA, 12AX7WA/7025, 12BZ7*, cont'd

Tube Type	Basing	Replacement
12DT7, cont'd		12DF7, 12DF7/12DT7, 12DM7*, 5751, 6681, 7025, 7025A, B339, B759, ECC83, ECC803
12DT8	9AJ	-
12DU7	9JX	-
12DV7	9JY	-
12DV8	9HR	-
12DW4	9HP	12AY3A, 12AY3A/12BS3A, 12BS3A, 12BS3A/12DW4A, 12CK3, 12CL3, 12DW4A
12DW4A	9HP	12AY3A, 12AY3A/12BS3A, 12BS3A, 12BS3A/12DW4A, 12CK3, 12CL3
12DW5	9CK	-
12DW7	9A	12DW7/7247, 7247
12DW7/7247	9A	12DW7, 7247
12DW8	9JC	-
12DY8	9JD	-
12DZ6	7BK	12DZ6/12EA6, 12EA6, 12EK6, 12EK6/12DZ6/12EA6
12DZ6/12EA6	7BK	12DZ6, 12EA6, 12EK6, 12EK6/12DZ6/12EA6
12DZ8	9JE	12FY8*
12E1	-	-
12E5	6Q	-
12E5GT	6Q	12J5, 12J5GT
12E13	-	6CA7, 6CA7/EL34, 6550#, EL34, EL34/6CA7
12EA6	7BK	12DZ6, 12DZ6/12EA6, 12EK6, 12EK6/12DZ6/12EA6
12EC8	9FA	-
12ED5	7CY	12CA5*, 12EH5*, 12FX5
12EF6	7S	-
12EG6	7CH	-
12EH5	7CV	12CA5, 12CA5/12EH5, 12CU5/12C5, 12ED5*, 12FX5*
12EH8	-	-
12EK6	7BK	12DZ6, 12DZ6/12EA6, 12EA6, 12EK6/12DZ6/12EA6
12EK6/12DZ6/12EA6	7BK	12DZ6, 12DZ6/12EA6, 12EA6, 12EK6
12EL6	7FB	-
12EM6	9HV	-
12EN6	7S	12L6GT, 12W6GT
12EQ7	9LQ	12KL8#
12EW6	-	-
12EX6	-	12DZ6, 12EA6, 12EK6, 12EK6/12DZ6/12EA6
12EZ6	7BK	12CX6
12F5	5M	12F5GT
12F5GT	5M	12F5
12F8	9FH	-
12F31	-	12BA6, 12GW6
12FA6	7CH	-
12FB5	9CV	30P12
12FG6	-	UM84
12FK6	7BT	12FM6
12FM6	7BT	12FK6
12FQ7	9LP	-
12FQ8	9KT	-
12FR8	9KU	-
12FS5	-	-
12FT6	7BT	12AE6, 12AE6A, 12BF6
12FV7	9A	-

Tube Type	Basing	Replacement
12FX5	7CV	12ED5, 12EH5*
12FX8	9KV	12FX8A
12FX8A	9KV	12FX8
12FY8	9EX	-
12G4	6BG	12G4/12H4, 12H4
12G4/12H4	6BG	12G4, 12H4
12G7	7V	12G7G, 12Q7G, 12Q7GT, 12Q7GT/G
12G7G	7V	12Q7G, 12Q7GT, 12Q7GT/G
12G8	9CZ	-
12G11	12BU	-
12GA6	7CH	-
12GB3	6AM	12BQ6GT, 12BQ6GTB/12CU6
12GB6	6AM	12AV6, 12GW6/12DQ6B
12GB7	6AM	12DQ6B, 12DQ6B/12GW6, 12GW6, 12GW6/12DQ6B, 13CM5
12GC6	8JX	-
12GE5	12BJ	-
12GJ5	9QK	12GJ5A
12GJ5A	9QK	12GJ5
12GK17	-	12D4
12GN6	7FW	-
12GN7	9BF	12GN7A, 12GN7A/12HG7, 12HG7*!, 12HG7/12GN7A
12GN7A	9BF	12GN7A/12HG7, 12HG7*!, 12HG7/12GN7A
12GN7A/12HG7	9BF	12GN7A, 12HG7, 12HG7/12GN7A
12GT5	9NZ	12GT5A
12GT5A	9NZ	12GT5
12GV5	12DR	-
12GW6	6AM	12DQ6B, 12DQ6B/12GW6, 12GW6/12DQ6B
12GW6/12DQ6B	6AM	12DQ6B, 12DQ6B/12GW6, 12GW6
12H4	7DW	12G4, 12G4/12H4
12H6	7Q	-
12H31	-	12BE6
12HB25	-	-
12HE7	12FS	-
12HG7	9BF	12GN7*, 12GN7A!, 12GN7A/12HG7, 12HG7/12GN7A
12HG7/12GN7A	9BF	12GN7A, 12GN7A/12HG7, 12HG7
12HL5	9QW	-
12HL7	9BF	-
12HU8	9NJ	PLL80
12J5	6Q	12J5GT, 12L5WGT
12J5GT	6Q	12J5, 12J5WGT
12J5WGT	6Q	12J5GT#
12J7	7R	12J7G, 12J7GT, 12J7GT/G
12J7G	7R	12J7, 12J7GT, 12J7GT/G
12J7GT	7R	12J7, 12J7G, 12J7GT/G
12J7GT/G	7R	12J7, 12J7G, 12J7GT
12J8	9GC	-
12JB6	9QL	12JB6A
12JB6A	9QL	12JB6
12JF5	12JH	-
12JN6	12FK	12JN6A
12JN6A	12FK	12JN6
12JN8	9FA	-
12JQ6	9RA	-
12JS6	12FY	-
12JT6	9QU	12JT6A
12JT6A	9QU	12JT6
12JZ8	-	-

Tube Type	Basing	Replacement
12K5	7FD	-
12K7	7R	12K7G, 12K7GT, 12K7GT/G, W76
12K7G	7R	12K7GT, 12K7GT/G, W76
12K7GT	7R	12K7G, 12K7GT/G, OF5, W76
12K7GT/G	7R	12K7G, 12K7GT, 0F5, W76
12K8	8K	12K8GT, 12K8Y, X71M, X76M
12K8GT	8K	12K8, 12K8Y, X71M, X76M
12K8Y	8K	12K8, 12K8GT, X71M, X76M
12KL8	9LQ	-
12L6	-	12EN6, 12L6GT, 12W6GT, 1632*
12L6GT	7AC	12EN6, 12W6GT, 1632*
12L8	-	12L8GT, 1644
12L8GT	8BU	1644
12MD8	9RQ	-
12NK7	-	-
12Q7	7V	12G7G, 12Q7G, 12Q7GT, 12Q7GT/G, DH74, DH76, DL74M
12Q7G	7V	12G7G, 12Q7,12Q7GT, 12Q7GT/G, DH74, DH76, DL74M
12Q7GT	7V	12G7G, 12Q7, 12Q7G, 12Q7GT/G, DH74, DH76, DL74M
12Q7GT/G	7V	12G7G, 12Q7, 12Q7G, 12Q7GT, DH74, DH76, DL74M
12R5	7CV	-
12RK19	9CB	12AF3/12BR3/12RK19, 12BR3, 12BR3/12RK19
12RLL3	-	12AV7
12RLL5	-	12FQ7
12S7	8GX	UAF42#, WD142#
12S8	8CB	12S8G, 12S8GT
12S8G	8CB	12S8, 12S8GT
12S8GT	8CB	12S8, 12S8G
12SA7	8R	12SA7G, 12SA7GT, 12SA7GT/G, 12SA7GTY, 12SA7Y, 12SY7, 12SY7GT
12SA7G	8R	12SA7, 12SA7GT, 12SA7GT/G, 12SA7GTY, 12SA7Y, 12SY7, 12SY7GT
12SA7GT	8R	12SA7, 12SA7G, 12SA7GT/G, 12SA7GTY, 12SA7Y, 12SY7, 12SY7GT
12SA7GT/G	8R	12SA7, 12SA7G, 12SA7GT, 12SA7GTY, 12SA7Y, 12SY7, 12SY7GT
12SA7GTY	8R	12SA7, 12SA7G, 12SA7GT, 12SA7GT/G, 12SA7Y, 12SY7, 12SY7GT
12SA7Y	8R	12SA7#, 12SA7G, 12SA7GT, 12SA7GT/G, 12SA7GTY, 12SA7GT
12SC7	8S	1634
12SF5	6AB	12SF5GT
12SF5GT	6AB	12SF5
12SF7	7AZ	12SF7GT, 12SF7Y
12SF7GT	7AZ	12SF7, 12SF7Y
12SF7Y	7AZ	12SF7, 12SF7GT
12SG7	8BK	12SG7GT, 12SG7Y
12SG7GT	8BK	12SG7, 12SG7Y
12SG7Y	8BK	12SG7#, 12SG7GT
12SH7	8BK	12SH7GT
12SH7GT	8BK	12SH7
12SJ7	8N	12SJ7GT
12SJ7GT	8N	12SJ7
12SK7	8N	12SK7G, 12SK7GT, 12SK7GT/G, 12SK7GTY, 12SK7Y, 5661
12SK7G	8N	12SK7, 12SK7GT, 12SK7GT/G, 12SK7GTY, 12SK7Y, 5661
12SK7GT	8N	12SK7, 12SK7G, 12SK7GT/G, 12SK7GTY, 12SK7Y, 5661
12SK7GT/G	8N	12SK7, 12SK7G, 12SK7GT, 12SK7GTY, 12SK7Y, 5661
12SK7GTY	8N	12SK7, 12SK7G, 12SK7GT, 12SK7GT/G, 12SK7Y, 5661
12SK7Y	8N	12SK7#, 12SK7G, 12SK7GT, 12SK7GT/G, 12SK7GTY, 5661
12SL7	8BD	2C52, 12SL7GT
12SL7GT	8BD	2C52*
12SN7	8BD	12SN7GT, 12SN7GTA, 12SX7GT, 13D2, B36
12SN7GT	8BD	12SN7GTA, 12SX7GT, 13D2, B36
12SN7GTA	8BD	12SN7GT, 12SX7GT, 13D2, B36
12SQ7	8Q	12SQ7G, 12SQ7GT, 12SQ7GT/G, OBC3
12SQ7G	8Q	12SQ7, 12SQ7GT, 12SQ7GT/G
12SQ7GT	8Q	12SQ7, 12SQ7G, 12SQ7GT/G
12SQ7GT/G	8Q	12SQ7, 12SQ7G, 12SQ7GT
12SR7	8Q	12SR7/12SW7, 12SR7GT, 12SW7, 12SW7GT
12SR7/12SW7	8Q	12SR7, 12SR7GT, 12SW7, 12SW7GT
12SR7GT	8Q	12SR7, 12SW7, 12SW7GT
12SS7	-	-
12ST7	-	-
12SV6	-	-
12SV7	-	-
12SW7	8Q	12SR7, 12SR7/12SW7, 12SR7GT, 12SW7GT
12SW7GT	8Q	12SR7, 12SR7/12SW7, 12SR7GT, 12SW7
12SX7	8BD	12SN7GT, 12SN7GTA, 12SX7GT, B36
12SX7GT	8BD	12SN7GT, 12SN7GTA, B36
12SY7	8R	12SA7, 12SA7G, 12SA7GT, 12SA7GT/G, 12SA7GTY, 12SA7Y, 12SY7GT
12SY7GT	8R	12SA7, 12SA7G, 12SA7GT, 12SA7GT/G, 12SA7GTY, 12SA7Y, 12SY7
12T10	12EZ	-
12TE8	-	-
12TE9	-	-
12U7	9A	-
12V6	7AC	12A6*#, 12A6G*#, 12A6GT*#, 12A6GTY*#, 12A6Y*#, 12V6GT
12V6GT	7AC	12A6*#, 12A6G*#, 12A6GT*#, 12A6GTY*#, 12A6Y*#
12W6	7AC	12EN6, 12L6GT, 12W6GT
12W6GT	7AC	12EN6, 12L6GT
12X3	-	-
12X4	5BS	HZ90
12Z3	4G	14Z3, H250, HZ90

Tube Type	Basing	Replacement
12Z5	6K	6Z5, 6Z5/12Z5
13	4C	5X3, 80, 83V, 88
13CM5	8GT	12GB7, XL36
13CW4	12AQ	-
13D1	-	6042
13D2	9HF	6SN7GT, 6SN7GTB
13D3	9HF	6158
13DE7	9HF	15EW7
13DR7	9HF	13FD7, 13FR7
13E1	-	-
13EA7	-	-
13EC7	9AQ	10F18, 12BX6*, W110, W119
13EM7	8BD	13EM7/15EA7, 15EA7
13EM7/15EA7	8BD	13EM7, 15EA7
13FD7	9HF	13FR7
13FM7	12EJ	13FM7/15FM7, 15FM7
13FM7/15FM7	12EJ	13FM7, 15FM7
13FR7	9HF	13FD7
13GB5	9NH	12B-B14, 13GB5/XL500, XL500
13GB5/XL500	9NH	13GB5, XL500
13GC8	9KZ	30L1, 30PL10, LN319
13GF7	9QD	13GF7A
13GF7A	9QD	13GF7
13J10	12BT	13J10/13Z10, 13Z10, 13Z10/13J10
13J10/13Z10	12BT	13J10, 13Z10, 13Z10/13J10
13JZ6	-	13JZ8
13JZ8	12DZ	13JZ8A
13JZ8A	12DZ	-
13V10	12EZ	12AE10
13Z10	12BT	13J10, 13J10/13Z10, 13Z10/13J10
13Z10/13J10	12BT	13J10, 13J10/13Z10, 13Z10
14	5E	-
14A4	5AC	-
14A5	6AA	-
14A7	8V	12B7, 12B7/14A7, 12B7ML, 14A7/12B7, 14A7ML, 14A7ML/12B7ML, 14H7
14A7/12B7	8V	12B7, 12B7/14A7, 12B7ML, 14A7, 14A7ML, 14A7ML/12B7ML, 14H7
14A7ML	8V	12B7, 12B7/14A7, 12B7ML, 14A7, 14A7/12B7, 14A7ML/12B7ML, 14H7
14A7ML/12B7ML	8V	12B7, 12B7/14A7, 12B7ML, 14A7, 14A7/12B7, 14A7ML, 14H7
14AF7	8AC	14AF7/XXD, 14N7*, XXD
14AF7/XXD	8AC	14AF7, XXD
14B6	8W	-
14B8	8X	-
14BL11	12GC	-
14BR11	12GL	-
14C5	6AA	-
14C7	8V	1280
14DA7	9EF	-
14E6	8W	-
14E7	8AE	14R7
14F7	8AC	-
14F8	8BW	-
14G6	9Z	10LD13, DH119, UBC81
14GT8	9KR	14GT8/14JG8, 14GT8A, 14JG8, 7724
14GT8/14JG8	9KR	14GT8, 14GT8A, 14JG8, 7724
14GT8A	9KR	14GT8!, 14JG8!
14GW8	-	14GW8/PCL86, PCL86
14GW8/PCL86	-	14GW8, PCL86
14H7	8V	12B7, 12B7ML, 14A7, 14A7/12B7, 14A7ML, 14A7ML/12B7ML
14J7	8BL	14S7
14JG8	9KR	14GT8, 14GT8/14JG8, 14GT8A
14K7	8GY	141TH, UCH42, UCH43, X142
14L7	8GZ	10LD3, 14L7/UBC41, 141DDT, DH118, DH142, UBC41
14L7/UBC41	8GZ	10LD3, 14L7, 141DDT, DH118, DH142, UBC41
14N7	8AC	14AF7*
14Q7	8AL	-
14R7	8AE	14E7
14S7	8BL	14J7
14V7	8V	-
14W7	8BJ	-
14X7	8BZ	-
14Y4	5AB	-
14Y7	9Q	-
14Z3	4G	12Z3, HZ90
15	5F	-
15A6	9AR	15A6/PL83, N153, N309, PL83
15A6/PL83	9AR	15A6, N153, N309, PL83
15A8	8GS	-
15AB9	10N	17AB9
15AF11	12DP	15BD11, 15BD11A
15AF11A	12DP	15BD11, 15BD11A
15BD7	-	-
15BD11	12DP	15AF11, 15BD11A
15BD11A	12DP	15AF11, 15BD11
15CW5	9CV	15CW5/PL84, 30P18, N379, PL84
15CW5/PL84	9CV	15CW5, 30P18, N379
15DQ8	9HX	15DQ8/PCL84, PCL84
15DQ8/PCL84	9HX	15DQ8/PCL84
15E	-	-
15EA7	8BD	13EM7, 13EM7/15EA7
15EM7	-	-
15EW6	7CM	-
15EW7	9HF	13DE7
15FM7	12EJ	13FM7, 13FM7/15FM7
15FM7/13FM7	12EJ	13FM7, 15FM7
15FY7	12EO	-
15HA6	-	-
15HB6	9NW	-
15KY8	9QT	15KY8A, 15KY8A/17DL8, 15MX8, 17DL8
15KY8A	9QT	15KY8, 15KY8A/17DL8, 15MX8, 17DL8
15KY8A/17DL8	9QT	15KY8, 15KY8A, 15MX8, 17DL8
15LE8	9QZ	-
15MF8	12DZ	-
15MX8	9QT	15KY8A
15R	-	4B30, 8022
16	4B	16B, 81
16A	4B	-
16B	4B	-
16A5	9BL	15CW5#, 16A5/PL82, 30P16, 30P18, 163PEN, N154, N329, N379, PL82, PL84
16A5/PL82	9BL	15CW5#, 16A5, 30P16, 30P18, 163PEN, N154, N329, N379, PL82, PL84
16A8	9EX	16A8/PCL82, 30PL12, cont'd

Tube Type	Basing	Replacement
16A8, cont'd		N369, PCL82
16A8/PCL82	9EX	16A8, 30PL12, N369, PCL82
16A	-	6AM5, 6AM5/EL91, EL91
16AK9	12GZ	-
16AQ3	9CB	16AQ3/XY88, XY88
16AQ3/XY88	9CB	16AQ3, XY88
16B	4B	16, 81
16BQ11	12DM	-
16BX11	12CA	-
16CB28	-	-
16GB16	-	-
16GK6	9GK	-
16GK8	9JE	30PL13, 30PL14, PCL88, PCL800
16GY5	12DR	-
16KA6	12GH	-
16LU8	12DZ	16LU8A, 16MY8
16LU8A	12DZ	16LU8, 16MY8
16MY8	12DZ	16LU8A
16Y9	10L	PFL200
17	5A	-
17A8	9DC	17A8/19EA8A, 19EA8, 19EA8A
17A8/19EA8A	9DC	17A8, 19EA8, 19EA8A
17AB9	10N	15AB9
17AB10	12BT	17AB10/17X10, 17X10
17AB10/17X10	12BT	17AB10, 17X10
17AV5	6CK	17AV5GA
17AV5GA	6CK	17AV5
17AX3	12BL	-
17AX4	4CG	17AX4GT, 17AX4GTA, 17AX4GTA/17DM4A, 17D4, 17D4A, 17DM4, 17DM4A, 17D4A, 17DQ4
17AX4GT	4CG	17AX4GTA, 17AX4GTA/17DM4A, 17D4, 17D4A, 17DM4, 17DM4A, 17D4A, 17DQ4
17AX4GTA	4CG	17AX4GTA/17DM4A, 17D4, 17D4A, 17DM4, 17DM4A, 17DQ4
17AX4GTA/17DM4A	4CG	17AX4GTA, 17D4, 17D4A, 17DM4, 17DM4A, 17DQ4
17AX10	-	-
17AY3	9HP	17AY3A, 17AY3A/17BS3A, 17BS3, 17BS3A, 17CK3, 17CL3, 17DW4A
17AY3A	9HP	17AY3, 17AY3A/17BS3A, 17BS3, 17BS3A, 17CK3, 17CL3, 17DW4A
17AY3A/17BS3A	9HP	17AY3, 17AY3A, 17BS3, 17BS3A, 17CK3, 17CL3, 17DW4A
17BB14	9NH	-
17BE3	12GA	17BE3/17BZ3, 17BE3A, 17BZ3
17BE3/17BZ3	12GA	17BE3, 17BE3A, 17BZ3
17BE3A	12GA	17BE3, 17BZ3
17BF11	12EZ	17BF11A
17BF11A	12EZ	17BF11
17BH3	9HP	17BH3A
17BH3A	9HP	17BH3
17BJ6	-	-
17BQ6	6AM	17BQ6GTB, 17DQ6, 17DQ6A, 17DQ6B, 17GW6, 17GW6/17DQ6B
17BQ6GTB	6AM	17DQ6, 17DQ6A, cont'd

Tube Type	Basing	Replacement
17BQ6GTB, cont'd		17DQ6B, 17GW6, 17GW6/17DQ6B
17BR3	9CB	17BR3/17RK19, 17RK19
17BR3/17RK19	9CB	17BR3, 17RK19
17BS3	9HP	17AY3A, 17AY3A/17BS3A, 17BS3A, 17BS3A/17DW4A, 17CK3, 17CL3, 17DW4A
17BS3A	9HP	17AY3A, 17AY3A/17BS3A, 17BS3, 17BS3A/17DW4A, 17CK3, 17CL3, 17DW4A
17BS3A/17DW4A	9HP	17AY3A, 17AY3A/17BS3A, 17BS3A, 17DW4A, 17CK3, 17CL3
17BW3	12FX	17AX3*, 17BE3*, 17BE3A*
17BZ3	12FX	17BE3, 17BE3/17BZ3, 17BE3A
17C5	7CV	17C5/17CU5, 17CU5, 17CU5/17C5, 17R5
17C5/17CU5	7CV	17C5, 17CU5, 17CU5/17C5, 17R5
17C8	9T	17C8/UBF80, 17N8, 171DDP, UBF80
17C8/UBF80	9T	17C8, 17N8, 171DDP, UBF80
17C9	10F	17C9A
17C9A	10F	17C9!
17CA5	7CV	-
17CA9	-	-
17CB24	-	-
17CK3	9HP	17CL3, 17DW4A
17CL3	9HP	17AY3A, 17AY3A/17BS3A, 17BS3A, 17BS3A/17DW4A, 17CK3, 17DW4A
17CQ4	4CG	17DE4
17CT3	9RX	-
17CU5	7CV	17C5, 17C5/17CU5, 17CU5/17C5, 17R5
17CU5/17C5	7CV	17C5, 17C5/17CU5, 17CU5, 17R5
17D4	4CG	17AX4GTA, 17AX4GTA/17DM4A, 17D4A, 17DM4, 17DM4A, 17DQ4
17D4A	4CG	17AX4GTA, A7AX4GTA/17DM4A, 17D4A, 17DM4, 17DM4A, 17DQ4
17DE4	4CG	17CQ4
17DL8	-	15KY8A, 15KY8A/17DL8
17DM4	4CG	17AX4GTA, 17AX4GTA/17DM4A, 17D4A, 17DM4A, 17DQ4
17DM4A	4CG	17AX4GTA, 17AX4GTA/17DM4A, 17D4A, 17DQ4
17DQ4	4CG	17AX4GTA, 17AX4GTA/17DM4A, 17D4, 17DM4A
17DQ6	6AM	17DQ6A, 17DQ6B, 17DQ6B/17GW6, 17GW6, 17GW6/17DQ6B
17DQ6A	6AM	17DQ6B, 17DQ6B/17GW6, 17GW6, 17GW6/17DQ6B
17DQ6B	6AM	17DQ6B/17GW6, 17GW6, 17GW6/17DQ6B
17DQ6B/17GW6	6AM	17DQ6B, 17GW6, 17GW6/17DQ6B
17DW4	9HP	17AY3A, 17AY3A/17BS3A, 17BS3A, 17BS3A/17DW4A, 17CK3, 17CL3, 17DW4A
17DW4A	9HP	17AY3A, 17AY3A/17BS3A, 17BS3A, 17BS3A/17DW4A, 17CK3, 17CL3

Tube Type	Basing	Replacement
17EM7	-	-
17EW8	9AJ	17EW8/HCC85, HCC85
17EW8/HCC85	9AJ	17EW8, HCC85
17GB3	6AM	-
17GE5	12BJ	-
17GJ5	9QK	17GJ5A
17GJ5A	9QK	17GJ5
17GT5	9NZ	17GT5A
17GT5A	9NZ	17GT5
17GV5	12DR	-
17GW6	6AM	17DQ6B, 17DQ6B/17GW6, 17GW6/17DQ6B
17GW6/17DQ6B	6AM	17GW6, 17DQ6B, 17DQ6B/17GW6
17H3	9FK	-
17HB25	17HB25	-
17HC8	9EX	-
17JB6	9QL	17JB6A
17JB6A	9QL	17JB6
17JF6	9QL	-
17JG6	9QU	17JG6A, 17KV6
17JG6A	9QU	17JG6, 17KV6
17JK8	9AJ	-
17JM6	12FJ	17JM6A
17JM6A	12FJ	17JM6
17JN6	12FK	17JN6A
17JN6A	12FK	17JN6
17JQ6	9RA	17JQ6A
17JQ6A	9RA	17JQ6
17JR6	9QU	17JG6, 17JG6A
17JR8	-	-
17JT6	9QU	17JT6A
17JT6A	9QU	17JT6
17JZ8	12DZ	17JZ8A
17JZ8A	12DZ	17JZ8
17KV6	9QU	17KV6A
17KV6A	9QU	17KV6
17KW6	-	-
17L6	7AC	17L6GT, 17W6GT
17L6GT	7AC	17L6, 17W6GT
17LD8	9QT	15KY8A
17N8	-	17C8, UBF80
17R5	7CV	-
17RHH2	-	-
17RK19	9CB	17BR3, 17BR3/17RK19
17W6	7AC	17L6GT, 17W6GT
17W6GT	7AC	17L6GT, 17W6
17X10	12BT	17AB10, 17AB10/17X10
17Y9	-	-
17Z3	9CB	17Z3/PY81, A61, PY81, PY83, PY800, PY801, U153, U193, U251, U349
17Z3/PY81	9CB	17Z3, A61, PY81, PY83, PY800, PY801, U153, U193, U251, U349
18	6B	-
18A5	6CK	-
18AJ10	12EZ	-
18AK5	-	6028
18AQ5	-	-
18D2	-	-
18DZ8	9JE	-
18FW6	7BK	18FW6A, 18GD6, 18GD6A, 19MR9
18FW6A	7BK	18FW6!, 18GD6!, 18GD6A, 19MR9
18FX6	7CH	18FX6A, X107#
18FX6A	7CH	18FX6!, X107#
18FY6	7BT	18FY6A, 18GE6, 18GE6A
18FY6A	7BT	18FY6!, 18GE6!, 18GE6A
18GB5	9NH	18GB5/LL500, LL500
18GB5/LL500	9NH	18GB5, LL500
18GD6	7BK	18FW6, 18FW6A, 18GD6A, 19MR9
18GD6A	7BK	18FW6!, 18FW6A, 18GD6!, 19MR9
18GE6	7BT	18FY6, 18FY6A, 18GE6A
18GE6A	7BT	18FY6!, 18FY6A, 18GE6!
18GV8	9LY	18GV8/PCL85, PCL85
18GV8/PCL85	9LY	18GV8, PCL85
18HB8	9ME	-
18J6	-	-
18RAL1	-	-
19	6C	-
19A3	-	-
19AJ8	-	19D8
19AQ5	7BZ	HL90
19AU4	4CG	17CQ4, 17DE4, 19AU4GT, 19AU4GTA
19AU4GT	4CG	17CQ4, 17DE4, 19AU4, 19AU4GTA
19AU4GTA	4CG	17CQ4, 17DE4, 19AU4, 19AU4GT
19BD	-	19X3
19BG6	5BT	19BG6G, 19BG6GA
19BG6G	5BT	19BG6, 19BG6GA
19BG6GA	5BT	19BG6, 19BG6G
19BR5	-	-
19BX6	-	UF80
19BY5	-	UF85
19BY7	-	-
19C8	9E	19T8, 19T8A, HABC80
19CG3	12HF	19CG3/19DQ3
19CG3/19DQ3	12HF	19CG3, 19DQ3
19CL8	9FX	19CL8A, 19CL8A/19JN8, 19CL8B, 19JN8, 19JN8/19CL8A
19CL8A	9FX	19CL8A/19JN8, 19CL8B, 19JN8, 19JN8/19CL8A
19CL8A/19JN8	9FX	19CL8A, 19CL8B, 19JN8, 19JN8/19CL8A
19CL8B	9FX	19CL8A!, 19JN8!
19CS4	8JT	PY301, U191, U339
19D8	9CA	10C14, 19AJ8, 19D8/UCH81, UCH81, X119
19D8/UCH81	9CA	10C14, 19AJ8, 19D8, UCH81, X119
19DC8	-	-
19DE3	12HX	-
19DE7	9HF	19EW7, 20EW7
19DK3	9SG	-
19DQ3	12HF	19CG3/19DQ3
19EA8	9AE	17A8, 17A8/19EA8A, 19EA8A
19EA8A	9AE	17A8!, 17A8/19EA8A, 19EA8!
19EW7	9HF	20EW7
19EZ8	9KA	-
19FK6	-	-
19FL8	-	-
19FX5	7CV	-
19GQ7	9RB	-
19HR6	7BK	-
19HS6	7BK	-
19HV8	9FA	-
19J6	7BF	-

Tube Type	Basing	Replacement
19JN8	9FA	19CL8A/19JN8, 19CL8A, 19CL8B, 19JN8/19CL8A
19JN8/19CL8A	9FA	19CL8A/19JN8, 19JN8, 19CL8A, 19CL8B
19KF6	-	N22LL
19KG8	9LY	-
19LF8	-	-
19MR9	7BK	18FW6A, 18GD6, 18GD6A
19MR10	-	18GD6, 18GD6A
19MR19	-	18FW6A
19Q9	10H	-
19RP11	-	-
19SU	-	19Y3
19T8	9E	19C8, 19T8A, HABC80
19T8A	9E	19T8!, 19C8!, HABC80
19U3	-	19X3
19V8	9AH	-
19W3	-	19X3
19X3	9BM	19BD, 19U3, 19W3, PY80, U152, U309
19X8	9AK	-
19Y3	9BM	19X3, 19SU, PY82, U154, U192, U319
20	4D	-
20A3	-	2D21#, 2D21/5727#, 2D21/PL51#, 5727#, 5727/2D21#, PL51#
20AQ3	9CB	20AQ3/LY88, LY88
20AQ3/LY88	9CB	20AQ3, LY88
20D3	-	12AH8
20D4	-	6AJ8, 6AJ8/ECH81, ECH81, ECH81/6AJ8
20EQ7	9LG	-
20EW7	9HF	19EW7
20EZ7	9MJ	-
20GF6	-	30P4, 30P19, N389, PL302
20J8	8H	-
20LF6	12GW	-
20LX6	-	-
20P18	-	-
21A6	9AS	21A6/21B6, 21A6/PL81, 21B6, 213PEN, N152, N339, N359, PL81, PL820
21A6/21B6	9AS	21A6, 21A6/PL81, 21B6, 213PEN, N152, N339, N359, PL81, PL820
21A6/PL81	9AS	21A6/21B6, 21A6, 21B6, 213PEN, N152, N339, N359, PL81, PL820
21A7	8AR	-
21B6	9AS	21A6/21B6, 21A6/PL81, 213PEN, N152, N339, N359, PL81, PL820
21EX6	5BT	25CD6GB, 25DN6
21GY5	12DR	-
21HB5	12BJ	21HB5A
21HB5A	12BJ	21HB5
21HD5	12ES	-
21HJ5	12FL	-
21JS6	12FL	21JS6A, 21JS6A/23JS6, 23JS6
21JS6A	12FY	21JS6A/23JS6, 23JS6A
21JS6A/23JS6	12FL	21JS6A, 23JS6, 23JS6A
21JV6	12FK	-
21JZ6	12GD	-
21KA6	12GH	-
21KE6	-	-
21KQ6	9RJ	LL521
21LG6	12HL	21LG6A
21LG6A	12HL	21LG6
21LR8	9QT	-
21LU8	12DZ	21MY8
21MY8	12DZ	21LU8
22	4K	-
22BH3	9HP	22BH3A
22BH3A	9HP	22BH3
22BW3	12FX	25CG3
22DE4	4CG	-
22JF6	9QL	22JF6/22KM6, 22KM6
22JF6/22KM6	9QL	22JF6, 22KM6
22JG6	9QU	22JG6A, 22JR6
22JG6A	9QU	22JG6, 22JR6
22JR6	9QU	-
22JU6	9QL	-
22KM6	9QL	22JF6, 22JF6/22KM6
22KV6	9QL	22KV6A
22KV6A	9QL	22KV6
23JS6	12FY	21JS6A, 21JS6A/23JS6, 23JS6A
23JS6A	12FY	21JS6A, 21JS6A/23JS6, 23JS6
23JZ8	12DZ	-
23MB6	12FY	-
23Z9	12GZ	-
24	-	-
24A	5E	24S, 35#, 35/51#, 35S#, 51S#, 35S/51S#
24AX4	-	-
24BF11	12EZ	-
24G	-	3C24, 3C24/24G
24GA7	12EB	-
24JE6	9QL	24JE6A, 24JE6A/24LQ6, 24JE6C, 24JE6C/24LQ6, 24LQ6, 24LQ6/24JE6C
24JE6A	9QL	24JE6A/24LQ6, 24JE6C, 24JE6C/24LQ6, 24LQ6, 24LQ6/24JE6C
24JE6B	9QL	24JE6A/24LQ6, 24JE6C, 24JE6C/24LQ6, 24LQ6, 24LQ6/24JE6C
24JE6C	9QL	24JE6A/24LQ6, 24JE6C/24LQ6, 24LQ6, 24LQ6/24JE6C
24JE6C/24LQ6	9QL	24JE6A/25LQ6, 24JE6C, 24LQ6, 24LQ6/24JE6C
24JZ8	12DZ	25JZ8
24LQ6	9QL	24JE6A, 24JE6A/24LQ6, 24JE6C, 24LQ6/24JE6C
24LQ6/24JE6C	9QL	24JE6A/24LQ6, 24JE6C, 24JE6C/24LQ6, 24LQ6
24LZ6	9QL	-
24S	5E	24A, 35#, 35/51#, 51#, 35S#, 51S#, 35S/51S#
25	6M	1B5, 1B5/25S, 25S
25A6	7S	25A6G, 25A6GT, 25A6GT/G, 25A6MG, KT33
25A6G	7S	25A6, 25A6GT, 25A6GT/G, 25A6MG
25A6GT	7S	25A6, 25A6G, 25A6GT/G, 25A6MG
25A6GT/G	7S	25A6, 25A6G, 25A6GT, 25A6MG
25A6MG	7S	25A6, 25A6G, 25A6GT, 25A6GT/G
25A7	8F	25A7G, 25A7GT, 25A7GT/G
25A7G	8F	25A7GT, 25A7GT/G
25A7GT	8F	25A7G, 25A7GT/G

Tube Type	Basing	Replacement
25A7GT/G	8F	25A7G, 25A7GT
25AC5	6Q	25AC5G, 25AC5GT, 25AC5GT/G
25AC5G	6Q	25AC5GT, 25AC5GT/G
25AC5GT	6Q	25AC5G, 25AC5GT/G
25AC5GT/G	6Q	25AC5G, 25AC5GT
25AQ8	-	-
25AU4	-	-
25AV5	6CK	25AV5GA, 25AV5GT
25AV5GA	6CK	25AV5GT
25AV5GT	6CK	25AV5GA
25AX4	4CG	25AX4GT, 25D4
25AX4GT	4CG	25AX4, 25D4
25B5	6D	-
25B6	7S	25B6G, 25B6GT
25B6G	7S	25B6, 25B6GT, 5824
25B6GT	7S	25B6, 25B6G
25B8	8T	25B8GT
25B8GT	8T	25B8
25BB14	-	-
25BK5	9BQ	-
25BQ5	-	-
25BQ6	6AM	25BQ6GA, 25BQ6GA/25CU6, 25BQ6GTB, 25BQ6GTB/25CU6, 25CU6, 25DQ6, 25DQ6A
25BQ6GA	6AM	25BQ6GA/25CU6, 25BQ6GTB, 25BQ6GTB/25CU6, 25CU6, 25DQ6, 25DQ6A
25BQ6GA/25CU6	6AM	25BQ6GA, 25BQ6GTA, 25BQ6GTB, 25BQ6GTB/25CU6, 25CU6, 25CU6A
25BQ6GT	6AM	25BQ6GA, 25BQ6GA/25CU6, 25BQ6GTA, 25BQ6GTB, 25BQ6GTB/25CU6, 25CU6, 25DQ6, 25DQ6A, 25G-B6
25BQ6GTA	6AM	25BQ6GTB, 25BQ6GTB/26CU6, 25CU6, 26DQ6, 25DQ6A
25BQ6GTB	6AM	25BQ6GTB/25CU6, 25CU6, 25DQ6, 25DQ6A
25BQ6GTB/25CU6	6AM	25BQ6GTB, 25CU6, 25CU6A
25BR3	9CB	-
25C5	7CV	-
25C6	7AC	-
25C6G	7AC	25C6GA
25C6GA	7S	25C6G
25C63	-	-
25CA5	7C	25C5, 25EH5
25CD6	5BT	25CD6G, 25CD6GA, 25CD6GB, 25DN6, 21EX6
25CD6G	5BT	25CD6GA, 25CD6GB, 25DN6, 21EX6
25CD6GA	5BT	21EX6, 25CD6GB, 25DN6
25CD6GB	5BT	21EX6
25CG3	12HF	-
25CK3	9HP	-
25CM3	9HP	-
25CR5	9HC	-
25CT3	9RX	-
25CU6	6AM	25BQ6GA/25CU6, 25BQ6GTB, 25BQ6GTB/25CU6, 25DQ6, 25DQ6A
25CU6A	6AM	25BQ6GA/25CU6, 25BQ6GTB, 25BQ6GTB/25CU6, 25DQ6A
25D4	4CG	-
25D6	-	-
25D8	8AF	25D8GT
25D8GT	8AF	25D8
25DK3	9SG	-
25DK4	5BQ	-
25DL3	-	-
25DN6	5BT	21EX6, 25CD6GB
25DQ6	6AM	25DQ6A
25DQ6A	6AM	25DQ6
25DT5	9HN	-
25E5	8GT	25E5/PL36, 30P4, 30P19, N308, PL36
25E5/PL36	8GT	25E5, 30P4, 30P19, N308, PL36
25EC6	5BT	21EX6, 25CD6GB
25EH5	7CV	25CA5
25F5	7CV	25F5A
25F5A	7CV	25F5!
25FY8	9EX	-
25GB6	-	25BQ6GT, 25BQ6GA/25CU6, 25BQ6GTB/25CU6, 25CU6
25GF6	6AM	-
25HX5	9SB	-
25JQ6	9RA	-
25JZ8	12DZ	24JZ8
25L6	7AC	25L6G, 25L6GT, 25L6GT/25W6GT, 25L6GT/G, 25W6GT, KT32
25L6G	7AC	25L6, 25L6GT, 25L6GT/25W6GT, 25L6GT/G, 25W6GT
25L6GT	7AC	25L6, 25L6G, 25L6GT/25W6, 25L6GT/25W6GT, 25L6GT/G, 25W6GT, 6046
25L6GT/25W6	7AC	25L6, 25L6G, 25L6GT, 25L6GT/25W6GT, 25L6GT/G, 25W6, 25W6GT, 6046
25L6GT/25W6GT	7AC	25L6GT, 25L6GT/25W6GT, 25L6GT/G, 25W6GT, 6046, KT32
25L6GT/G	7AC	25L6G, 25L6GT, 25L6GT/25W6GT, 25W6GT
25LQ6	-	-
25MK15	-	-
25MP20	-	-
25N6	7W	25N6G
25N6G	7W	25N6
25RK19	-	25BR3
25S	6M	1B5, 1B5/25S, 25
25SN7	-	-
25T	-	25T/HK24, HK24
25T/HK24	-	25T, HK24
25U4	4CG	25AX4GT, 25D4, 25U4, 25W4GT
25U4GT	4CG	25AX4GT, 25D4, 25W4GT
25W4	-	25AX4GT, 25D4, 25U4GT, 25W4GT
25W4GT	4CG	25AX4GT, 25D4, 25U4GT
25W6	7AC	25L6, 25L6G, 25L6GT/25W6GT, 25L6GT/G, 25W6GT, 6046, KT32
25W6GT	7AC	25L6, 25L6G, 25L6GT/25W6GT, 25L6GT/G, 6046, KT32
25X6	7Q	25X6GT
25X6GT	7Q	25X6

Tube Type	Basing	Replacement
25Y4	5AA	25Y4GT, PY31, U31
25Y4GT	5AA	25Y4, PY31, U31
25Y5	6E	25Z5, 25Z5
25Z3	4G	-
25Z4	5AA	25Z4GT, U31
25Z4GT	5AA	25Z4, U31
25Z5	6E	25Y5, 25Z5GT, 25Z5MG
25Z5GT	6E	25Y5, 25Z5, 25Z5MG
25Z5MG	6E	25Y5, 25Z5, 25Z5GT
25Z6	7Q	25Z6G, 25Z6GT, 25Z6GT/G, 25Z6MG, 25Z6WGT
25Z6G	7Q	25Z6, 25Z6GT, 25Z6GT/G, 25Z6MG
25Z6GT	7Q	25Z6, 25Z6G, 25Z6GT/G, 25Z6MG
25Z6GT/G	7Q	25Z6, 25Z6G, 25Z6GT, 25Z6MG
25Z6MG	7Q	25Z6, 25Z6G, 25Z6GT, 25Z6GT/G
25Z6WGT	7Q	25Z6GT#, 25Z6GT/G#
26	4D	-
26A6	7BK	26CG6
26A7	8BU	26A7GT
26A7GT	8BU	26A7#
26AQ8	-	-
26B6	7CH	26D6
26BK6	7BT	-
26C6	7BT	-
26CG6	7BK	26A6
26D6	7CH	26B6
26E6	7S	26E6G, 26E6WG, 26E6WGB
26E6G	7S	26E6WG, 26E6WGB
26E6WG	7S	26E6WGB
26E6WGB	7S	26E6WG#
26FZ6	-	-
26HU5	8NB	-
26LW6	8NC	-
26LX6	12JA	-
26Z5	9BS	26Z5W
26Z5W	9BS	26Z5#
27	5A	27A*, 27S, 56, 56S
27A	5A	27*, 27S*
27BL7	-	-
27BL8	-	-
27GB5	9NH	27GB5/PL500, 27GB5/PL504, 28GB5, PL500, PL504
27GB5/PL500	9NH	27GB5, 27GB5/PL504, 28GB5, PL500, PL504
27GB5/PL504	9NH	27GB5, 27GB5/PL500, 28GB5, PL500, PL504
27KG6	9RJ	LL505
27S	5A	27, 27A*, 56, 56S
28AK8	-	-
28D7	8BS	28D7W, 1238
28D7W	8BS	28D7#, 1238
28EC4	-	LY500
28GB5	9NH	27GB5, 27GB5/PL504, 27GB5/PL500, PL500, PL504
28HA6	9NW	-
28HD5	12ES	30HD5
28Z5	6BJ	-
29GK6	9GK	-
29HA6	-	-
29KQ6	9RJ	29KQ6/PL521, 29LE6
29KQ6/PL521	9RJ	29KQ6, 29LE6, PL521
29LE6	9RJ	29KQ6
30	4D	-
30A5	7CV	35C5, 35C5A, 35EH5, HL94
30AE3	9CB	30AE3/PY88, PY88
30AE3/PY88	9CB	30AE3, PY88
30AG3	-	-
30AG11	12DA	-
30C1	-	9A8, 9A8/PCF80, PCF80
30C13	-	-
30C15	-	9EN7
30C18	-	7GV7
30CW5	9CV	-
30F5	-	7ED7
30FL1	-	9GB8
30HD5	12ES	28HD5
30HJ5	12FL	-
30JZ6	12GD	-
30KD6	12GW	-
30L1	-	7AN7
30L15	-	7EK7
30MB6	12FY	-
30MP23	-	-
30MP27	-	-
30P4	-	25E5, 25E5/PL36, 25GF6, PL36
30P12	-	12FB5
30P16	-	16A5
30P18	-	15CW5, 15CW5/PL84, PL84
30P19	-	25E5, 25E5/PL36, 25GF6, PL36
30PL1	-	13GC8
30PL10	-	13GC8
30PL12	-	16A8, 16A8/PCL82, PCL82
30PL13	-	16GK8
30PL14	-	16GK8
30S	-	30S/VT67, VT67
30S/VT67	-	30S, VT67
31	4D	-
31A3	8HB	31A3/UY41, 311SU, U118, U142, U145, U404, UY41, UY42, V311, V312
31A3/UY41	8HB	31A3, 311SU, U118, U142, U145, U404, UY41, UY42, V311, V312
31AL10	12HR	-
31AV3	-	-
31BX7GT	8BD	-
31JS6	12FY	31JS6A, 31JS6C
31JS6A	12FY	31JS6#,31JS6C
31JS6C	12FY	31JS6A
31LQ6	9QL	-
31LR8	9QT	-
31LZ6	9QL	-
31Z	-	1231Z
32	4K	1A4, 1A4P, 1A4T, 1B4, 1B4G, 1B4P, 1B4P/951, 1B4T, 1C4*, 1K4*, 34, 951
32A5	6AA	-
32A8	9EX	-
32ET5	7CV	32ET5A, 34GD5, 34GD5A
32ET5A	7CV	32ET5!, 34GD5A
32GA7	12EB	-
32HQ7	12HT	-
32L7	8Z	32L7GT
32L7GT	8Z	32L7
33	5K	-
33GT7	12FC	-
33GY7	12FN	33GY7A
33GY7A	12FN	33GY7
33HE7	12FS	-
33HK7	-	-

Tube Type	Basing	Replacement
33JR6	9QU	-
33JV6	12FK	-
34	4M	1A4, 1A4P, 1A4T, 1B4, 1B4G, 1B4P, 1B4P/951, 1B4T, 1C4*, 1K4*, 32, 951
34CD3	12FX	34CD3/34CE3, 34CE3, 34CE3/34CD3
34CD3/34CE3	12FX	34CD3, 34CE3, 34CE3/34CD3
34CE3	12GK	34CD3, 34CD3/34CE3, 34CE3/34CD3
34CE3/34CD3	12GK	34CD3, 34CD3/34CE3, 34CE3
34CM3	9HP	-
34DK3	9SG	-
34GD5	7CV	32ET5, 32ET5A, 34GD5A
34GD5A	7CV	32ET5A, 32ET5!, 34GD5!
34R3	9CB	-
35	5E	24A#, 24S#, 35/51, 35S, 35S/51S, 51, 51S
35/51	5E	24A#, 24S#, 35, 35S, 35S/51S, 51, 51S
35A3	7ET	35C3
35A5	6AA	35A5LT
35A5LT	6AA	35A5
35B5	7BZ	-
35C3	7ET	35A3
35C5	7CV	30A5, 35C5A, HL94
35C5A	7CV	30A5!, 35C5!, HL94
35CD6	5BT	35CD6GA
35CD6GA	5BT	35CD6#
35D5	9FU	-
35DZ8	9JE	-
35EH5	7CV	35EH5A
35EH5A	7CV	35EH5!
35FM5	-	-
35FN5	8GN	-
35GL6	7FZ	-
35HB8	9ME	-
35L6	7AC	35L6G, 35L6GT, 35L6GT/G
35L6G	7AC	35L6GT, 35L6GT/G
35L6GT	7AC	35L6G, 35L6GT/G
35L6GT/G	7AC	35L6G, 35L6GT
35LR6	12FY	-
35QL6	-	-
35S	5E	24A#, 24S#, 35, 35/51, 35S/51S, 51, 51S
35S/51S	5E	24A#, 24S#, 35, 35/51, 35S, 51, 51S
35T	-	35TG
35TG	-	35T
35W4	5BQ	35W4A, HY90
35W4A	5BQ	35W4!, HY90
35W4GT	5BQ	35W4#, 35W4A, HY90#
35W5	-	-
35X4	-	-
35Y4	5AL	35Y4GT
35Y4GT	5AL	35Y4
35Y5	-	-
35Z3	4Z	35Z3LT
35Z3LT	4Z	35Z3
35Z4	5AA	35Z4GT, U74, U76
35Z4GT	5AA	35Z4, U74, U76
35Z5	6AD	35Z5G, 35Z5GT, 35Z5GT/G
35Z5G	6AD	35Z5, 35Z5GT, 35Z5GT/G
35Z5GT	6AD	35Z5, 35Z5G, 35Z5GT/G
35Z5GT/G	6AD	35Z5, 35Z5G, 35Z5GT
35Z6	6AD	35Z6G, 35Z6GT
35Z6G	6AD	35Z6GT

Tube Type	Basing	Replacement
35Z6GT	6AD	35Z6, 35Z6G
36	5E	36A, 64*, 64A*
36A	5E	36, 64*, 64A*
36AM3	5BQ	36AM3A, 36AM3B
36AM3A	5BQ	36AM3, 36AM3B
36AM3B	5BQ	36AM3!, 36AM3A!
36KD6	12GW	36KD6/40KD6, 40KD6!
36KD6/40KD6	12GW	36KD6, 40KD6
36LW6	-	-
36MC6	9QL	-
37	5A	37A, 56A*, 56AS*, 67*, 67A*, 76
37A	5A	37, 56A*, 56AS*, 67*, 67A*, 76
38	5F	38A
38A	5F	38
38A3	9BM	U119, U381, UY85
38HE7	12FS	38HE7/38HK7, 38HK7
38HE7/38HK7	12FS	38HE7, 38HK7
38HK7	12FS	38HE7, 38EH7/38HK7
39	5F	39/44, 44
39/44	5F	39, 44
40	4D	-
40A1	8ES	-
40B2	8ES	-
40FR5	7CV	-
40KD6	12GW	36KD6/40KD6
40KG6	9RJ	40KG6A, 40KG6A/PL509, 40KG6A/PL519, PL509, PL519
40KG6A	9RJ	40KG6A/PL509, 40KG6A/PL519, PL509, PL519
40KG6A/PL509	9RJ	40KG6A, 40KG6A/PL519, PL509, PL519
40KG6A/PL519	9RJ	40KG6A, 49KG6A/PL509, PL509, PL519
40SUA	-	1D5
40Z5	6AD	40Z5GT, 40Z5/45Z5GT, 45Z5GT
40Z5/45Z5GT	6AD	40Z5, 40Z5GT, 45Z5GT
40Z5GT	6AD	40Z5, 40Z5/45Z5GT, 45Z5GT
41	6B	42*
42	6B	41*
42EC4	6EC4	42EC4/PY500A, 42EC4A, 42EC4A/PY500, PY500, PY500A
42EC4/PY500A	6EC4	42EC4, 42EC4A, 42EC4A/PY500, PY500, PY500A
42EC4A	6EC4	42EC4, 42EC4/PY500A, 42EC4A/PY500, PY500, PY500A
42EC4A/PY500	6EC4	42EC4, 42EC4/PY500A, 42EC4A, PY500, PY500A
42EL4	-	-
42KN6	12GU	-
43	6B	43MG
43MG	6B	43
44	5F	39, 39/44
44/39	5F	39, 39/44, 44
45	4D	45A
45A	4D	45
45A5	8GW	10P14, 45A5/UL41, 451PT, BF451, N142, UL41, UL46
45A5/UL41	8GW	10P14, 45A5, 451PT, BF451, N142, UL41, UL46
45B5	9CV	10P18, 45B5/UL84, N119, UL84

Tube Type	Basing	Replacement
45B5/UL84	9CV	10P18, 45B5, N119, UL84
45Z3	5AM	-
45Z5	6AD	40Z5, 40Z5/45Z5GT, 40Z5GT, 45Z5GT
45Z5GT	6AD	40Z5, 40Z5/45Z5GT, 40Z5GT
46	5C	-
47	5B	-
48	6A	-
48A8	-	50BM8/UCL82
49	5C	-
49KG6A	9RJ	49KG6A/PL509
49KG6A/PL509	9RJ	49KG6A, PL509
50	4D	-
50A1	9CM	50A1/D5TF30, D5TF30
50A1/D5TF30	9CM	50A1, D5TF30
50A5	6AA	-
50AX6	7Q	50AX6G, 50AX6GT
50AX6G	7Q	50AX6GT
50AX6GT	7Q	50AX6G
50B5	7BZ	WTT126
50BK5	9BQ	-
50BM8	9EX	10PL12, 48A8, 50BM8/UCL82, LN119, UCL82
50BM8/UCL82	9EX	10PL12, 48A8, 50BM8, LN119, UCL82
50C5	7CV	50C5A, HL92
50C5A	7CV	50C5!
50C6	7AC	50C6G, 60C6GA
50C6G	7AC	50C6GA
50C6GA	7S	50C6G
50CA5	7CV	50EH5, 50EH5A
50CD6	5BT	50CD6G
50CD6G	5BT	50CD6
50DC4	5BQ	-
50E5	8GT	-
50EH5	7CV	50CA5, 50EH5A
50EH5A	7CV	50CA5!, 50EH5!
50FA5	7CV	50FA5/50FK5, 50FK5
50FA5/50FK5	7CV	50FA5, 50FK5
50FE5	8KB	-
50FK5	7CV	50FA5, 50FA5/50FK5
50FY8	9EX	-
50GY7	12FN	50GY7A
50GY7A	12FN	50GY7
50HC6	7FZ	-
50HC8	-	-
50HK6	7FZ	50HC6
50HN5	9QW	-
50JY6	8MG	-
50L6	7AC	50L6G, 50L6GT
50L6G	7AC	50L6GT
50L6GT	7AC	50L6G, KT71
50X6	7DX	-
50Y6	7Q	50Y6G, 50Y6GT, 50Y6GT/G
50Y6G	7Q	50Y6GT, 50Y6GT/G
50Y6GT	7Q	50Y6G, 50Y6GT/G
50Y6GT/G	7Q	50Y6G, 50Y6GT
50Y7	8AN	50Y7GT, 50Z7G, 50Z7GT
50Y7GT	8AN	50Z7G, 50Z7GT
50Z6	7Q	50AX6G, 50AX6G/GT, 50Z6G
50Z6G	7Q	50AX6G, 50AX6G/GT
50Z7	8AN	50Y7GT, 50Z7G, 50Z7GT
50Z7G	8AN	50Y7GT, 50Z7GT
50Z7GT	8AN	50Y7GT, 50Z7G
51	5E	24A#, 24S#, 35, 35/51, 35S, 35S/51S, 51S
51S	5E	24A#, 24S#,35, cont'd
51S, cont'd		35/51, 35S, 35S/51S, 51
52	5C	-
52KU	-	5Y3, 5Y3/6087, 5Y3G, 5Y3G/GT, 5Y3GT, 5Y3GT/G, 5Z4G, 6087
53	7B	-
53HK7	12FS	58HE7
53KU	-	5Y3, 5Y3/6087, 5Y3G, 5Y3G/GT, 5Y3GT, 5Y3GT/G, 6087
54KU	-	5Y3, 5Y3/6087, 5Y3G, 5Y3G/GT, 5Y3GT, 5Y3GT/G, 6087
55	6G	55S
55A3	-	-
55N3	9BM	UY82
55S	6G	55
56	5A	56S, 27, 27S
56A	5A	37*, 37A*, 56AS, 67, 67A, 76*
56AS	5A	37*, 37A*, 56A, 67, 67A, 76*
56R9	12EN	-
56S	5A	27, 27S, 56
57	6F	57S
57A	6F	6C6*, 57AS, 77*
57AS	6F	6C6*, 57A, 77*
57S	6F	57
58	6F	58AS, 58S
58A	6F	58S
58AS	6F	58A, 6D6*, 78*
58HE7	12FS	53HK7
58S	6F	58
59	7A	-
60E3	7ET	-
60EH5	7CV	-
60FX5	7CV	-
60HL5	9QW	-
62DDT	-	6CV7, 6CV7/EBC41, EBC41
62TH	-	6CU7, 6CU7/ECH42, ECH42
62VP	-	6CJ5, 6CJ5/EF41, EF41
63T1	-	-
63TP	-	6AB8, 6AB8/ECL80, ECL80
63VP	-	6AL3, 6AL3/EY88, 6BR3, 6BR3/6RK19, 6R3, 6R3/EY81, 6RK19, EY81, EY88
64	5E	36*, 36A*, 64A, 65, 65A
64A	5E	36*, 36A*, 64, 65, 65A
64ME	-	6CD7, 6CD7/EM34, 6CD7/EM34-35, EM34, EM34/6CD7, EM34-35, EM34-35/6CD7
64SPT	-	6BX6, 6BX6/EF80, EF80
65	5E	64, 64A, 65A
65A	5E	65, 64, 64A
65ME	-	6BR5, 6BR5/EM80, EM80, EM80/6BR5
65W7	8BD	6SN7
66KU	-	6BT4
67	5A	37*, 37A*, 56A, 56AS, 67A, 76*
67A	5A	37*, 37A*, 56A, 56AS, 67, 76*
67PT	-	6CK5, 6CK5/EL41, BF61, EL41, N150
68	5E	68A
68A	5E	68
70A7	8AB	70A7GT

Tube Type	Basing	Replacement
70A7GT	8AB	70A7
70L7	8AA	70L7GT
70L7GT	8AA	70L7
71	4D	71A*, 71B*
71A	4D	71*, 71B*
71B	4D	71*, 71A*
75	6G	75S
75C1	-	-
75S	6G	75
75TL	-	-
76	5A	37, 37A, 56A*, 56AS*, 67*, 67A*
77	6F	6C6, 57A*, 57AS*
78	6F	6D6, 58AS
79	6H	-
80	4C	5X3, 13, 83V, 88, 213B, RE1, WT270
81	4B	16, 16B, RE2
82	4C	82V
82V	4C	82
83	4C	5Z3, WT301
83V	4AD	-
84	5D	6Z4, 84/6Z4, 98
84/6Z4	5D	6Z4, 98
85	6G	85S
85A1	-	0E3, 0E3/85A1, 5TV85-10
85A2	-	0G3, 0G3/85A2
85A3	-	5783#
85AS	6G	-
85S	6G	85
88	4C	82, 82V
89	6F	89Y
89Y	6F	89
90C1	-	-
95	6B	2A5, 2A5/KR65, KR65
96	4G	-
98	5D	6Z4, 84/6Z4
99	-	-
100TH	-	-
101F	-	-
105	-	-
108C1	-	0B2, 0B2WA, 0C3#, 0C3/VR105#, 0C3A#, 6074, 6074/0B2, 6627, 6627/0B2WA
108C2	-	0B2, 0B2WA, 0C3#, 0C3/VR105#, 0C3A#, 6074, 6074/0B2, 6627, 6627/0B2WA
112A	4D	12A, 12A/112A
113HY	-	123HY
114B	-	-
115HY	-	145HY
117L7	8AO	117M7
117L7/117M7GT	8A0	117L7, 117M7GT
117L7/M7	8AO	117L7, 117M7
117L7/M7GT	8AO	117L7GT, 117M7GT
117L7GT	8AO	117L7/117M7GT, 117M7GT
117M7	8AO	117L7
117M7GT	8AO	117L7GT, 117L7/117M7GT
117N7	8AV	117N7GT, 117P7GT, WTT115
117N7GT	8AV	117P7GT, WTT115
117P7	8AV	117N7GT, 117P7GT
117P7GT	8AV	117N7GT
117Z3	4CB	-
117Z4	5AA	117Z4GT
117Z4GT	5AA	117Z4
117Z6	7Q	117Z6G, 117Z6GT, 117Z6GT/G, WT377
117Z6G	7Q	117Z6GT, 117Z6GT/G, WT377
117Z6GT	7Q	117Z6G, 117Z6GT/G, WT377
117Z6GT/G	7Q	117Z6G, 117Z6GT, WT377
120	-	-
121VP	-	12AC5
123HY	-	113HY
124A	-	-
128A	-	2523
141DDT	-	14L7
141TH	-	14K7
144	-	-
145HY	-	115HY
150B2	-	6354
150C1	-	0A2, 0A2WA, 6073, 6073/0A2, 6626, 6626/0A2WA
150C2	-	0A2, 0A2WA, 6073, 6073/0A2, 6626, 6626/0A2WA
150C3	-	0D3, 0D3/VR150, 0D3A, VR150, VR150/0D3
150C4	-	0A2#, 0A2WA, 6073, 6073/0A2, 6626, 6626/0A2WA
163PEN	-	16A5
160L	-	572B, 572B/160L
171	-	-
171A	-	-
171DDP	-	17C8
172	-	WTT119
180C1	-	0B2, 0B2WA, 6074, 6074/0B2, 6627, 6627/0B2WA
182B	4D	182B/482B, 482B
182B/482B	4D	71*, 71A*, 71B*, 182B, 482B
183	4D	183/483, 483
183/483	4D	183, 483
183B	4D	183B/483, 483
183B/483	4D	183B, 483
184	-	-
200A	-	-
203	-	2RA15, 215
203A	-	-
203H	-	HF125
205	-	2RA5
205D	-	-
206	-	2RA6, 619
210	-	-
210-T	4D	-
211	-	-
213	-	2RA3
213A	-	2RA3
213B	4C	80
213PEN	-	21A6
215	-	2RA15, 203
222	-	-
224	-	-
226	-	-
227	-	-
230	-	-
231	-	-
232	-	-
233	-	-
234	-	-
235	-	235/251, 251
235/251	-	235, 251
236	-	-
237	-	-

Tube Type	Basing	Replacement
238	-	-
239	-	239, 244
239/244	-	239/244, 244
239A	-	-
240	-	-
244	-	239, 239/244
245	-	884, 884/6Q5, 884W
246	-	-
247	-	-
249B	-	-
250	-	-
250TH	-	-
250TL	-	-
251	-	235, 235/251
253A	-	-
256	-	-
257B	-	4E27, 4E27/257B
259A	-	259B
259B	-	259A
262A	-	262B
262B	-	262A
264	-	264C, 864
264C	-	264
271A	-	-
272A	-	-
274	5L	5V4G, 5V4GA
274A	-	5R4GB, 5R4GYB, 2076, 2076/5R4GB, 2076/5R4GYB
274B	5T	5R4GB, 5R4GY, 5R4GYB, 2076, 2076/5R4GB, 2076/5R4GYB
275A	-	-
280	-	-
281	-	-
282	-	-
285	-	-
287A	-	-
300B	-	-
301A	-	83
304TH	-	-
304TL	-	-
307A	-	-
310A	6F	6C6, 310B, 1620#
310B	6F	310A, 1620#
311A	-	311B
311B	-	311A
311SU	-	31A3
312A	-	-
313C	-	1C21#
313CA	-	-
316A	-	-
323A	-	-
328A	-	6C6#
329A	-	-
331A	-	805, 805/331A
332A	-	-
336A	-	-
337A	-	-
338A	-	-
345A	-	-
347A	-	-
348A	-	1620#
349A	-	6F6#, 6F6GT#, 6K6#, 6K6GT
350A	-	-
350B	-	-
351A	-	6X5GT#
352A	-	-
354A	-	-
356	-	5771
359A	-	1C21#
393A	-	-
394A	-	-
395A	-	5823#
396A	8CJ	2C51, 2C51W*, 1219*, 5670*, 5670/2C51W*, 5670WA*, 6185, 6385*
398A	-	5603
399A	-	-
399B	-	-
400A	-	400B
400B	-	400A
401	-	401A, 401B, 5590, 5590/401A, 5590/401B
401A	-	401B, 5590, 5590/401A, 5590/401B
401B	-	401A, 5590, 5590/401A, 5590/401B
403	7BD	6AK5, 6AK5/5654, 6AK5/EF95, 403A, 403B, 5591, 5591/403B, 5654, EF95
403A	7BD	6AK5, 6AK5/5654, 6AK5/EF95, 403B, 5591, 5591/403B, 5654, EF95
403B	-	6AK5, 6AK5/5654, 6AK5/EF95, 5654, 5591/403B, EF95
404A	9X	5847, 5847/404A
407A	407A	-
408A	7BD	408A/6028, 6028
408A/6028	7BD	408A, 6028
409A	7CM	6AS6, 6AS6W, 5725, 5725/6AS6W, 6187, 6187/6AS6
412A	-	6754
413B	-	-
415	-	415A, 5550, 5550/415
415A	-	415, 5550, 5550/415
416B	-	416B/6280, 6280
416B/6280	-	416B, 6280
417A	9V	417B, 5842, 5842/417A
417B	9V	417A, 5842, 5842/417A
418A	-	-
420A	-	5755
421	-	6AS7G#, 6AS7GA#, 6080#, 6080W#
421A	8BD	6AS7G#, 6AS7GA#, 5998, 6080#, 6080W#
422A	-	-
423A	9BY	5651A, 5651WA, 6140
426A	-	-
427A	-	-
429A	-	-
442A	-	-
443A	-	-
451PT	-	45A5
471A	-	-
482	-	-
482B	-	182B
483	4D	183, 183/483, 183B
484	-	RIT
485	5A	RIR
501R	-	5759
502A	-	2050, 2050A
502X	-	-
510AX	-	-

Tube Type	Basing	Replacement
512AX	-	-
518AX	-	-
522AX	-	6088, 6088/522AX
527AX	-	-
534AX	-	-
547DX	-	-
548DX	-	6418, 6418/548DX
572B	-	160L, 572B/160L
572B/160L	-	160L, 572B
573AX	-	6029, 6029/573AX
575A	4AT	-
579B	7O	-
586	-	-
604	4BS	604/7014, 7014, NL604, NL604/7014
604/7014	4BS	604, 7014, NL604, NL604/7014
606L	-	606L/7017
606L/7017	-	606L, 7017
610	-	610/7723
610/7723	-	610, 7723
615	4AU	615/7018
615/7018	4AU	615, 7018
618	-	618P, 5892, EL6B
618P	-	618, 5892, EL6B
619	-	2RA6, 206
627	-	-
630	-	2050, 2050A
630A	-	2050, 2050A
631P1	-	631P1/SN4, SN4
631P1/SN4	-	631P1, SN4
632B	4CD	-
635	-	635/7019, 7019
635/7019	-	635, 7019
635L	-	635L/7020, 7020
635L/7020	-	635L, 7020
651	-	651/5552A, 5552A, 5552A/651
651/5552A	-	651, 5552A, 5552A/651
660L	-	660L/7786, 7786
660L/7786	-	660L, 7786
672A	4CE	-
673	2P	-
676	-	6T7, 676
677	-	-
684	-	2Z2, 2Z2/G84, G84
702A	-	-
704A	-	-
705A	-	-
707B	-	2K28, 2K28/707B
710	4CF	710/6011, 6011
710/6011	4CF	710, 6011
710L	-	710L/7509A, 7509A
710L/7509A	-	710L, 7509A
714	3G	714/7021, 7021
714/7021	3G	714, 7021
715B	-	-
715C	-	-
716	4D	716/6855, 6855, NL716
716/6855	4D	716, 6855, NL716
717A	-	-
723A	-	-
723B	-	-
726A	-	-
726B	-	-
731A	7BD	6AK5, 6AK5/5654, 6AK5/EF95, 5654, EF95
734	-	734/5544, 5544
734/5544	-	734, 5544
740	-	6856, 6856/740
740P	-	740P/6857, 6857
740P/6857	-	740P, 6857
760	4BZ	760/6858, 6858
760/6858	4BZ	760, 6858
760L	-	760L/7023, 7023
760L/7023	-	760L, 7023
760P	-	760P/6859, 6859
760P/6859	-	760P, 6859
800	-	-
801	-	-
801A	-	-
802	-	-
803	-	-
804	-	-
805	-	331A, 805/331A
805/331A	-	331A, 805
806	-	-
807	5AW	4Y25, 5S1, 807W, 807WA, 5933, 5933/807W, 5933WA, 8018, HY61, P17A, QE06/50, QE06140, QV05-25, RK39
807W	5AW	807WA, 5933, 5933/807W, 5933WA
807WA	5AW	807W, 5933, 5933/807W, 5933WA
808	-	-
809	-	-
810	-	-
811	-	-
811A	3G	-
812	-	-
812A	3G	-
813	-	-
814	-	-
815	-	-
816	4P	-
826	-	-
828	-	-
829A	-	3E29, 3E29/829A
829B	-	-
830B	-	-
832A	-	-
833A	-	-
834	-	-
835	-	-
836	-	-
837	-	-
838	-	-
840	-	840/8754, 8754
840/8754	-	840, 8754
841	-	-
842	-	-
843	-	-
845	-	-
849	-	-
850	-	-
857-B	-	-
860	-	-
861	-	-
864	4D	264
865	-	-
866	4P	866/866A, 866A
866/866A	4P	866, 866A

Tube Type	Basing	Replacement
866A	4P	866, 866/866A
866JR	-	-
868	2K	-
869B	-	-
872	4AT	872A
872A	4AT	872
874	-	-
876	-	-
877	-	-
878A	-	-
879	4AB	2X2, 2X2/879, 2X2A
884	6Q	6Q5, 6Q5G, 884/6Q5, 884W
884/6Q5	6Q	6Q5, 6Q5G, 884, 884W
884(THYRATRON)	-	-
885	5A	2B4
885(THYRATRON)	-	884#
886	-	-
889RA	-	5667
902A	-	-
913	-	-
917	1A	-
918	2K	918/CE1C, CE1C
918/CE1C	2K	918, CE1C
919	1B	-
920	4BG	-
921	2AQ	-
922	2AQ	-
923	2K	-
924	2AR	1P41, 1P41/924
925	3J	-
926	2AQ	-
927	2F	-
928	-	-
929	3J	-
930	3J	-
931A	-	-
931B	-	-
932	4K	1229, 1229/932
934	2F	-
935	1C	-
936	-	-
941	-	-
950	5K	-
951	4K	1A4, 1A4P, 1A4T, 1B4P, 1B4P/951, 1B4G, 1B4T, 1C4*, 1K4*, 32, 34
954	-	9001#, E1F, UN954, ZA2*
955	5BC	1650, 4671, E1C, HA2, UN955
956	-	9003#, E2F
957	-	D1C
958	-	958A, 9002#, D2C
958A	-	9002#, D2C
959	5BE	D3F
991	991	-
1003	4R	0Z4, 0Z4A, 0Z4A/0Z4, 0Z4G, 1003, WTT114
1005	-	-
1006	-	-
1007	-	-
1026	-	-
1036	-	1036/6436, 6436
1036/6436	-	1036, 6436
1038	-	-
1089	-	-
1201	8BN	7E5, 7E5/1201, 1201A
1201A	8BN	7E5, 7E5/1201, 1201
1203	4AH	7C4, 7C4/1203A, 1203A

Tube Type	Basing	Replacement
1203A	4AH	7C4, 7C4/1203A, 1203
1203A/7C4	6AA	7C4, 7C4/1203A, 1203A
1204	8BO	7AB7, 7AB7/1204
1206	8BV	7G8
1216	7BF	5844
1217	7CH	6BE6, 6BE6/5750, 6BE6/EK90, 6BE6W, 5750, 5750/6BE6W, 5915#, 5915A, 7036, EK90
1218A	-	8334, 8334/1218A
1219	8CJ	2C51W, 5670, 5670/2C51W, 5670WA, 6185, 6385*
1220	-	6AK5, 6AK5/5654, 6AK5/EF95, 6AK5W, 6AK5WA, 6AK5WB, 6F32, 403A, 5654, 5654/6AK5W, 6096, 6096/6AK5W, 6968, DP61, EF95, EF905, F95F,
1221	6F	6C6#
1222	1222	1222A
1222A	1222	1222
1223	7R	6J7, 1620#, 7000
1225	7T	6L7#
1229	4K	1229/932, 932
1229/932	4K	1229, 932
1230	4D	-
1231	8V	7V7
1231Z	-	31Z
1232	8V	7G7, 7G7/1232, 7V7
1236	1236A	1236A
1236A	1236A	1236
1238	8BS	28D7, 28D7W
1247	1247	-
1252	-	6AX8, 6EA8, 6KD8, 6KD8/6U8A, 6LN8, 6U8A, 6U8A/6KD8, 6U8A/6AX8/6KD8, 6678, 7731, ECF80, ECF82
1265	4AJ	-
1266	4AJ	5823#
1267	4V	0A4, 0A4/PL1267, 0A4G, 1267, 1267/0A4, PL1267, WTT132, Z300T
1267/0A4	4V	0A4, 0A4/PL1267, 0A4G, 1267, PL1267, WTT132, Z300T
1273	8V	7AJ7
1274	6S	7Y4*
1275	4C	5Z3
1276	4D	-
1280	8V	14C7
1282	8BJ	7W7
1284	8V	-
1288	7BE	3B7, 1291, 1292
1291	7BE	3B7, 1288, 1292
1292	7BE	3B7, 1288, 1291
1293	4AA	-
1294	4AH	1R4, 1R4/1294
1299	6BA	3D6, 3D6/1299, 1299
1381HQ	7BD	6AK5, 6AK5/5654, 6AK5/EF95, 5654, EF95
1602	-	-
1603	6F	6C6#, 7700
1608	-	-
1609	-	-
1610	-	-
1611	7S	6F6#, 6F6GT#, 1621
1612	7T	6L7#

Tube Type	Basing	Replacement
1613	7S	6F6#, 6F6GT#, 1621
1614	7AC	6L6#, 6L6GC#
1616	-	-
1619	7AW	-
1620	7R	6J7#, 1223, 7000
1621	7S	6F6#, 6F6GT#
1622	7AC	6L6#, 6L6GC#
1624	-	-
1625	5AZ	-
1626	6Q	-
1629	7AL	6E5#
1631	-	6L6#, 6L6GC#, 1614#, 1622#
1632	7S	12L6
1633	-	-
1634	8R	12SC7
1635	8B	-
1641	-	1641/RK60, RK60
1641/RK60	-	1641, RK60
1642	7BH	2C21, 2C21/1642
1644	8BU	12L8GT
1649	8N	6AC7#, 6AC7/1852#, 6AC7/1852/6134#, 6AC7/EL34#, 1852#, 6134#, EL34#
1650	5BC	955
1654	-	-
1655	8S	6SC7#
1657	6BS	2050#, 2050A#, 2050W
1659	6G	2A6
1662	7BB	3A4, 3A4/DL93, DL93
1664	8E	12C8
1665	6BS	2050, 2050A, 2050W
1679	-	-
1723	-	PY81
1851	-	-
1852	8N	6AC7#, 6AC7/1852#, 6AC7/1852/6134#, 6AC7/EL34#, 1852#, 6134, EL34
1853	8N	6AB7, 6AB7/1853
1919	-	-
1946	-	-
1947	-	-
1949	-	-
2013	9A	6211
2014	9BV	6CL6#, 6CL6/6677#, 6197, 6677/6CL6#
2020	-	-
2033	-	-
2050	6BS	1657, 1665, 2050/2051, 2050A#, 2050W, 2051, EN32, ME1501
2050/2051	6BS	1657, 1665, 2050, 2050A#, 2050W, 2051, EN32, ME1501
2050(THYRATRON)	-	-
2050A	6BS	1657, 1665, 2050#, 2050W, EN32, ME1501
2050A(THYRATRON)	-	-
2050W	6BS	1657, 1665, 2051, EN32, ME1501
2051	6BS	2050#, 2050A#, EN32, ME1501
2054	-	-
2057	7Q	6H6, 2057/6H6
2057/6H6	7Q	6H6, 2057
2060	12AE	-
2061	14AA	-
2062	-	-
2063	-	-
2064B	-	-
2065	-	-
2076	5T	5R4GB, 5R4GY, 5R4GYA, 5R4GYB, 5R4GTY, 2076/5R4GB, 2076/5R4GYB
2076/5R4GB	5T	5R4GB, 5R4GY, 5R4GYA, 5R4GYB, 5R4GTY, 2076, 2076/5RYGB
2076/5R4GYB	5T	5R4GB, 5R4GY, 5R4GYB, 5R4GTY, 2076, 2076/5R4GB
2081	9DX	6AW8A, 6AU8A, 6AU8A/6BH8, 6BA8A, 6BH8, 6JV8, 6KS8, 6LF8, 2081/6AW8A
2081/6AW8A	9DX	6AW8A, 6AU8A, 6AU8A/6BH8, 6BA8A, 6BH8, 6JV8, 6KS8, 6LF8, 2081
2082	9A	12AY7#, 2082/12AY7
2082/12AY7	9A	12AY7#, 2082
2101	5K	1F4
2523	-	128A
2935A	4B	81
3107	-	5V4GA
3841	4C	80
3871	4D	50
3872	4D	30
3873	4K	32
3921	4D	45
3924	5E	24A
4028A	-	-
4037	-	-
4037A	-	-
4041	-	-
4046	-	-
4046A	-	-
4053	-	-
4054	-	-
4055	-	-
4058	-	-
4060-4061	-	-
4062A	-	-
4068	-	-
4070-4071	-	-
4072	-	-
4438	-	-
4439	-	-
4440	12AE	-
4441	-	-
4441A	-	-
4449A	-	-
4459	20E	-
4460	12BG	-
4461	-	-
4463	14AA	-
4464	14AM	-
4465	5E	35
4465(PHOTOTUBE)	14AM	-
4471	-	-
4472	-	-
4473	-	-
4478	-	-
4492	-	4536
4492V1	-	4536
4492V2	-	4536
4493	8LN	-
4494	8LN	-

Tube Type	Basing	Replacement
4495	8LN	-
4500	-	-
4503A	-	-
4516	-	-
4517	-	-
4518	14AA	-
4521	-	-
4522	-	-
4523	14AA	-
4524	14AA	-
4525	14AA	-
4526	-	-
4536	-	-
4542	-	-
4589	-	-
4600A	-	-
4604	7CL	-
4610	5E	36
4611	5A	37
4612	5F	38
4616	-	-
4616V1	-	-
4617	-	-
4618	-	-
4621	-	-
4624	-	-
4626	-	-
4628	-	-
4630	-	-
4631	-	-
4632	-	-
4634	-	-
4635	-	-
4636	-	-
4637	-	-
4638	-	-
4651	-	-
4652	-	4652/8042, 8042
4652/8042	-	4652, 8042
4657	-	-
4658	-	-
4659	-	-
4660	-	-
4661	-	-
4662	-	-
4671	-	955, 1650, E1C, HA2, UN955
4707	-	6X4
4832	-	-
4848	-	-
4855	-	-
5441	-	-
5514	-	-
5516	5516	-
5517	-	1B48, 5517/CX1013, CX1013
5517/CX1013	5BU	1B48, 5517, CX1013
5528	-	5528/C6L, C6L
5528/C6L	-	5528, C6L
5541	-	-
5544	-	734, 734/5544
5545	-	-
5550	-	415, 5550/415
5550/415	-	415, 5550
5551A	-	-
5552A	-	651, 651/5552A, 5552A/651
5552A/651	-	651, 651/5552A, 5552A
5556	-	PJ8
5557	3G	5557/FG17, FG17

Tube Type	Basing	Replacement
5557/FG17	3G	5557, FG17
5557A	-	-
5558	-	-
5559	4BL	5559/FG57, FG57
5559/FG57	4BL	5559, FG57
5560	4CD	5560/FG95, FG95
5560/FG95	4CD	5560, FG95
5561	-	5561/FG104, FG104
5561/FG104	-	5561, FG104
5563A	3X	-
5581	3J	-
5582	2AQ	-
5583	2F	-
5588	-	6161
5590	7BD	401A, 401B, 5590, 5590/401A, 5590/401B
5590/401A	7BD	401A, 401B, 5590, 5590/401B
5590/401B	7BD	6AK5#, 6AK5/5654#, 6AK5/EF95#, 401A, 401B, 5590, 5590/401A, 5654#, EF95#
5591	7BD	403B, 5591/403B, DP61, E95F, EF95, EF905, M8180
5591/403B	7BD	6AK5#, 6AK5/5654#, 6AK5/EF95#, 403B, 5591, 5654#, EF95#
5603	6BD	398A
5608	7B	5608A, DP61, E95F, EF95, EF905, M8100, M8180
5608A	7B	5608, DP61, E95F, EF95, EF905, M8100, M8180
5610	-	-
5618	7CU	-
5624	-	E82M
5632	-	5632/C3J, C3J
5632/C3J	-	5632, C3J
5633	5633	-
5634	-	EN70, ME1500
5635	8DB	-
5636	8DC	EF730
5636A	-	5636#
5637	8DK	-
5638	5638	-
5639	8DL	-
5640	8DL	-
5641	-	EY70
5642	5642	DY70
5643	-	EN70
5644	4CN	-
5645	5646	-
5646	5646	-
5647	-	EA71
5651	5BO	0G3, 0G3/85A2, 85A2, M8098, STV85/10, 5651A, 5651WA
5651A	5BO	5651WA
5651WA	5BO	5651A#
5652	2AB	-
5653	3J	-
5654	7BD	6AK5#, 6AK5/5654, 6AK5/EF95#, 6AK5W, 6AK5WB, 1220, 5654/6AK5W, 6096, 6096/6AK5W, DP61, E95F, EF95, EF905, M8100, M8180, PM05
5654/6AK5W	-	6AK5#, 6AK5/EF95#, 6AK5W, 5654, 6096, 6096/6AK5W, EF95#

Tube Type	Basing	Replacement
5656	-	-
5659	7S	12A6#
5660	8E	12C8, VP12D
5661	8N	12SK7
5662	-	-
5663	6CE	5696#, 5696A#
5665	-	5665/C16J, C16J, C16J/5665
5665/C16J	-	5665, C16J, C16J/5665
5667	-	889RA
5670	8CJ	2C51, 2C51W, 1219, 5670/2C51W, 5670WA, 6185, 6385*
5670/2C51W	8CJ	2C51, 2C51W, 1219, 5670, 5670WA, 6185, 6385*
5670WA	-	2C51W, 5670#, 6185
5671	-	
5672	5672	DL69, DL75, DL620, DL652, XFY14
5675	-	-
5676	-	XFR3
5677	-	FA6
5678	5678	DF60, DF654, XFR2
5679	7CX	-
5680	-	7C23, 5680/7C23
5680/7C23	-	7C23, 5680
5684	-	5684/C3JA, C3JA
5684/C3JA	-	5684, C3JA
5685	-	5685/C6J/K, C6J, C6K
5685/C6J/K	-	5685, C6J, C6K
5686	9G	-
5687	9H	5687WA
5687WA	9H	5687#
5690	-	-
5691	8BD	6SL7#, 6SL7GT#
5692	8BD	6SN7#, 6SN7GTB#, B65, ECC33
5693	8N	6SJ7#
5694	8CS	-
5695	-	-
5696	7BN	5696A#, EN92
5696A	7BN	5696#, EN92
5702	5702	5702WA, 5702WB
5702WA	5702	5702#, 5702WB
5702WB	5702	5702#, 5702WA
5703	5703	5703WA, 5703WB
5703WA	5703	5703#, 5703WB
5703WB	5703	5703#, 5703WA
5704	5704	-
5718	8DK	5718A, EC70
5718A	8DK	5718, EC70
5719	8DK	-
5721	9A	5721/QK209, 6057, 7494, 7729, QK209
5721/QK209	9A	5721, 6057, 7494, 7729, QK209
5722	5CB	-
5725	7CM	6AS6#, 6AS6W, 5725/6AS6W, 6187, 6187/6AS6, M8196
5725/6AS6W	7CM	6AS6, 6AS6W, 5725, 6187, 6187/61S6, M8196
5726	6BT	6AL5#, 6AL5/6EB5#, 6AL5/5726/6663, 6AL5/EAA91, 6AL5W, 6D2, 6EB5#, 5726/6AL5W, 5726/6AL5W/6097, 6097, 6663, 6663/6AL5#, D2M9, cont'd
5726, cont'd		D77, D152, D717, DD6, DD6G, E91AA, EAA91, EAA91/6AL5, EAA901S, EB91, M8079, M8212, QA2404, QD77
5726/6AL5W	6BT	6AL5W, 5726, 5726/6AL5W/6097, 6097
5726/6AL5W/6097	6BT	6AL5W, 5726, 5726/6AL5W, 6097
5727	7BN	2D21#, 2D21/5727, 2D21/PL21#, 2D21W, 5727/2D21, 5727/2D21W, E91N, EN91, M8204, PL21
5727/2D21	7BN	2D21#, 2D21/5727, 2D21/PL21#, 2D21W, 5727, 5727/2D21W, E91N, EN91, M8204, PL21
5727/2D21W	7BN	2D21#, 2D21/5727#, 2D21W, 5727, 5727/2D21#
5731	5BC	955, E1F, ZA2
5732	-	6K7
5734	5734	-
5742	-	PJ7
5743	-	PJ21
5744	5744	5744/6151, 6151
5744/6151	5744	5744, 6151
5749	7BK	6BA6#, 6BA6/5749/6660, 6BA6/EF93#, 6BA6W, 6BA6WA, 5749/6BA6W, 6660, 6660/6BA6, EF93, M8101, PM04, W727
5749/6BA6W	7BK	6BA6/5749/6660, 6BA6/EF93#, 6BA6W, 6BA6WA, 5749, 6660, 6660/6BA6, EF93#
5750	7CH	6BE6#, 6BE6/5750, 6BE6/EK90, 6BE6W, 5750/6BE6W, EK90, HM04, MH04, X77, X727
5750/6BE6W	7CH	6BE6#, 6BE6/5750, 6BE6/EK90, 6BE6W, 5750, EK90, HM04, MH04, X77, X727
5751	9A	12AX7#, 12AX7A#, 12AX7A/ECC83#, 12AX7A/ECC83/7025#, 12AX7WA#, 12AX7WA/7025#, 12DT7, 5751WA, 6681, 6681/12AX7A#, 7025, B759, ECC83, B339, M8137
5751WA	9A	12AX7A#, 12AX7A/ECC83#, 12AX7A/ECC83/7025#, 12AX7WA#, 12AX7WA/7025#, 5751#, 6681, 6681/12AX7A#, 7025#, ECC83#
5754	-	8196
5755	9J	420A
5757	-	E91N, EN91, M8204, PL21
5759	-	501R
5762	-	7C24, 5762/7C24, 5762A
5762/7C24	-	7C24, 5762, 5762A
5762A	-	7C24, 5762, 5762/7C24,
5763	9K	6062, QE03/10, QV03-12
5767	-	6481
5770	-	-
5771	-	356
5783	5783	85A3, M8190

Tube Type	Basing	Replacement
5784	5784	-
5785	5785	-
5787	5783	-
5794	-	5794A, 6562, 6562/5794A
5794A	-	6562, 6562/5794A
5796	-	-
5802	-	ME1401
5812	-	5763#
5814	9A	12AU7#, 12AU7A/ECC82#, 5814A, 5814WA, 6189#, 6680, B749, ECC82#, M8136
5814A	9A	12AU7A/ECC82#, 5814WA, 6189#, ECC82#
5814WA	9A	5814A#, 6189#
5819	14M	-
5820	-	5820A, 5820A/L, 8775
5820A	-	5820A/L, 8775
5820A/L	-	5820A, 8775
5823	4CK	Z900T
5824	7AC	25B6G
5825	4P	-
5829	-	-
5835	-	5835/NL653, NL653, NL653/5835
5835/NL653	-	5835, NL653, NL653/5835
5838	-	EZ35, U70, U147
5839	6S	-
5840	8DE	5840/6625, 5840A, 5840W, 6225, EF732
5840/6625	8DE	5840, 5840A, 5840W, 6225, EF732
5840A	-	5840#, 5840/6225#, 5840W, 6225#
5840W	-	5840#, 5840/6225#, 5840A#, 6225#
5842	9V	417A, 417B, 5842/417A
5842/417A	9V	417A, 417B, 5842
5844	7BF	6J6A#, 1216, 5964#
5845	5CA	-
5847	9X	6RR8, 6RR8C, 404A, 5847/404A, E182F
5847/404A	9X	6RR8, 6RR8C, 404A, 5847, E182F
5851	6CL	-
5852	-	EZ35, U70, U147
5853	-	-
5854	-	-
5857	-	-
5861	-	EC55
5868	-	AX9902, AX9902/5868
5869	-	5869/9950, 9950
5869/9950	-	5869, 9950
5870A	-	-
5871	7AC	6V6GT#
5876A	-	-
5879	9AD	-
5880	-	-
5881	7AC	6L6G#, 6L6GC#, 6L6WGB, 5881/6L6WGB, 5881/6L6WGC, 5932, 7581, 7581A, EL37, KT66
5881/6L6WGB	7AC	6L6G#, 6L6GC#, 6L6WGB, 6L6WGC, 5881, 5881/6L6WGC, 5932, 7581, 7581A, EL37, KT66
5881/6L6WGC	7AC	6L6G#, 6L6GC#, cont'd
5881/6L6WGC, cont'd		6L6WGB, 6L6WGC, 5881, 5881/6L6WGB, 5932, 7581, 7581A, EL37, KT66
5886	-	DF703
5889	-	ME1403
5890	-	-
5892	-	618, 618P, EL6B
5893	-	-
5894	-	9903, AX9903, AX9903/5894, QQC04/15, QQE06/40, QQV06-40
5894B	-	-
5896	8DJ	-
5897	8DK	5718
5898	8DK	-
5899	8DE	EF731
5899A	-	5899
5900	-	5899, EF71
5901	-	5840, 5840/6225, 5840W, 5901/5840, 5915#, 6225#, EF72, EF732, M8121
5901/5840	-	5840, 5840/6225, 5840W, 5901, 5915#, 6225#, EF72, EF732, M8121
5902	8DE	5902A, EL71
5902A	8DE	5902, EL71
5903	8DJ	-
5904	8DK	5904/6055, 6055
5904/6055	8DK	5904, 6055
5905	8DL	-
5906	8DL	-
5907	8DL	-
5910	6AR	1U4#, 1U4WA, DF904
5911	-	DF67
5913	-	DL67
5915	7CH	6BE6, 6BE6/5750, 6BE6/EK90, 6BE6W, 6BY6, 1217, 5750, 5750/6BE6W, 5915A, 7036, E91H, EH900S, EK90
5915A	7CH	6BE6, 6BE6/5750, 6BE6/EK90, 6BE6W, 1217, 5750, 5750/6BE6W, 5915, 7036, EK90
5916	8DC	-
5920	-	6J6A#, 5920/E90CC, 5964#, 6101#, E90C, E90CC
5920/E90CC	-	6J6A#, 5920, 5964#, 6101#, E90C, E90CC
5923	-	9904
5928	-	-
5930	4D	2A3, 2A3/5930, 2A3W, 5930/2A3W
5930/2A3W	4D	2A3, 2A3/5930, 2A3W, 5930
5931	5T	5AS4A#, 5AS4A/5U4GB#, 5U4#, 5U4GB#, 5U4GB/5AS4A#, 5U4WGB, GZ31, U52
5932	7S	6L6, 6L6WGA, 6L6WGB, 6L6WGC, 5881, 5881/6L6WGC, 5932/6L6WGB, 7027A#, EL37, KT66
5932/6L6WGB	7S	6L6, 6L6WGA, 6L6WGB, 6L6WGC, 5881, 588A/6L6WGC, 5932, 7027A#, EL37, KT66
5933	5AW	807, 807W, 807WA, 5933/807W, 5933WA, QE06140
5933/807W	5AW	807, 807W, 807WA, cont'd

Tube Type	Basing	Replacement
5933/807W, cont'd		5933, 5933WA, QE06140
5933WA	25AW	807, 807W, 807WA, 5933
5934	-	-
5946	-	-
5960	-	S340
5962	-	6085
5963	9A	12AU7A, 12AU7A/ECC82#, 5814A#, 5963/12AU7A, 6680/12AU7A#
5963/12AU7A	9A	12AU7A, 12AU7A/ECC82#, 5814A#, 5963, 6680/ 12AU7A#
5964	7BF	6J6WA#, 6101#
5965	9A	12AV7, 6829
5965A	-	5965#
5968	8DQ	-
5969	8DR	-
5970	8DS	-
5971	-	-
5972	-	-
5977	8DK	-
5987	8DM	-
5991	-	NL8422, NL8422/5991
5992	7AC	6V6GT*, 6V6GTA
5993	-	6Z31, EZ90, M8138, QA2407, QU78, U78, U707, V2M70*
5998	-	-
6000	-	-
6004	-	U50
6005	7BZ	6AQ5#, 6AQ5/6005/6669, 6AQ5A#, 6AQ5A/6HG5#, 6AQ5W, 6HG5#, 6005/ 6AQ5W, 6005/6AQ5W/ 6095, 6095, 6669, 6669/ 6AQ5A, EL90, M8245, N727
6005/6AQ5W	7BZ	6AQ5#, 6AQ5/6005/6669, 6AQ5A#, 6AQ5A/6HG5#, 6AQ5W, 6HG5#, 6005, 6005/6AQ5W/6095, 6095, 6669#, 6669/6AQ5A#
6005/6AQ5W/6095	7BZ	6AQ5#, 6AQ6/6005/6669, 6AQ5A#, 6AQ5A/6HG5#, 6AQ5W, 6HG5#, 6005, 6005/6AQ5W, 6095, 6669#, 6669/6AQ5A#
6006	-	6SG7
6007	-	DL67
6008	-	DF67
6011	-	710, 710/6011
6012	6CO	5727#
6013	-	EL3B
6014	-	C1K, C1K/6014
6021	8DG	-
6024	-	6F12, 8D3, EF91, HP6, M8083, PM07, QZ77, QA2403, SP6, Z77
6026	-	-
6028	7BD	408A, 6028/408A
6028/408A	-	408A, 6028
6029	-	573AX, 6029/573AX
6029/573AX	-	573AX, 6029
6030	-	ECC91, M8081, T2M05
6042	-	13D1
6045	-	6J6
6046	-	-
6049	8DL	-
6050	-	XFR3
6052	8DJ	-
6053	8DJ	-
6055	8DK	5904, 5904/6055
6056	8DL	-
6057	-	12AX7A, 12AX7A/ECC83, 12AX7A/ECC83/7025, 12AX7WA, 12AX7WA/ 7025, 7025, B339, B759, ECC83, ECC803, M8137
6058	6BT	6AL5#, 6AL5/6EB5#, 6AL5/ 5726/6663#, 6AL5/ EAA91#, 6AL5W, 6D2, 6EB5#, 5726#, 5726/ 6AL5W, 5726/6AL5W/ 6097, 6097, 6663#, 6663/ 6AL5#, D77, D152, D717, D2M9, DD6, DD6G, EAA91, EAA91/6AL5, EB91, QA2404, QD77
6059	-	8D5
6060	-	6201#, ECC801
6061	-	6BW6
6062	-	5763#, M8096
6063	5BS	6X4#, 6X4W#, 6X4WA, 6Z31, EZ90, EZ90Z, EZ900, M8138, QA2407, QU78, U78, U707, V2M70
6064	-	6F12, 8D3, EF91, HP6, M8083, PM07, QA2403, QZ77, SP6, Z77
6065	-	6F21, 9D6, EF92, M8161, QA2400, QW77, VP6, W77
6066	-	6AT6, 6AV6
6067	-	12AU7A/ECC82#, 5814A#, 6680/12AU7A#, B749, M8136
6072	9A	12AY7#, 2082/12AY7#, 6072A
6072A	9A	12AY7#, 2082/12AY7#, 6072#
6073	5BO	0A2#, 0A2WA#, 150C2, 150C4, 6073/0A2, 6626, 6626/0A2WA, HD51, M8223, STV150/30
6073/0A2	5BO	0A2#, 0A2WA#, 150C2, 150C4, 6073, 6626, 6626/ 0A2WA, HD51, M8223, STV150/30
6074	5BO	0B2#, 0B2WA, 180C1, 6074/ 0B2, 6627, 6627/0B2WA, HD52, M8224, STV108/30
6074/0B2	5BO	0B2#, 0B2WA, 180C1, 6074, 6627, 6627/0B2WA, HD52, M8224, STV108/30
6075	-	9907
6076	-	6076/QBL5, 9907R, QBL5
6076/QBL5	-	6076, 9907R, QBL5
6077	-	9906
6078	-	9906R
6079	-	9908
6080	8BD	6AS7G#, 6AS7GA#, 6AS7GT#, 6080WA, 6080WB, A1834, ECC230
6080W	8BD	6AS7G#, 6AS7GA#, 6080#, 6080WA, 6080WB
6080WA	8BD	6AS7G#, 6AS7GA#, 6080#, 6080W#, 6080WB

Tube Type	Basing	Replacement
6080WB	8BD	6AS7G#, 6AS7GA#, 6080#, 6080W#, 6080WA
6082	8BD	6082A
6082A	8BD	6082#
6083	-	9909
6084	-	5879#, 6084/E80F, E80F
6084/E80F	-	5879#, 6084, E80F
6085	-	5962#, 6085/E80CC, E80CC
6085/E80CC	-	6SN7GTB#, 5962#, 6085, E80CC
6086	-	18042, TS53
6087	5L	5Y3#, 5Y3/6087, 5Y3G#, 5Y3G/GT#, 5Y3GT#, 5Y3GT/G#, 5Y3WGT, 5Y3WGTB, 6087/5Y3WGTB, 6106, 6106/5Y3WGT, 6853, U50
6087/5Y3WGTB	5L	5Y3#, 5Y3/6087, 5Y3G#, 5Y3G/GT#, 5Y3GT#, 5Y3GT/G#, 5Y3WGT, 5Y3WGTB, 6087, 6106, 6106/5Y3WGT, 6853, U50
6088	-	522AX, 6088/522AX
6088/522AX	-	522AX, 6088
6094	-	6AQ5#, 6AQ5/6005/6669#, 6AQ5A#, 6AQ5A/6HG5#, 6HG5#, 6005#, 6669#, 6669/6AQ5A#
6095	7BZ	6AQ5#, 6AQ5/6005/6669#, 6AQ5A#, 6AQ5A/6HG5#, 6AQ5W, 6HG5#, 6005, 6005/6AQ5W, 6005/6AQ5W/6095, 6095, 6669#, 6669/6AQ5A#
6096	7BD	6AK5, 6AK5/5654, 6AK5/EF95, 6AK5W, 5654#, 5654/6AK5W, 6096/6AK5W, DP61, E95F, EF95, EF905, M8100, M8180, PM05
6096/6AK5W	7BD	6AK5, 6AK5/5654, 6AK5/EF95, 6AK5W, 5654#, 5654/6AK5W, 6096, DP61, E95F, EF95, EF905, M8100, M8180, PM05
6097	6BT	6AL5#, 6AL5/6EB5#, 6AL5/5726/6663, 6AL5/EAA91, 6AL5W, 6EB5#, 5726#, 5726/6AL5W, 5726/6AL5W/6097, 6663#, 6663/6AL5#, D717, EAA91, EAA91/6AL5, M8212
6098	6BQ	6AR6, 6AR6WA, 6098/6AR6WA, 6384, 6384/6098
6098/6AR6WA	6BQ	6AR6, 6AR6WA, 6098, 6384, 6384/6098
6099	7BF	6J6, 6J6WA#, 5964#, 6099/6J6, 6101#, ECC91, M8081, T2M05
6099/6J6	7BF	6J6, 6J6WA#, 5964#, 6099, 6101#, ECC91, M8081, T2M05
6100	6BG	6C4#, 6C4/6135#, 6C4/EC90#, 6C4WA, 6100/6C4WA, 6135*, cont'd
6100, cont'd		EC90, L77, M8080, QA2401, QL77
6100/6C4WA	6BG	6C4#, 6C4/6135#, 6C4/EC90#, 6C4WA, 6100, 6135*, EC90, L77, M8080, QA2401, QL77
6101	7BF	6J6WA, 5964#, 6099, 6101/6J6WA, ECC91, M8081, T2M05
6101/6J6WA	7BF	6J6WA, 5964#, 6099, 6101, ECC91, M8081, T2M05
6105	-	U50
6106	5T	5Y3, 5Y3/6087, 5Y3G#, 5Y3G/GT#, 5Y3GT#, 5Y3GT/G#, 5Y3WGT, 5Y3WGTA, 5Y3WGTB, 6087, 6087/5Y3WGTB, 6106/5Y3WGT, 6853, ECC91, M8081, T2M05
6106/5Y3WGT	5T	5Y3, 5Y3/6087, 5Y3G#, 5Y3G/GT#, 5Y3GT#, 5Y3GT/G#, 5Y3WGT, 5Y3WGTA, 5Y3WGTB, 6087, 6087/5Y3WGTB, 6106, 6853, ECC91, M8081, T2M05
6110	8DJ	-
6111	8DG	-
6112	8DG	-
6113	8BD	6SL7W#, 6SL7GT#, 6SL7WGT
6118	7V	-
6125	-	EC90, L77, M8080, QA2401, QL77
6130	4BL	3C45, 3C45/6130, 6130/3C45
6130/3C45	4BL	3C45, 3C45/6130, 6130
6132	-	7D10, EF82, EL821, EL822, M8135
6134	8N	6AC7#, 6AC7/1852#, 6AC7/1852/6134, 6AC7/EL34#, 6AC7W, 6AC7WA, 1852#, 6134#, EL34#
6135	6BG	6C4#, 6C4/6135#, 6C4/EC90#, 6C4WA*, 6100*, 6100/6CWA*, 88080, EC90, L77, QA2401, QL77
6136	7BK	6AU6#, 6AU6/6136, 6AU6WA#, 6AU6WB#, 6136/6AU6WA, EF94
6136/6AU6WA	7BK	6AU6#, 6AU6WA, 6AU6WB, 6136, EF94
6137	-	6SK7, 6137/6SK7WA
6137/6SK7WA	-	6SK7, 6137
6140	-	423A, 5651A#, 6140
6140/423A	-	423A, 5651A#, 6140
6143	-	-
6145	8V	-
6146	7CK	QE05/40, QV06-20, 6146B, 6146B/8192A, 6146B/8298A, 8192A, 8298A
6146A	7CK	6146B, 6146B/8192A, 6146B/8298A, 8192A, 8298A
6146B	7CK	6146B/8192A, 6146B/8298A, 8192A, 8298A
6146B/8192A	7CK	6146B, 6146B/8298A, 8192A, 8298A
6146B/8298A	7CK	6146B, 6146B/8192A, cont'd

Tube Type	Basing	Replacement
6146B/8298A, cont'd		8192A, 8298A
6147	6CL	-
6151	-	5744, 5744/6151
6155	-	4-125A, 4-125A/4D21, 4D21
6156	-	4-250, 4-250/5D22, 4-250A, 4-250A/5D22, 5D22
6157	-	R17
6158	-	13D3
6159	-	6159A, 6159B, 6159W, 6159W/7357, 7357, QE05/40H
6159A	-	6159B, 6159W, 6159W/7357, 7357, QE05/40H
6159B	-	6159A, 6159W, 6159W/7357, 7357
6159W	-	6159W/7357, 7357
6159W/7357	7CK	6159W, 7357
6161	-	5588
6166	-	6166A, 6166A/7007, 7007
6166A	-	6166A/7007, 7007
6166A/7007	-	6166A, 7007
6173	-	-
6180	-	6SN7GTB#, 5692
6185	-	2C51W, 1219, 5670, 5670/2C51W, 5670WA
6186	7BD	6AG5#, 6AG5/EF96#, 6AG5WA, 6186/6AG5WA, EF96
6186/6AG5WA	7BD	6AG5#, 6AG5/EF96#, 6AG5WA, 6186, EF96#
6187	7CM	6AS6, 6AS6W, 5725*, 5725/6AS6W, 6187/6AS6
6187/6AS6	7CM	6AS6, 6AS6W, 5725*, 5725/6AS6W, 6187
6188	8BD	6SU7, 6SU7WGT, 6SU7GTY, 6188/6SU7GTY
6188/6SU7GTY	8BD	6SU7, 6SU7WGT, 6SU7GTY, 6188
6189	9A	12AU7#, 12AU7A, 12AU7A/ECC82#, 12AU7WA, 5814#, 5814A#, 5963#, 6189/12AU7A, 6189/12AU7WA, 6680#, 6680/12AU7A#, B749, EB2CC, ECC82, ECC802, M8136
6189/12AU7A	9A	12AU7#, 12AU7A, 12AU7A/ECC82#, 12AU7WA, 5814#, 5814A#, 5963#, 6189, 6189/12AU7WA, 6680#, 6680/12AU7A#, B749, EB2CC, ECC82, ECC802, M8136
6189/12AU7WA	-	12AU7A#, 12AU7A/ECC82#, 12AU7WA, 5814A#, 5963#, 6189, 6680/12AU7A#
6189W	9A	12AU7W, 12AU7WA, 6189#
6195	-	DL700
6197	9BV	6CL6#, 6CL6/6677#, 2014, 6677/6CL6#
6199	12AE	-
6201	9A	12AT7#, 12AT7WA#, 12AT7WB#, 6201/12AT7WA, 6679#, 6679/12AT7#, B152, B309, E81CC, ECC81, ECC801S, M8162
6201/12AT7WA	9A	12AT7#, 12AT7WA, 12AT7WB#, 6201, 6679#, 6679/12AT7#, B152, B309, E81CC, ECC81, ECC801S, M8162
6202	5BS	6X4#, 6X4W#, 6X4WA, 6Z31, 6202/6X4WA, EZ90, M8138, QA2407, QU78, U78, U707, V2M70
6202/6X4WA	5BS	6X4#, 6X4W#, 6X4WA, 6Z31, 6202, EZ90, M8138, QA2407, QU78, U78, U707, V2M70
6203	-	6Z31, EZ90, M8138, QA2407, QU78, U78, U707, V2M70
6205	-	EF734
6206	8DC	-
6211	9A	2013, 6211A
6211A	-	6211#
6213	-	-
6215	-	-
6216	-	-
6217	14M	-
6218	-	E80T
6225	-	5840, 5840/6225, 6225
6227	-	E80L
6236	-	-
6247	-	-
6252	-	9910, QQE03/20, QQV03-20
6263A	-	-
6264A	-	-
6265	-	6BH6, 6BH6/6265/6661, 6265, 6661, 6661/6BH6
6267	9CQ	6BQ5, 6BQ5/EL84, 6F22, 8D8, 6267/EF86, EF86, EF87, EF806S, EL84, EL84/6BQ5, M8195, Z729
6267/EF86	9CQ	6BQ5, 6BQ5/EL84, 6F22, 8D8, 6267, EF86, EF87, EF806S, EL84, EL84/6BQ5, M8195, Z729
6268	-	-
6269	-	-
6277	-	-
6278	-	ELC5F14
6280	-	416B, 6280/416B
6280/416B	-	416B, 6280
6281	-	CK512AX, CK574
6286	-	-
6287	9CT	-
6293	7CK	-
6297	-	6CL6, 6CL6/6677, 6677
6299	-	-
6305	-	R10
6308	8EX	-
6325	-	-
6327	-	-
6328	-	-
6336	8BD	6336A, 6336B
6336A	8BD	6336, 6336B
6336B	8BD	6336, 6336A
6337	-	6336A, 6336B
6342A	14AA	-
6350	9CZ	-
6351	-	Z319
6352	8EY	-
6354	-	150B2, M8163

Tube Type	Basing	Replacement
6360	6360A	6360A, QQE03/12, QQV03-10
6360A	6360A	6360#, QQE03/12, QQV03-10
6370	-	E1T
6373	-	DL70
6374	-	6374/EY84, EY84
6374/EY84	-	6374, EY84
6375	-	DC70
6384	-	6098, 6384/6098
6384/6098	-	6098, 6384
6385	-	5670#
6386	8CJ	-
6391	-	EF74
6394	8BD	6082#, 6394A
6394A	8BD	6394
6397	-	-
6399	-	-
6405	2K	1640, 6405/1640
6405/1640	2K	1640, 6405
6414	-	5965#
6417	9K	7551#
6418	-	548DX, 6418/548DX
6418/548DX	-	548DX, 6418
6436	-	1036, 1036/6436
6438	-	6438/CK1038, CK1038
6438/CK1038	-	6438, CK1038
6442	-	-
6443	-	EY84, M8091, R18
6448	-	-
6463	-	CC863, ECC86, ECC813
6472	-	-
6476A	-	-
6481	-	5767
6483	-	-
6484	-	4B24, EL3C, EL3C/4B24
6485	7BK	6AH6WA, 6485/6AH6WA
6485/6AH6WA	7BK	6AH6WA, 6485
6486	-	6AS6#, 6AS6W#, 5725#, 5725/6AS6W, 6187#, 6187/6AS6#
6486A	9DV	6AS6#, 5AS6W#, 5725#, 5725/6AS6W, 6187#, 6187/6AS6#
6487	-	EF70, M8125
6488	-	EF73, M8122
6489	-	6489/EA76, EA76, M8123
6489/EA76	-	6489, EA76, M8123
6516	-	M8082
6520	8BD	6AS7G#, 6AS7GA#
6521	-	-
6524	-	-
6525	-	-
6528	8BD	-
6528A	-	-
6533	-	-
6535	-	ECC91, M8081, T2M05
6542	-	-
6550	7S	7D11, 12E13, 6550A, 7027A#, KT88
6550A	-	7D11, 12E13, 6550, 7027A#, KT88
6562	-	5794A, 6562/5794A
6562/5794A	-	5794A, 6562
6570	-	-
6574	-	EN32, ME1501
6582	9EJ	TE35
6582A	9EJ	-
6611	-	DF61
6626	5BO	0A2#, 0A2A#, 0A2WA, 150C2, 150C4, 6073#, 6073/0A2#, 6626/0A2WA, HD51, M8223, STV150/30
6626/0A2WA	5BO	0A2A#, 0A2WA, 150C2, 150C4, 6073#, 6073/0A2#, 6626, HD51, M8223, STV150/30
6627	5BO	0B2#, 0B2WA#, 108C1, 108C2, 6074#, 6074/0B2#, 6627/0B2WA, HD52, M8224, STV108/30
6627/0B2WA	5BO	0B2#, 0B2WA#, 108C1, 108C2, 6074#, 6074/0B2#, 6627, HD52, M8224, STV108/30
6655A	14AA	-
6660	7CC	6BA6#, 6BA6/5749/6660, 6BA6/EF93#, 6BA6W, 6BA6WA, 5749#, 5749/6BA6W, 6660/6BA6, EF93#, M8101, PM04, W727
6660/6BA6	7BK	6BA6#, 6BA6/5749/6660, 6BA6/EF93#, 6BA6W, 6BA6WA, 5749#, 5749/6BA6W, 6660, EF93#, M8101, PM04, W727
6661	7CM	6BH6#, 6BH6/6265/6661, 6265, 6661/6BH6, 7693, E90F
6661/6BH6	7CM	6BH6#, 6BH6/6265/6661, 6265, 6661, 7693, E90F
6662	7CM	6BJ6#, 6BJ6/6662, 6BJ6A, 6662/6BJ6, 7694, N78, E99F
6662/6BJ6	7CM	6BJ6#, 6BJ6/6662, 6BJ6A, 6662, 7694, N78, E99F
6663	6BT	6AL5#, 6AL5/6EB5#, 6AL5/5726/6663, 6AL5/EAA91, 6AL5W, 6EB5#, 5726#, 5726/6AL5W, 5726/6AL5W/6097, 6058, 6097, 6663/6AL5, EAA91, EAA91/6AL5, EB91
6663/6AL5	6BT	6AL5#, 6AL5/6EB5#, 6AL5/5726/6663, 6AL5/EAA91, 6AL5W, 6EB5#, 5726#, 5726/6AL5W, 5726/6AL5W/6097, 6058, 6097, 6663, EAA91, EAA91/6AL5
6664	5CE	6AB4#, 6AB4/EC92, 6664/6AB4, EC92, EC92/6AB4
6664/6AB4	5CE	6AB4#, 6AB4/EC92, 6664, EC92, EC92/6AB4
6669	7BZ	6AQ5#, 6AQ5/6005/6669, 6AQ5A#, 6AQ5A/6HG5#, 6AQ5W, 6HG5#, 6005#, 6005/6AQ5W, 6005/6AQ5W/6095, 6095, 6669/6AQ5A, EL90, N727
6669/6AQ5A	7BZ	6AQ5A#, 6AQ5A/6HG5#, 6AQ5W, 6HG5, 6005, 6005/6AQ5W, 6005/6AQ5W/6095, 6095, 6669
6670	-	12AU7, 12AU7A, 12AU7A/ECC82, 12AU7W, cont'd

Tube Type	Basing	Replacement
6670, cont'd		12AU7WA, 12AX7#, 12AX7A#, 12AX7A/ECC83#, 12AX7A/ECC83/7025#, 12AX7WA#, 12AX7WA/7025#, 6670/12AU7A, 7025#, ECC82, ECCC82/12AU7A, ECC83#
6670/12AU7A	-	12AU7, 12AU7A, 12AU7A/ECC82, 12AU7W, 12AU7WA, 12AX7#, 12AX7A#, 12AX7A/ECC83#, 12AX7A/ECC83/7025#, 12AX7WA#, 12AX7WA/7025#, 6670, 7025#, ECC82, ECCC82/12AU7A, ECC83#
6671	-	12AT7, 12AT7/ECC81, 12AT7WA, 12AT7WB, 12AZ7*, 12AZ7A*, 6060, 6201, 6679, 7492, 7728, A2900, B152, B309, B739, CV4024, E81CC, ECC81, ECC801, ECC801S, M8162, QA2406, QB309
6676	7CM	6CB6#, 6CB6A#, 6CB6A/6CF6#, 6CF6, 6676/6CB6A
6676/6CB6A	7CM	6CB6A#, 6CB6A/6CF6#, 6CF6#, 6676
6677	9BV	6CL6#, 6CL6/6677, 6L43, 2014, 6197, 6297, 6677/6CL6
6677/6CL6	9BV	6CL6#, 6CL6/6677, 6L43, 6677, 2014, 6197, 6297, 6677
6678	9AE	6AX8#, 6EA8, 6KD8#, 6LN8, 6U8#, 6U8A#, 6U8A/6KD8#, 6U8A/6AX8/6KD8#, 6678/6U8A, ECF82
6678/6U8A	9AE	6AX8#, 6EA8, 6KD8#, 6LN8, 6U8A#, 6U8A/6KD8#, 6U8A/6AX8/6KD8#, 6678
6679	9A	12AT7#, 12AT7WA#, 12AT7WB#, 6201, 6679/12AT7, B152, B309, B739, ECC81
6679/12AT7	9A	12AT7#, 12AT7WA#, 12AT7WB#, 6201, 6679, B739
6680	9A	12AU7#, 12AU7A#, 12AU7A/ECC82#, 12AU7WA, 5814, 6189, 6680/12AU7A, 7730, B749, ECC82#
6680/12AU7A	9A	12AU7A#, 12AU7A/ECC82#, 12AU7WA, 6189, 6680, 7730, B749
6681	9A	12AX7#, 12AX7A#, 12AX7A/ECC83#, 12AX7A/ECC83/7025#, 12AX7WA, 12AX74WA/7025, 12DT7, 5721, 5751, 6057, 6681/12AX7A, 7025, 7494, 7729, B339, B759, E83CC, ECC83#, M8137
6681/12AX7A	9A	12AX7#, 12AX7A#, 12AX7A/ECC83#, 12AX7A/ECC83/7025#, 12AX7WA, 12AX7WA/7025, 5721, 6057, 6681, 7025, cont'd
6681/12AX7A, cont'd		7494, 7729, B339, B759, E83CC, ECC83#, M8137
6686	-	6686/E81L, E81L
6686/E81L	-	6686, E81L
6687	-	5915#, E91H
6688	-	6688/E180F, E180F
6688/E180F	-	6688, E180F
6688A	9EQ	-
6689	-	6689/E83F, E83F
6689/E83F	-	6689, E83F
6690	8GQ	-
6754	9ET	412A, TE36
6761	-	-
6771	-	-
6775	-	4-400C, 4-400C/6775
6778	-	EC70
6779	-	Z803U
6788	-	-
6792	-	-
6806	-	-
6807	-	-
6808	-	-
6810A	20D	-
6814	8DK	-
6816	-	-
6829	-	12AV7, 5965#
6832	8DG	-
6833	-	-
6840	9CZ	-
6844A	-	-
6850	-	-
6851	9A	-
6853	5T	5Y3, 5Y3/6087, 5Y3G#, 5Y3G/GT#, 5Y3GT#, 5Y3GT/G#, 5Y3WGT, 5Y3WGTA, 5Y3WGTB, 6087, 6087/5Y3WGTB, 6106, 6106/5Y3WGT
6854	9FV	-
6855	-	NL716, NL716/6855
6856	-	740, 6856/740
6856/740	-	740, 6856
6857	-	740P, 740P/6857
6858	-	760, 760/6858
6859	-	760P, 760P/6859
6861	-	-
6870	9BF	-
6877	9GB	-
6883	7CK	6883B, 6883B/8552, 6883B/8032A/8552, 8032A, 8552
6883B	7CK	6883B/8552, 6883B/8032A/8552, 8032A, 8552
6883B/8552	7CK	6883B, 6883B/8552, 6883B/8032A/8552, 8032A, 8552
6883B/8032A/8552	7CK	6883B, 6883B/8552, 8032A, 8552
6884	-	-
6887	6BT	-
6888	8N	TE40
6889	-	-
6893	7CK	2E26, 6893/2E26
6893/2E26	7CK	2E26, 6893
6894	4AT	-
6895	-	-
6897	-	-
6900	6900	-

Tube Type	Basing	Replacement
6903	14AA	-
6907	-	-
6909	-	-
6910	-	-
6913	9A	-
6914	-	-
6914A	-	-
6919	6BT	-
6922	9AJ	6922/E88CC, CV2492, E88CC, ECC88
6922/E88CC	9AJ	6922, CV2492, E88CC, ECC88
6923	-	6923/EA52, EA52
6923/EA52	-	6923, EA52
6927	-	6J6
6928	-	-
6929	-	-
6931	-	-
6939	9HL	7645, QQE02/5, QQV02-6
6943	8DC	-
6944	8DC	-
6945	8DL	-
6946	8DK	-
6947	8DG	-
6948	8DG	-
6949	-	-
6952	-	-
6953	3J	-
6954	7CM	-
6955	9A	-
6957	-	-
6968	7ED	6AK5#, 6AK5/5654#, 6AK5/EF95#, 5654#, EF95#
6973	9EU	6C25
6977	6977	6977/DM160, DM160
6977/DM160	6977	6977, DM160
6979	-	4X250B
6989	-	6989/C6J/KL, 6989/C6KL, C6J, C6KL
6989/C6J/KL	-	6989, 6989/C6KL, C6J, C6KL
6989/C6KL	-	6989, 6989/C6J/KL, C6KL
6998	-	-
7000	7R	6J7, 1223, 1620#
7001	-	M8167
7007	-	6166A
7008	-	-
7014	-	NL604, NL604/7014
7015	-	NL604L, NL604L/7015
7016	-	NL606, NL606/7016
7017	-	606L, 606L/7017
7018	-	615, 615/7018
7019	-	635, 635/7019
7020	-	635L, 635L/7020
7021	-	N714, N714/7021
7023	-	760L, 760L/7023
7025	9A	12AD7*, 12AX7#, 12AX7A#, 12AX7A/ECC83, 12AX7A/ECC83/7025, 12AX7WA, 12AX7WA/7025, 12BZ7*, 12DF7, 12DM7*, 12DT7, 5751, 6681, 7025/12AX7A, 7025A, B339, B759, ECC83#, M8137
7025/12AX7A	9A	12AD7*, 12AX7#, 12AX7A#, 12AX7A/ECC83, 12AX7A/ECC83/7025, 12AX7WA, 12AX7WA/7025, 12BZ7*, 12DF7, 12DM7*, cont'd
7025/12AX7A, cont'd		12DT7, 5751, 6681, 7025, 7025A, B339, B759, ECC83#, M8137
7025A	9A	12AX7#, 12AX7A#, 12AX7A/ECC83#, 12AX7A/ECC83/7025#, 12AX7WA, 12AX7WA/7025, 12DF7, 12DT7, 12AD7*, 12BZ7*, 12DM7*, 7025#, ECC83#
7027	8HY	7027A, KT66
7027A	8HY	7027, KT66
7032	7CH	-
7034	-	4X150A, 4X150A/7034, 7034/4X150A
7034/4X150A	-	4X150A, 4X150A/7034, 7034
7035	-	4X150D, 7035/4X150D, 7609
7035/4X150D	-	4X150D, 7035, 7609
7036	7CH	6BE6, 6BE6/5750, 6BE6/EK90, 6BE6W, 6BY6, 1217, 5750, 5750/6BE6W, 5915#, 5915A, EK90
7038	8HM	-
7043	-	7043/CE881, CE881
7043/CE881	-	7043, CE881
7044	9H	-
7054	9GK	7054/8077, 8077, 8077/7054
7054/8077	9GK	7054, 8077, 8077/7054
7055	6BT	-
7056	7CM	-
7057	9AJ	-
7058	9A	-
7059	9AE	-
7060	9DX	-
7061	9EU	-
7062	-	5965#, 7062/E180CC, E180CC
7062/E180CC	-	5965#, 7062, E180CC
7077	7077	6BY4
7094	-	-
7102	12AE	-
7105	8BD	6AS7G#, 6AS7GA#, 6080#, 6080W#, 6080WA#
7111	-	-
7117	11K	-
7118	-	E181CC
7119	-	7119/E182CC, E182CC
7119/E182CC	-	7119, E182CC
7136	-	-
7137	7BQ	-
7163	-	-
7167	7EW	-
7184	9CV	6V6#, 6V6GTA#
7189	9CV	7189A
7189A	9LE	7189
7191	-	-
7193	-	2C22, 7193/2C22
7193/2C22	-	2C22, 7193
7194	-	-
7196	-	2C22, 2C22/7193, 7193
7199	9JT	-
7200	11K	-
7203	-	4CX250B, 4CX250B/7203, 7203/4CX250B
7203/4CX250B	-	4CX250B, 4CX250B/7203, 7203
7204	-	4CX250F, 7204/4CX250F
7204/4CX250F	-	4CX250F, 7204

Tube Type	Basing	Replacement
7211	-	-
7212	8EC	-
7213	-	-
7214	-	-
7224	-	-
7227	9BA	-
7233	-	-
7235	-	-
7236	7236	-
7239	-	-
7241	7241	-
7242	7241	-
7244	7BF	6J6WA#, 5964#, 6101#
7244A	7BF	6J6WA#, 5964#, 6101#
7245	7BQ	6J4#, 8532#
7245A	-	6J4#, 8532#
7247	9A	12DW7, 12DW7/7247, B759
7258	9DA	-
7262A	8HM	-
7263A	-	-
7264	20D	-
7265	20C	-
7271	-	-
7292	-	JN2-25W
7293A	-	-
7295	-	7295A, 7295B, 7295C, 8749
7295A	-	7295B, 7295C, 8749
7295B	-	7295C, 8749
7295C	-	8749
7308	9AJ	7308/E188CC, E188CC
7308/E188CC	9AJ	7308, E188CC
7316	-	ECC186
7318	9A	12AU7A/ECC82#, 5814A#, 6680/12AU7A#
7320	-	7320/E84L, E84L
7320/E84L	-	7320, E84L
7326	14AM	-
7327	8DG	-
7348	-	-
7355	8KN	-
7357	-	6159W, 6159W/7357, 7357
7358	8EC	-
7360	9KS	-
7370	-	5687#
7377	-	7377, 7377/QQE04/5
7377/QQE04/5	-	7377, QQE04/5
7378	-	QE08/200, QV08-100
7389	-	7389A, 7389B, 7389C, 8748
7389A	-	7389B, 7389C, 8748
7389B	-	7389C, 8748
7389C	-	8748
7408	7AC	6V6, 6V6G, 6V6GT, 6V6GTA, 6V6GT/G, 6V6GTY, 6V6GTX, 7408/6V6GT, OSW3106
7408/6V6GT	7AC	6V6, 6V6G, 6V6GT, 6V6GTA, 6V6GT/G, 6V6GTY, 6V6GTX, 7408, OSW3106
7410	-	-
7412	-	-
7452	-	7452/QK324, QK324
7452/QK324	-	7452, QK324
7457	-	-
7486	-	-
7489	-	12AU7
7492	-	12AT7
7494	9A	12AX7A#, cont'd

Tube Type	Basing	Replacement
7494, cont'd		12AX7A/ECC83#, 12AX7A/ECC83/7025#, 12AX7WA, 12AX7WA/7025, 5721, 6057, 6681, 6681/12AX7A#, 7025#, 7729, ECC83#
7496	-	6BA6
7498	-	6AM6, 6AM6/EF91, EF91
7499	-	6CH6, 6CH6/EL821, EL821
7502	-	6BE6, 6BE6/5750, 6BE6/EK90, 5750, EK90
7509A	-	710L, 710L/7509A
7527	-	4-400B, 4-400B/7527
7533	-	-
7534	8KE	E180L
7536	-	-
7543	7BK	6AU6#, 6AU6/6136, 6AU6A#, 6AU6A/EF94#, 6AU6WA#, 6AU6WB#, 6136, 6136/6AU6WA, 7543/E130LF, 8425A#, 8425A/6AU6A#, E130LF, EF94
7543/E130LF	7BK	6AU6#, 6AU6/6136, 6AU6A#, 6AU6A/EF94#, 6AU6WA#, 6AU6WB#, 6136, 6136/6AU6WA, 7543, 8425A#, 8425A/6AU6A#, E130LF, EF94
7548	-	-
7550	8DG	-
7551	9LK	-
7554	-	-
7556A	-	C3J/AL/7556A, C3J/AL
7558	9LK	-
7581	7AC	6L6GC, 7581A, 7581A/KT66, KT66, KT77
7581A	7AC	6L6GC, 7581, 7581A/KT66, KT66
7581A/KT66	7AC	6L6GC, 7581A, KT66
7586	12AQ	-
7587	12AS	-
7588	-	-
7591	8KQ	7591A, 7591AP
7591A	8KQ	7591
7591AP	8KQ	7591A
7607	-	-
7609	-	4X150D, 7035, 7035/4X150D
7623	-	-
7624	-	-
7631	-	6AL5
7642	-	-
7643	-	7643/E80CF, E80CF
7643/E80CF	-	7643, E80CF
7645	9HL	6939
7650	-	-
7651	-	-
7683	-	-
7687	9AE	-
7693	-	E90F
7694	-	E99F
7695	9MQ	-
7699	-	-
7700	6F	6C6#, 1603
7701	-	7551#
7709	-	Z70W
7710	-	7710/Z70U, Z70U

Tube Type	Basing	Replacement
7711	-	Z71U
7713	-	Z804U
7714	-	Z805U
7716	-	-
7717	7EW	6CY5#, 6EA5, 6EV5, 7717/6CY5, 8113
7717/6CY5	7EW	6CY5#, 6EA5, 6EV5, 7717, 8113
7719	-	-
7721	-	-
7722	-	7722/E280F, E280F
7722/E280F	-	7722, E280F
7723	-	610, 610/7723
7724	9KR	14GT8#, 14GT8A#, 14JG8, 7724/14GT8
7724/14GT8	9KR	14GT8#, 14GT8A, 14JG8, 7724
7728	-	12AT7, 6201, 7728/12AT7
7728/12AT7	-	12AT7, 6201, 7728
7729	9A	12AX7A#, 12AX7A/ECC83#, 12AX7A/ECC83/7025#, 12AX7WA, 12AX7WA/7025, 5721, 6057, 6681#, 6681/12AX7A#, 7025, 7494, ECC83
7730	9A	12AU7A#, 12AU7WA, 5814A#, 6189#, 6680#, 6680/12AU7A#, 7730/12AU7WA
7730/12AU7WA	9A	12AU7A#, 12AU7WA, 5814A#, 6189#, 6680#, 6680/12AU7A#, 7730
7731	-	6AX8#, 6KD8#, 6U8A#, 6U8A/6KD8#, 6U8A/6AX8/6KD8#, 6678#, 6678/6U8A#
7732	-	6CB6A#, 6CB6A/6CF6#, 6CF6#, 6676#, 6676/6CB6A#
7733	-	12BV7, 12BV7/12BY7A, 12BY7A, 12BY7A/12BV7/12DQ7#, 7733/12BY7A
7733/12BY7A	-	12BV7, 12BV7/12BY7A, 12BY7A, 12BY7A/12BV7/12DQ7#, 7733
7734	-	6GE8, 6GE8/7734
7735	8HM	7735A, 7735B
7735A	8HM	7735, 7735B
7735B	8HM	7735, 7735A
7737	-	7737/E186F, E186F
7737/E186F	-	7737, E186F
7738	-	-
7746	14AV	-
7751	-	-
7752	7CM	6AS6#, 6AS6W#, 5725#, 5725/6AS6W#, 6187#, 6187/6AS6#
7754	9MQ	-
7755	7BD	6AJ5
7756	6BQ	6AR6
7763	-	-
7764	9NG	-
7767	-	-
7786	-	660L, 660L/7786
7788	-	7788/E810F, E810F
7788/E810F	-	7788, E810F
7801	-	-
7802	-	-
7803	-	-
7835	-	-
7842	-	-
7843	-	-
7850	20E	-
7854	-	-
7861	-	-
7867	5BT	6EX6*
7868	9RW	7868P
7868P	9RW	7868
7870	-	-
7892	-	-
7894	-	-
7895	12AQ	-
7898	9EP	-
7905	9PB	-
7911	-	-
7963	-	-
7983	-	-
7984	-	-
8000	-	-
8005	-	-
8008	2P	-
8012	-	8012A
8012A	-	8012
8013	-	8013A
8013A	4P	8013
8016	3C	1AU3, 1B3, 1B3/8016, 1B3GT, 1B3GT/1G3GTA, 1G3GT, 1G3GT/1B3GT, 1G3GTA, 1G3GTA/1B3GT, 1J3, 1J3/1K3A, 1J3A, 1K3, 1K3/1J3, 1K3A, 1K3A/1J3, 1N2, 1N2A, 1N2A/1AU3, 8016, DY30, U41
8018	5AW	4Y25, 5S1, 807, 807W, 807WA, 5933, 5933WA, HY61, P17A, QE06/50, QE06140, QV05-25, RK39
8020	-	-
8022	-	4B30, 15R
8025	-	8025A
8025A	-	8025
8032	-	8032A, 8032A/8552, 8552
8032A	-	8032A/8552, 8552
8032A/8552	-	8032A, 8552
8042	-	4652, 8042/4652
8042/4652	-	4652, 8042
8051	8LB	-
8053	14AA	-
8054	14AA	-
8055	14AA	-
8056	12AQ	-
8058	12CT	-
8068	-	-
8072	-	-
8077	9GK	7054#, 8077/7054
8077/7054	9GK	7054#, 8077
8084	-	-
8092A	-	-
8101	-	EC157
8102	-	-
8106	9PL	-
8113	-	6CY5, 8113/6CY5
8113/6CY5	-	6CY5, 8113
8117A	-	-

Tube Type	Basing	Replacement
8121	-	-
8122	-	-
8134	8LN	-
8134/V1	-	-
8136	5C	6DK6#
8149	-	-
8150	-	-
8156	-	-
8162	-	12AT7#, 12AT7/ECC81#, 12AT7WA, ECC81#
8163	-	3-400Z, 3-400Z/8163
8165	-	4-65A, 4-65A/8165
8167	-	4CX300A, 4CX300A/8167
8184	-	-
8187	-	4PR65A, 4PR65A/8187
8192A	-	6146B, 6146B/8192A, 6146B/8298A, 8298A
8196	-	6AS6#, 5725#, 5725/6AS6W#, 5754, 6187#, 6187/6AS6#
8203	12AQ	-
8204	-	2D21#, 2D21/5727#, 2D21/PL21#, 5727#, 5727/PL21#, PL21#
8212	-	-
8223	-	8223/E288CC, E288CC
8223/E288CC	-	8223, E288CC
8226	-	-
8228	-	8228/ZZ1000, ZZ1000
8228/ZZ1000	-	8228, ZZ1000
8233	-	8233/E55L, E55L
8233/E55L	-	8233, E55L
8236	-	-
8245	-	4CX250K, 4CX250K/8245
8254	-	EC1000
8255	-	E88C, EC88, 8255/E88C
8255/E88C	-	E88C, EC88, 8255
8270	-	-
8278	9QB	EL503, F1EL
8298	-	6146B, 6146B/8298A, 8298A
8298A	-	6146B, 6146B/8298A
8321	-	4CX350A, 4CX350A/8321
8323	-	-
8327	-	-
8334	-	-
8344	-	1218A, 8344/1218A
8344/1218A	-	1218A, 8344
8353	-	-
8379	-	-
8380	-	7587
8382	-	7586
8393	12AQ	-
8408	-	-
8417	7S	8417P
8417P	7S	8417
8422	-	8422/B5991, B5991
8422/B5991	-	8422, B5991
8425	7BK	6AU6A, 6AU6A/EF94, 8425A, 8425A/6AU6A, EF94
8425A	7BK	6AU6A, 6AU6A/EF94, 8425, 8425A/6AU6A, EF94
8425A/6AU6A	7BK	6AU6A, 6AU6A/EF94, 8425, 8425A, EF94
8426	7BK	12AU6, 12AU6A, 12BA6, 12BA6A, 12F31, 8426A, 8426A/12AU6, HF93, HF94
8426A	7BK	12AU6, 12AU6A, cont'd
8426A, cont'd		12BA6, 12BA6A, 12F31, 8426, 8426A/12AU6, HF93, HF94
8426A/12AU6	7BK	12AU6, 12AU6A, 12AW6#, 12BA6, 12BA6A, 12F31, 8426A, HF93, HF94
8431	-	-
8437	-	-
8438	-	4-400A, 4-400A/8438
8441	-	7895
8445	-	-
8446	-	-
8447	-	12BR7
8448	-	12BY7
8449	-	-
8453	-	Z550M
8456	-	-
8457	-	-
8458	-	-
8462	-	-
8463	-	-
8480	8MD	-
8489	-	-
8501	-	-
8507A	8ME	-
8509	-	-
8521	8LB	-
8532	7BQ	6J4#, 6J4WA, 8532/6J4WA
8532/6J4WA	7BQ	6J4#, 6J4WA, 8532
8536	-	-
8538	-	-
8541	-	8541A
8541A	-	8541
8552	-	6883B, 6883B/8552, 8032A, 8032A/8552
8556	-	8556/EC8010, EC8010
8556/EC8010	-	8556, EC8010
8560A	-	8560AS
8560AS	-	8560A
8562	-	8562/E88C, E88C
8562/E88C	-	8562, E88C
8567	8LN	-
8568	-	-
8571	12FZ	-
8572A	8ME	-
8573A	-	-
8575	-	-
8587	-	-
8596	-	-
8605/V1	-	-
8605/V2	-	-
8606	-	-
8608	-	-
8624	-	-
8627	12CT	8627A#
8627A	-	8627#
8628	12AQ	-
8643	-	-
8644	-	-
8645	-	-
8664	-	-
8664/V1	-	-
8673	-	-
8674	-	-
8675	-	-
8683	-	-
8684	-	-
8727	-	-

Tube Type	Basing	Replacement
8748	-	7389B
8749	-	7295B
8753	-	-
8754	-	840, 840/8754, NL840, NL840/8754
8765	-	-
8775	-	-
8791/V1	-	-
8792/V1	-	-
8793	-	-
8794	-	-
8806	-	-
8807	-	-
8808	8808	-
8828	-	-
8844	-	-
8850	-	-
8851	-	-
8857/V1	-	-
8857/V2	-	-
8858	-	-
8874	-	-
8890	-	-
8891	-	-
8904	-	-
8906AL	-	-
8908	-	-
8950	-	-
9001	7BD	-
9002	7BS	-
9003	7BD	-
9004	4BJ	-
9005	5BG	-
9006	6BH	-
9903	-	5894
9904	-	5923
9905	-	QQC04/15
9906	-	6077
9906R	-	6078
9907	-	6075
9907R	-	6076, 6076/QBL5
9908	-	6079
9909	-	6083
9910	-	QQE03/20, QQV03-20
9950	-	5869, 5869/9950
18042	-	6086
88080	-	6135
A59RX	-	-
A61	4G	17Z3, 17Z3/PY81
A677	-	6C6
A863	7R	6J7, 6J7GT
A1834	-	6AS7G#, 6AS7GA, 6080, 6080W, 6080WA, 6080WB
A2521	-	6CR4
A2900	-	12AT7, 12AT7/ECC81
AA91E	-	6AL5#, 6AL5/6EB5#, 6AL5/5726/6663#, 6AL5/EAA91#, 6AL5W, 6EB5#, 5726, 5726/6AL5W, 5726/6AL5W, 6097, 6663#, 6663/6AL5#, EAA91#, EAA91/6AL5#
ABC91	-	12A6
AD17	6AU	-
AD	4G	1V, 1V/6Z3, 1V/KR1, 6Z3, KR1
AFX212		
AG5210	-	0B2, 0B2WA, 6074, cont'd
AG5210, cont'd		6074/0B2, 6627, 6627/0B2WA
AG5211	-	0A2, 0A2WA, 6073, 6073/0A2, 6626, 6626/0A2WA
ARS25A	-	4Y25, 5S1, 807, 807W, 807WA, 5933, 5933WA, 8018, HY61, P17A, QE06/50, QE06140, QV05-25, RK39
ASG512	-	2D21, 2D21/5727, 2D21/PL21, 5727, 5727/2D21, PL21
ASG5121	-	2D21, 2D21/5727, 2D21/PL21, 5727, 5727/2D21, PL21
ATS225A	-	4Y25, 5S1, 807, 807W, 807WA, 5933, 5933WA, 8018, HY61, P17A, QE06/50, QE06140, QV05-25, RK39
AX9902	-	5868, AX9902/5868
AX9902/5868	-	5868, AX9902
AX9903	-	5894, AX9903/5894
AX9903/5894	-	5894, AX9903
AZ11	-	-
AZ41	-	-
B36	8BD	12SN7, 12SN7GT
B63	-	6A6
B65	8BD	6SN7, 6SN7GT, 6SN7GTB, ECC33
B109	-	-
B139	-	7AN7
B152	9A	12AT7, 12AT7/ECC81, 6201, 6679, B309, ECC81
B309	9A	12AT7, 12AT7/ECC81, 6201, 6679, B152, ECC81
B317	-	-
B319	-	7AN7
B329	9A	12AU7, 12AU7A/ECC82
B339	9A	12AX7#, 12AX7A#, 12AX7A/ECC83#, 12AX7A/ECC83/7025#, 12AX7WA#, 12AX7WA/7025#, 5751#, 6681/12AX7A#, 7025#, ECC83#
B349	-	7EK7
B719	9AJ	6AQ8, 6AQ8/ECC85, ECC85, ECC85/6AQ8
B729	-	6GA8
B739	9A	12AT7#, 12AT7/ECC81#, 12AT7WA#, 6679#, 6679/12AT7#
B749	9A	12AU7#, 12AX7A#, 12AX7A/ECC83#, 12AX7A/ECC83/7025#, 12AX7WA#, 12AX7WA/7025#, 12AU7A/ECC82#, 5814#, 6189#, 6680#, 6680/12AU7A#, 7025#, ECC83#
B759	9A	12AX7#, 12AX7A#, 12AX7A/ECC83#, 12AX7A/ECC83/7025#, 12AX7WA#, 12AX7WA/7025#, 5751#, 7025#, ECC83#
B5991	-	8422, 8422/B5991
BA2	-	2050#
BF61	-	6CK5, 6CK5/EL41, cont'd

Tube Type	Basing	Replacement
BF61, cont'd		67PT, EL41, N150
BF451	-	45A5
BH	-	-
BPM04	7BZ	6AQ5, 6AQ5/6005/6669, 6AQ5A, 6AQ5A/6HG5, 6HG5, 6005, 6669, 6669/6AQ5A
BR	-	-
BVA264	-	6AG6G, 6AG6G/EL33, EL33
BVA265	-	6AG6G, 6AG6G/EL33, EL33
C1K	4D	C1K/6014, 6014
C1K/6014	4D	C1K, 6014
C3J	4CF	5632, 7556A, C3J/5632, C3J/AL/7556A
C3J/5632	4CF	5632, 7556A, C3J, C3J/AL/7556A
C3J/AL	4CF	7556A, C3J/AL/7556A
C3J/AL/7556A	4CF	7556A, C3JAL
C3J/L	4CF	7556A, C3J/AL/7556A
C3JA	4CF	5684, 7556A, C3JA/5684, C3J/AL/7556A
C3JA/5684	4CF	5684, 7556A, C3JA, C3J/AL/7556A
C6J	4BZ	5C21, C6J/5C21, C6K, 5685, 5685/C6J/K
C6J/5C21	4BZ	5C21, C6J
C6J/K	-	5685, 5685/C6J/K
C6J/KL	-	6989, 6989/C6J/KL
C6JA	4BZ	5685, 5685/C6J/K, C6JA/5685
C6JA/5685	4BZ	5685, 5685/C6J/C6K, C6J, C6JA, C6K
C6K	-	5685, 5685/C6J/C6K, C6JA
C6KL	-	6989, 6989/C6J/KL
C6L	-	5528, 5528/C6L
C10B	-	-
C16J	-	C16J/5665, 5665, 5665/C16J
C16J/5665	-	C16J, 5665, 5665/C16J
C610	-	7J7
C1106	-	6AZ6
CC81E	-	12AT7/ECC81#12AT7WA, 12AT7WB, 6201, 6679#, 6679/12AT7#, ECC81#
CC86E	-	6GM8
CC863	-	ECC86, ECC813
CCA	-	6922/E88CC
CE1C	-	918, 918/CE1C
CE881	-	7043, 7043/CE881
CFX80	9AE	-
CK108	6F	77
CK501AX	-	-
CK502AX	-	-
CK503AX	-	-
CK505AX	-	-
CK506AX	-	-
CK507AX	-	-
CK511AX	-	-
CK512AX	-	-
CK521AX	-	-
CK522AX	-	-
CK527AX	-	-
CK532DX	-	-
CK533AX	-	-
CK536AX	-	-
CK551AXA	-	-
CK553AXA	-	-
CK556AX	-	-
CK558AX	-	-
CK569AX	-	-
CK573AX	-	-
CK574	-	6281
CK605CX	-	-
CK606BX	-	-
CK608CX	-	-
CK619CX	-	-
CK1003	4R	0Z4A, 0Z4A/0Z4
CK1009	-	CK1009A, CK1009BA
CK1009A	-	CK1009BA
CK1009BA	-	-
CK1013	-	5517, 5517/CK1013
CK1038	-	6438, 6438/CK1038
CK1917	-	NL842, NL842/CK1917
CK5517	-	-
CK5608	-	-
CK5676AX	-	-
CK5678	-	-
CK5694	-	-
CK5829	-	-
CK5854	-	-
CSF80	-	4BL8, 4BL8/XCF80
CV131	-	-
CV133	-	-
CV136	-	-
CV138	-	-
CV140	-	-
CV216	-	0D3, 0D3/VR150, 0D3A, VR150, VR150/0D3
CV281	-	-
CV283	-	-
CV303	-	-
CV346	-	-
CV378	-	GZ37, GZ37/CV378
CV417	-	-
CV426	-	-
CV452	-	-
CV453	-	-
CV454	-	-
CV455	-	-
CV484	-	-
CV491	-	-
CV492	-	-
CV493	-	-
CV500	-	-
CV503	-	-
CV509	-	-
CV510	-	-
CV511	-	-
CV512	-	-
CV515	-	-
CV522	-	-
CV525	-	-
CV526	-	-
CV529	-	-
CV531	-	-
CV534	-	-
CV535	-	-
CV537	-	-
CV538	-	-
CV540	-	-
		-
CV543	-	-
CV544	-	-
CV546	-	-
CV547	-	-

Tube Type	Basing	Replacement
CV550	-	-
CV551	-	-
CV552	-	-
CV553	-	-
CV561	-	-
CV562	-	-
CV568	-	-
CV569	-	6SL7GT, ECC35, ECC35/ CV569, CV569
CV571	-	-
CV574	-	-
CV578	-	-
CV580	-	-
CV581	-	-
CV582	-	-
CV583	-	-
CV586	-	-
CV587	-	-
CV588	-	-
CV589	-	-
CV590	-	-
CV591	-	-
CV592	-	-
CV593	-	-
CV594	-	-
CV595	-	-
CV618	-	83#
CV660	-	-
CV661	-	-
CV686	-	0C3, 0C3/VR105, 0C3A, VR105
CV694	-	-
CV698	-	-
CV700	-	-
CV703	-	-
CV705	-	-
CV706	-	-
CV728	-	-
CV729	-	-
CV731	-	-
CV741	-	-
CV747	-	-
CV753	-	-
CV752	-	0A4G
CV753	-	-
CV755	-	-
CV756	-	-
CV760	-	-
CV765	-	-
CV766	-	-
CV768	-	-
CV770	-	-
CV771	-	-
CV772	-	-
CV773	-	-
CV774	-	-
CV775	-	-
CV776	-	-
CV777	-	-
CV778	-	-
CV779	-	-
CV780	-	-
CV781	-	-
CV782	-	-
CV783	-	-
CV784	-	-
CV785	-	-

Tube Type	Basing	Replacement
CV786	-	-
CV797	-	2D21, 2D21/5727, 2D21/ PL21, 5727, 5727/2D21, PL21
CV807	-	3A4, 3A4/DL93, DL93
CV808	-	-
CV815	-	-
CV818	-	-
CV819	-	-
CV820	-	-
CV837	-	-
CV844	-	-
CV845	-	-
CV846	-	-
CV848	-	-
CV849	-	-
CV850	-	-
CV851	-	-
CV852	-	-
CV858	-	-
CV859	-	-
CV860	-	-
CV861	-	-
CV862	-	-
CV864	-	-
CV865	-	-
CV866	-	-
CV867	-	-
CV870	-	-
CV872	-	-
CV873	-	-
CV876	-	-
CV877	-	-
CV878	-	-
CV879	-	-
CV880	-	-
CV882	-	-
CV883	-	-
CV885	-	-
CV886	-	-
CV887	-	-
CV888	-	-
CV890	-	-
CV891	-	-
CV892	-	-
CV893	-	-
CV894	-	-
CV895	-	-
CV896	-	-
CV897	-	-
CV898	-	-
CV899	-	-
CV900	-	-
CV901	-	-
CV902	-	-
CV911	-	-
CV916	-	-
CV917	-	-
CV918	-	-
CV919	-	-
CV920	-	-
CV921	-	-
CV922	-	-
CV923	-	-
CV924	-	-
CV925	-	-
CV930	-	-

Tube Type	Basing	Replacement
CV937	-	-
CV938	-	-
CV939	-	-
CV940	-	-
CV945	-	-
CV946	-	-
CV948	-	-
CV995	-	-
CV1067	-	-
CV1074	-	-
CV1075	-	-
CV1100	-	-
CV1195	-	-
CV1280	-	-
CV1285	-	-
CV1286	-	-
CV1287	-	-
CV1301	-	-
CV1347	-	-
CV1376	-	-
CV1377	-	-
CV1535	-	-
CV1633	-	-
CV1741	-	-
CV1753	-	-
CV1758	-	1L4, 1L4/DF92, DF92
CV1762	-	-
CV1763	-	-
CV1770	-	-
CV1777	-	-
CV1784	-	-
CV1800	-	-
CV1802	-	-
CV1803	-	-
CV1805	-	-
CV1806	-	-
CV1811	-	-
CV1812	-	-
CV1815	-	-
CV1817	-	-
CV1818	-	-
CV1819	-	-
CV1820	-	-
CV1821	-	-
CV1823	-	-
CV1824	-	-
CV1826	-	-
CV1829	-	-
CV1832	-	0A2, 0A2WA, 6073, 6073/0A2, 6626/0A2WA
CV1833	-	0B2, 0B2WA, 6074, 6074/0B2, 6627, 6627/0B2WA
CV1834	-	6AS7G, 6AS7GA, 6080, 6080W, 6080WA
CV1854	-	-
CV1856	-	-
CV1862	-	-
CV1865	-	-
CV1873	-	-
CV1878	-	-
CV1882	-	-
CV1886	-	-
CV1887	-	-
CV1888	-	-
CV1893	-	-
CV1894	-	-
CV1896	-	-
CV1901	-	-
CV1902	-	-
CV1908	-	-
CV1909	-	-
CV1910	-	-
CV1911	-	-
CV1912	-	-
CV1917	-	-
CV1918	-	-
CV1926	-	-
CV1928	-	-
CV1929	-	-
CV1930	-	-
CV1931	-	-
CV1932	-	-
CV1933	-	-
CV1934	-	-
CV1935	-	-
CV1936	-	-
CV1937	-	-
CV1938	-	-
CV1940	-	-
CV1941	-	-
CV1942	-	-
CV1943	-	-
CV1944	-	-
CV1945	-	-
CV1946	-	-
CV1947	-	-
CV1948	-	-
CV1949	-	-
CV1950	-	-
CV1951	-	-
CV1956	-	-
CV1957	-	-
CV1958	-	-
CV1959	-	-
CV1961	-	-
CV1962	-	-
CV1963	-	-
CV1964	-	-
CV1966	-	-
CV1967	-	-
CV1969	-	-
CV1970	-	-
CV1971	-	-
CV1972	-	-
CV1973	-	-
CV1974	-	-
CV1975	-	-
CV1978	-	-
CV1981	-	-
CV1982	-	-
CV1984	-	-
CV1985	-	-
CV1986	-	-
CV1988	-	-
CV1990	-	-
CV1991	-	-
CV1992	-	0A4G
CV1993	-	-
CV1995	-	-
CV1996	-	-
CV2004	-	-
CV2005	-	-
CV2007	-	-
CV2009	-	-

Tube Type	Basing	Replacement
CV2010	-	-
CV2011	-	-
CV2013	-	-
CV2014	-	-
CV2016	-	-
CV2020	-	-
CV2021	-	-
CV2022	-	-
CV2023	-	-
CV2024	-	-
CV2026	-	-
CV2005	-	-
CV2127	-	-
CV2128	-	-
CV2129	-	5763
CV2135	-	-
CV2136	-	-
CV2195	-	-
CV2240	-	3B4WA#
CV2241	-	5642
CV2300	-	-
CV2361	-	-
CV2370	-	-
CV2382	-	-
CV2390	-	3A4, 3A4/DL93, DL93
CV2466	-	6939
CV2492	-	6922, 6922/E88CC, E88CC, ECC88
CV2500	-	-
CV2507	-	-
CV2522	-	6AS6, 6AS6W, 5725, 5725/6AS6W, 6187, 6187/6AS6
CV2523	-	-
CV2524	-	-
CV2526	-	-
CV2527	-	-
CV2530	-	-
CV2534	-	-
CV2556	-	-
CV2557	-	-
CV2558	-	-
CV2573	-	5651A, 5751WA
CV2578	-	-
CV2642	-	5842/417A
CV2704	-	-
CV2706	-	-
CV2707	-	-
CV2709	-	-
CV2710	-	-
CV2714	-	-
CV2716	-	-
CV2721	-	-
CV2726	-	-
CV2729	-	-
CV2742	-	1L4, 1L4/DF92, DF92
CV2748	-	-
CV2769	-	-
CV2795	-	1L4, 1L4/DF92, DF92
CV2798	-	-
CV2842	-	-
CV2844	-	-
CV2854	-	-
CV2876	-	2D21#, 2D21/5727, 2D21/PL21#, 5727, 5727/2D21, PL21#
CV2877	-	-
CV2882	-	-
CV2883	-	-
CV2884	-	-
CV2901	-	-
CV2940	-	-
CV2975	-	-
CV2983	-	-
CV2984	-	6AS7G#, 6AS7GA#, 6080, 6080W, 6080WA
CV3508	-	12AT7WA#, 12AT7WB#, 6201, 6679/12AT7#
CV3512	-	5696, 5696A
CV3523	-	-
CV3526	-	-
CV3789	-	417A, 5842, 5842/417A
CV3798	-	0A3, 0A3/VR75, 0A3A, VR75
CV3908	-	-
CV3912	-	-
CV3928	-	5636, 5840#, 5840/6225#, 5840W#, 6225#
CV3930	-	5718
CV3986	-	6021
CV3990	-	-
CV3995	-	-
CV3998	-	-
CV4003	-	12AU7A, ECC82, ECC82/12AU7A
CV4004	-	12AX7, 12AX7A, 12AX7A/ECC83, 12AX7A/ECC83/7025, 12AX7WA, 12AX7WA/7025, 7025, ECC83
CV4007	-	-
CV4009	-	6BA6, 6BA6/5749/6660, 6BA6/EF93#, 5749, 6660, 6660/6BA6, EF93#
CV4010	-	-
CV4011	-	6AS6#, 6AS6W, 5725, 5725/6AS6W, 6187#, 6187/6AS6#
CV4012	-	-
CV4014	-	-
CV4015	-	-
CV4016	-	-
CV4017	-	12AX7A#, 12AX7A/ECC83#, 12AX7A/ECC83/7025#, 12AX7WA#, 12AX7WA/7025#, 5751, 6681/12AX7A#, 7025#, ECC83#
CV4018	-	2D21#, 2D21/5727, 2D21/PL21#, 5727, 5727/2D21, PL21#
CV4019	-	-
CV4020	-	0A2#, 0A2WA, 6073, 6073/0A2, 6626, 6626/0A2WA
CV4022	-	-
CV4023	-	6AU6A#, 6AU6A/EF94#, 6AU6WB, 8425A#, 8425A/6AU6A#, EF94#
CV4024	-	12AT7#, 12AT7WA, 12AT7WB, 6679/12AT7#, ECC81#
CV4025	-	6AL5#, 6AL5/6EB5#, 6AL5/5726/6663#, 6AL5/EAA91#, 6AL5W, 6EB5#, 5726#, 5726/6AL5W, 5726/6AL5W/6097, 6097, 6663#, 6663/6AL5#, EAA91#, EAA91/6AL5#

Tube Type	Basing	Replacement
CV4026	-	-
CV4028	-	0B2#, 0B2WA, 6074#, 6074/ 0B2#, 6627, 6627/0B2
CV4031	-	6J6WA#, 6101
CV4039	-	5763
CV4048	-	5651A, 5651WA
CV4055	-	-
CV4058	-	-
CV4100	-	0A2#, 0A2WA, 6073#, 6073/ 0A2#, 6626, 6626/0A2WA
CV4101	-	0B2#, 0B2WA, 6074#, 6074/ 0B2#, 6627, 6627/0B2
CV4108	-	-
CV5008	-	-
CV5021	-	-
CV5032	-	-
CV5034	-	-
CV5036	-	-
CV5037	-	-
CV5040	-	-
CV5042	-	-
CV5055	-	-
CV5065	-	-
CV5071	-	-
CV5072	-	-
CV5074	-	-
CV5077	-	-
CV5094	-	-
CV5122	-	5823
CV5156	-	-
CV5172	-	-
CV5181	-	-
CV5186	-	5651A, 5651WA
CV5189	-	-
CV5190	-	-
CV5192	-	-
CV5212	-	12AT7WA#, 12AT7WB#, 6201, 6679/12AT7#
CV5212	-	-
CV5214	-	-
CV5215	-	-
CV5216	-	-
CV5220	-	-
CV5231	-	-
CV5281	-	-
CV5311	-	-
CV5331	-	-
CV5354	-	-
CV5358	-	-
CV5365	-	-
CV5404	-	-
CV5427	-	-
CV5434	-	-
CV5724	-	-
CV5724	-	-
CV5817	-	-
CV5831	-	-
CV5843	-	-
CV5893	-	-
CV5895	-	-
CV5896	-	-
CV5905	-	-
CV5989	-	-
CV8017	-	-
CV8020	-	-
CV8045	-	-
CV8047	-	-
CV8048	-	-
CV8065	-	-
CV8068	-	-
CV8069	-	-
CV8070	-	-
CV8071	-	-
CV8073	-	-
CV8076	-	-
CV8154	-	-
CV8155	-	-
CV8156	-	-
CV8159	-	-
CV8160	-	-
CV8189	-	-
CV8190	-	-
CV8191	-	-
CV8192	-	-
CV8200	-	-
CV8201	-	-
CV8202	-	-
CV8203	-	-
CV8204	-	-
CV8205	-	-
CV8206	-	-
CV8208	-	-
CV8209	-	-
CV8210	-	-
CV8211	-	-
CV8215	-	-
CV8216	-	-
CV8218	-	-
CV8221	-	-
CV8222	-	-
CV8223	-	-
CV8224	-	-
CV8225	-	-
CV8226	-	-
CV8227	-	-
CV8229	-	-
CV8231	-	-
CV8232	-	-
CV8237	-	-
CV8246	-	-
CV8248	-	-
CV8249	-	-
CV8280	-	-
CV8287	-	-
CV8297	-	-
CV8310	-	-
CV8311	-	-
CV8312	-	-
CV8403	-	-
CV8430	-	-
CV8431	-	-
CV8433	-	-
CV8458	-	-
CV8470	-	-
CX1013	-	5517, 5517/CX1013
CXF80	9AE	-
D1C	-	957
D2C	-	958A
D2M9	6BT	6AL5, 6AL5/6EB5, 6AL5/5726/ 6663, 6AL5/EAA91, 6AL5W, 6EB5, 5726#, 5726/6AL5W, 5726/ 6AL5W/6097, 6097, 6663, 6663/6AL5, cont'd

Tube Type	Basing	Replacement
D2M9, cont'd		EAA91, EAA91/6AL5
D3F	-	959
D5TF30	-	50A1, 50A1/D5TF30
D27	6BT	6AL5, 6AL5/6EB5, 6AL5/5726/ 6663, 6AL5/EAA91, 6AL5W, 6EB5, 5726, 5726/ 6AL5W, 5726/6AL5W/ 6097, 6097, 6663, 6663/ 6AL5, EAA91, EAA91/6AL5
D61	-	6CT7#, 6CT7/EAF42#, EAF42#
D63	6BT	6H6
D77	-	6AL5, 6AL5/6EB5, 6AL5/5726/ 6663, 6AL5/EAA91, 6AL5W, 6EB5, 5726#, 5726/6AL5W, 5726/ 6AL5W/6097, 6097, 6663, 6663/6AL5, D152, DD6, EAA91, EAA91/6AL5, EB91
D152	6BT	6AL5, 6AL5/6EB5, 6AL5/5726/ 6663, 6AL5/EAA91, 6AL5W, 6EB5, 5726#, 5726/6AL5W, 5726/ 6AL5W/6097, 6097, 6663, 6663/6AL5, D77, DD6, EAA91, EAA91/6AL5, EB91
D193	-	-
D717	-	6AL5, 6AL5/6EB5, 6AL5/5726/ 6663, 6AL5/EAA91, 6AL5W, 6EB5, 5726#, 5726/6AL5W, 5726/ 6AL5W/6097, 6097, 6663, 6663/6AL5, D77, D152, DD6, EAA91, EAA91/6AL5, EB91
DA42	-	-
DA90	5AP	1A3, 1A3/DA90/1D13, 1D13, DA90
DAC21	-	1S5, 1S5/DAF91, DAF91
DAC32	5Z	1H5, 1H5/HD14/DAC32, 1H5G, 1H5GT, 1H5GT/G, HD14
DAE32	-	-
DAF90	-	1A3, 1A3/DA90/1D13, 1D13, DA90
DAF91	6AU	1FD9, 1S5, 1S5/DAF91, ZD17
DAF92	6BW	1U5
DAF96	6AP	1AF5, 1AF5/1AH5, 1AH5
DAF97	-	1AN5, 1AN5/DF97, DF97
DC70	-	6375
DC80	-	1E3
DC90	-	-
DCC90	7BC	3A5, 3A5/DCC90
DCF60	-	1V6
DD6	-	6AL5, 6AL5/6EB5, 6AL5/6726/ 6663, 6AL5/EAA91, 6AL5W, 6EB5, 5726#, 5726/6AL5W, 5726/ 6AL5W/6097, 6097, 6663, 6663/6AL5, D77, D152, EAA91, EAA91/6AL5, EB91
DD6G	-	5726#
DD7	-	6AM5, 6AM5/EL91, EL91
DD77	-	5726
DDR7	-	6AM5, 6AM5/EL91, EL91
DF26	-	1S5, 1S5/DAF91, DAF91
DF33	5Y	1N5GT, 1N5GT/1P5GT, 1N5GT/G, 1P5GT
DF60	-	5678
DF61	-	6611
DF62	-	1AD4, 1AD4/DF62
DF67	-	5911, 6008
DF91	6AR	1T4, 1F3, W17
DF92	6AR	1F2, 1L4, 1L4/DF92, DF92
DF96	-	1AF4, 1AF4/1AJ4, 1AJ4
DF97	-	1AN5, 1AN5/DF97
DF652	-	1AD4, 1AD4/DF62, DF62
DF654	-	5678, DF60, XFR2
DF668	-	1AD4, 1AD4/DF62, DF62
DF703	-	5886
DF904	6AR	1U4
DG7-5	-	-
DH63	7V	6B6, 6B6G, 6Q7, 6Q7G, 6Q7GT, 6Q7MG, 6T7, 6T7G*, 6118
DH74	-	12Q7GT
DH76	-	12Q7GT
DH77	7BT	6AT6, 6AT6/6BK6, 6BK6, EBC90
DH81	-	7B6
DH109	-	-
DH118	-	14L7
DH119	-	14G6
DH142	-	14L7
DH147	-	-
DH149	-	7C6
DH150	-	6CV7, 6CV7/EBC41, EBC41
DH718	-	6CV7, 6CV7/EBC41, EBC41
DH719	-	6AK8, 6AK8/EABC80, EABC80
DH817	-	6CV7, 6CV7/EBC41, EBC41
DK32	-	1A7G, 1A7GT, 1A7GT/G
DK91	7AT	1C1, 1R5, 1R5/DK91, X17
DK92	-	1AC6, 1AC6/DK92, DK92
DK96	7DH	1AB6, 1AB6/DK96, 1C3, 1C3/ DK96/1AB6, 1H35, DK96, X25
DK97	-	1AB6, 1AB6/DK96, 1C3, 1C3/ DK96/1AB6, DK96
DL012	9E	6T8A
DL29	6BA	3D6, 3D6/1299, 1299
DL31	6X	1A5G, 1A5GT, 1A5GT/G
DL33	7AP	3Q5GT, 3Q5GT/G
DL35	6X	1C5, 1C5/DL35/N14, 1C5GT, 1C5GT/G, DL35, N14
DL36	-	1Q5GT, 1Q5GT/G
DL37	-	6L6GC
DL63	-	-
DL67	-	5913, 6007
DL69	-	5672, DL75, DL620, DL652, XFY14
DL70	-	6373
DL74M	-	12Q7
DL75	-	5672, DL69, DL75, DL620, DL652, XFY14
DL82	-	7B6
DL91	7AV	1S4
DL92	7BA	1P10, 3S4, 3S4/DL92, N17
DL93	7BB	3A4, 3A4/DL93, DL93
DL94	6BX	1P11, 3V4, 3V4/DL94
DL95	7BA	3Q4, N18
DL96	-	3C4, 3C4/DL96
DL98	7CY	3B4#, 3B4/DL98, 3B4WA#, HD30
DL193	-	-
DL620	-	5672, DL69, DL75, cont'd

Tube Type	Basing	Replacement
DL620, cont'd		DL652, XFY14
DL652	-	5672, DL69, DL75, DL620, XFY14
DL700	-	6195
DM70	8EM	1M3, 1M3/DM70, 1N3, 1N3/DM70/1M3, 1N3/DM71, DM70, DM70/1N3, DM71, Y25
DM70/1N3	-	1M3, 1M3/DM70, 1N3, 1N3/DM70/1M3, 1N3/DM71, DM70, DM71, Y25
DM71	-	1M3, 1M3/DM70, 1N3, 1N3/DM70/1M3, 1N3/DM71, DM70, DM70/1N3, DM71
DM160	-	6977#, 6977/DM160
DP51	-	-
DP61	7BD	6AK5, 6AK5/5654, 6AK5/EF95, 5654#, EF95, PM05
DR2000	-	-
DR2010	-	-
DR2020	-	-
DR2100	-	-
DR2110	-	-
DR2120	-	-
DR2130	-	-
DR2200	-	-
DR2210	-	-
DY30	3C	1B3GT, 1B3GT/1G3GTA, 1G3GT, 1G3GT/1B3GT, 1G3GTA, 1B3GTA/1B3GT
DY51	-	1BG2
DY70	-	5642
DY80	9Y	1BX2, 1X2A, 1X2A/1X2B, 1X2B, 1X2B/1X2A, 1X2C, 1C2C/1BX2
DY86	9DT	1RK23, 1S2, 1S2A, 1S2A/DY87, 1S2A/DY87/DY86, DY87
DY87	9DT	1S2, 1S2A, 1S2A/DY87, 1S2A/DY87/DY86, 1S2B, DY86
DY802	-	1BQ2, 1BQ2/DY802
E1C	-	1650, 4671, HA2, UN955
E1F	-	954
E1T	-	6370
E2D	-	-
E2F	-	956
E55L	-	8233#, 8233/E55L
E80CC	-	6085, 6085/E80CC
E80CF	-	7643, 7643/E80CF
E80F	-	6084, 6084/E80F
E80L	-	6227
E80T	-	6218
E81CC	9A	12AT7#, 12AT7/ECC81#, 12AT7/ECC81#, 12AT7WA, 6201, 6679/12AT7#, ECC81#
E81L	-	6686#, 6686/E81L
E82CC	9A	12AU7#, 12AU7A/ECC82#, 5814A, 6189, 6680/12AU7A#, ECC82#
E82CF	-	-
E82M	-	5624
E83CC	9A	12AX7A#, 12AX7A/ECC83#, 12AX7A/ECC83/7025#, 12AX7WA#, 12AX7WA/7025#, 6681#, 6681/12AX7A#, 7025#, ECC83#
E83F	-	6689, 6689/E83F
E84L	-	7320, 7320/E84L
E86C	-	6CM4, 6CM4/E86C, EC86
E88C	-	6DL4, 8255, 8562, 8255/E88C, 8562/E88C
E88CC	9AJ	6922, 6922/E88CC, E88CC/6922
E88CC/6922	9AJ	6922, 6922/E88CC, E88CC
E89F	-	6DG7
E90C	-	5920
E90CC	-	5920, 5920/E90CC
E90F	-	6BH6, 6BH6/6265/6661, 6265, 6661, 6661/6BH6
E90Z	-	6X4
E91	-	2D21, 2D21/5727, 2D21/PL21, 2D21W, 20A3, 5727, 5727/2D21, 5727/2D21W, EN91, EN91N, M8204, PL21
E91AA	-	6AL5#, 6AL5/6EB5#, 6AL5/5726/6663#, 6AL5/EAA91#, 6AL5W, 6EB5#, 5726, 5726/6AL5W, 5726/6AL5W/6097, 6097, 6663#, 6663/6AL5#, EAA91#, EAA91/6AL5#
E91H	-	5915#, 6687
E91N	-	2D21#, 2D21/5727, 2D21/PL21#, 5727, 5727/2D21, PL21#
E92CC	-	-
E95	-	6AG5*, 6AG5/EF96#, 6AK5/5654, 6AK5/EF95, 6AK5W, 6AK5WA, 6AK5WB, 6BC5*, 6BC5/6CE5*, 6CE5*, 6CE5/6BC5*, 6CY5*#, 6EA5*#, 6EV5*#, 6F32, 403A, 1220, 5654, 5591*, 5654, 5654/6AK5W, 6096, 6096/6AK5W, 6968, DP61, E95#, E95F, EF95, EF96#, EF905, M8100#, PM05#
E95F	-	6AK5#, 6AK5/5654#, 6AK5/EF95#, 5654, EF95#
E99F	-	6BJ6, 6BJ6/6662, 6662, 6662/6BJ6
E130L	-	7543, 7543/E130LF, E130LF
E130LF	-	7543, 7543/E130LF
E180CC	-	7062, 7062/E180CC
E180F	-	6688#, 6688/E180F, 6688A#
E180L	-	7534
E181CC	-	7118
E182CC	-	7119, 7119/E182CC
E182F	-	5847#, 5847/404A#
E186F	-	7737, 7737/E186F
E188CC	-	7308#, 7308/E188CC
E280F	-	7722, 7722/E280F
E288C	-	-
E288CC	-	8223, 8223/E288CC
E810F	-	7788#, 7788/E810F
E902	-	6X4
E1485	-	3A4, 3A4/DL93, DL93
E1955	-	2D21, 2D21/5727, 2D21/PL21, 5727, 5727/2D21, PL21
E2016	-	6CQ6, 6CQ6/EF92, EF92
E2157	-	12AT7, 12AT7/ECC81

Tube Type	Basing	Replacement
E2163	-	12AU7, 12AU7A/ECC82
E2164	-	12AX7, 12AX7A, 12AX7A/ ECC83, 12AX7A/ECC83/ 7025, 12AX7WA, 12AX7WA/7025, 7025, ECC83
EA41	-	6CT7, 6CT7/EAF42, EAF42
EA50	-	2B35, 2B35/EA50
EA52	-	6923, 6923/EA52
EA53	-	-
EA71	-	5647
EA76	-	6489, 6489/EA76
EAA91	6BT	6AL5, 6AL5/6EB5, 6AL5/5726/ 6663, 6AL5/EAA91, 6AL5W, 6B32, 6EB5, 6D2, 5726, 5726/6AL5W, 5726/ 6AL5W/6097, 6058, 6097, 6663, 6663/6AL5, 7631, D2M9, D27, D77, D152, D717, DD6, EAA91/6AL5, EAA901, EAA901S, EB91
EAA91/6AL5	6BT	6AL5, 6AL5/6EB5, 6AL5/5726/ 6663, 6AL5W, 6B32, 6D2, 6EB5, 5726, 5726/6AL5W, 5726/6AL5W/6097, 6058, 6097, 6663, 6663/6AL5, 7631, D2M9, D27, D77, D152, D717, DD6, EAA91, EAA901, EAA901S, EB91
EAA901	-	6AL5#, 6AL5/6EB5#, 6AL5/ 5726/6663#, 6AL5/ EAA91#, 6AL5W, 6EB5#, 5726, 5726/6AL5W, 5726/ 6AL5W/6097, 6097, 6663, 6663/6AL5, EAA91#, EAA91/6AL5#
EAA901S	-	6AL5#, 6AL5/6EB5#, 6AL5/ 5726/6663#, 6AL5/ EAA91#, 6AL5W, 6EB5#, 5726, 5726/6AL5W, 5726/ 6AL5W/6097, 6097, 6663, 6663/6AL5, EAA91#, EAA91/6AL5#
EAB80	-	-
EABC80	9E	6AK8, 6AK8/EABC80, 6LD12, 6T8, 6T8A, DH719, EABC80/6T8
EABC80/6T8	9E	6AK8, 6AK8/EABC80, 6LD12, 6T8, 6T8A, DH719, EABC80
EAF41	-	6CT7#, 6CT7/EAF42#, EAF42#
EAF42	-	6CT7, 6CT7/EAF42
EAF801	-	-
EAM86	-	6GX8
EB2CC	-	6189
EB34	7Q	6H6*
EB41	-	-
EB80CC	-	-
EB91	6BT	6AL5, 6AL5/6EB5, 6AL5/5727/ 6663, 6AL5/EAA91, 6AL5W, 6EB5, 5726, 5726/ 6AL5W, 5726/6AL5W/ 6097, 6097, 6663, 6663/ 6AL5, D77, D152, DD6, EAA91, EAA91/6AL5
EBC3	-	6BD7A, 6BD7A/EBC81, EBC81
EBC33	-	-
EBC41	8GZ	6CV7, 6CV7/EBC41, cont'd

Tube Type	Basing	Replacement
EBC41, co nt'd		6LD3, 62DDT, DH150, DH718
EBC80	9E	6BD7
EBC81	-	6BD7, 6BD7A, 6BD7A/EBC81
EBC90	7BT	6AT6, 6AT6/6BK6, 6BK6, DH77
EBC91	7BT	6AV6, 6AV6/EBC91
EBF32	-	6B8*
EBF41	-	6CJ5, 6CJ5/EF41, EF41
EBF80	9T	6AD8, 6AD8/6DC8, 6DC8, 6N8,6N8/EBF80, EBF81, WD709, ZD152
EBF81	-	6AD8, 6AD8/6DC8, 6DC8
EBF83	-	6DR8, 6DR8/EBF83
EBF85	9HE	6DC8
EBF89	9HE	6AD8, 6AD8/6DC8, 6DC8, 6DC8/EBF89, 6FD12, 6N8, EBF85
EC22	-	6R4
EC55	-	5861
EC70	-	5718
EC71	-	5718
EC80	-	6Q4
EC81	-	6R4
EC84	-	6AJ4
EC86	-	6CM4, 6CM4/EC86
EC88	9NY	6DL4, 6DL4/EC88, 8255
EC90	6BG	6C4, 6C4/6135, 6C4/EC90, 6135
EC91	-	6AQ4, 6AQ4/EC91
EC92	5CE	6AB4, 6AB4/EC92, 6664, 6664/6AB4, EC92/6AB4
EC92/6AB4	5CE	6AB4, 6AB4/EC92, 6664, 6664/6AB4, EC92
EC93	-	6BS4
EC94	7DK	6AF4, 6AF4A, 6AF4A/6DZ4, 6DZ4, 6DZ4/6AF4A
EC95	7FP	6ER5
EC97	7FP	6FY5, 6FY5/EC97, 6ER5
EC157	-	8101
EC806	-	6CM4*, 6CM4/E86C*, E86C*, EC806S
EC806S	-	6CM4*, 6CM4/E86C*, E86C*
EC900	7GM	6HA5, 6HM5/6HA5, 6HM5
EC903	-	-
EC1000	-	8254
EC8010	-	8556, 8556/EC8010
EC8020	-	-
ECC32	8BD	6SN7GT, 6SN7GTB
ECC33	-	6SN7, 6SN7, B65
ECC35	8BD	6SL7*, 6SL7GT, CV569, ECC35/CV569
ECC35/CV569	8BD	6SL7GT, CV569, ECC35
ECC40	-	-
ECC70	-	6021
ECC81	9A	12AT7, 12AT7/ECC81, 12AT7WA, 12AT7WB, 6060, 6201, 6671, 6679, 7492, 7728, A2900, B152, B309, B739, CV4024, E81CC, ECC81/12AT7, ECC81/M8162, ECC801, ECC801S, M8162, QA2406, QB309
ECC81/12AT7	9A	12AT7, 12AT7WA, 12AT7WB, 6060, 6201, 6671, 6679, 7492, 7728, A2900, B152, B309, B739, CV4024, E81CC, ECC81, cont'd

Tube Type	Basing	Replacement
ECC81/12AT7, cont'd		ECC81/M8162, ECC801, ECC801S, M8162, QA2406, QB309
ECC81/M8162	9A	12AT7, 12AT7/ECC81, 12AT7WA, 12AT7WB, 6060, 6201, 6671, 6679, 7492, 7728, A2900, B152, B309, B739, CV4024, E81CC, ECC81, ECC81/12AT7, ECC801, ECC801S, M8162, QA2406, QB309
ECC82	9A	12AU7, 12AU7A, 12AU7A/ECC82, 5814, 6189, 6680, CV4003, ECC82, ECC82/12AU7, ECC82/12AU7A
ECC82/12AU7	9A	12AU7, 12AU7A, 12AU7A/ECC82, 5814, 6189, 6680, CV4003, ECC82, ECC82/12AU7A
ECC82/12AU7A	9A	12AU7, 12AU7A, 12AU7A/ECC82, 5814, 6189, 6680, CV4003, ECC82, ECC82/12AU7
ECC83	9A	6L13, 12AX7, 12AX7A, 12AX7A/ECC83, 12AX7A/ECC83/7025, 12AX7WA, 12AX7WA/7025, 12DF7, 12DT7, 5751, 6057, 6681, 7025, 7025A, 7494, 7729, B339, B759, CV4004, E83CC, ECC83/12AX7, ECC803, M8137
ECC83/12AX7	9A	6L13, 12AX7, 12AX7A, 12AX7A/ECC83, 12AX7A/ECC83/7025, 12AX7WA, 12AX7WA/7025, 12DF7, 12DT7, 6057, 6681, 7025, 7025A, 7494, 7729, B339, B759, E83CC, ECC83, ECC803, M8137
ECC84	-	6CW7, 6CW7/ECC84
ECC85	9DE	6AQ8, 6AQ8/ECC85, 6L12, B719, ECC85, ECC85/6AQ8
ECC85/6AQ8	9DE	6AQ8, 6AQ8/ECC85, 6L12, B719, ECC85
ECC86	9DE	6GM8, 6GM8/ECC86
ECC88	9AJ	6DJ8, 6DJ8/ECC88, 6ES8, 6ES8/ECC189, 6922, CV2492, ECC88/6DJ8, ECC189
ECC88/6DJ8	9AJ	6DJ8, 6DJ8/ECC88, 6ES8, 6ES8/ECC189, 6922, CV2492, ECC88, ECC189
ECC89	9DE	6FC7
ECC91	7BF	6J6, 6J6A, 6J6A/ECC91, 6101
ECC180	9AJ	6BC8, 6BC8/6BQ7A, 6BQ7A, 6BQ7A/6BZ7, 6BQ7A/6BZ7/6BS8, 6BS8, 6BS8/6BK7B, 6BZ7, 6BZ7/6BQ7A
ECC186	9A	12AU7, 12AU7A/ECC82
ECC189	9AJ	6DJ8, 6ES8, 6ES8/ECC189, ECC88
ECC230	-	6AS7G, 6AS7GA, 6080, 6080W, 6080WA
ECC801	9A	12AT7/ECC81, 6060, 6201, ECC801S
ECC801S	9A	12AT7/ECC81, 6060, 6201, ECC801
ECC802	9A	12AU7A/ECC82#, 6189, 6680/12AU7A, ECC802S
ECC802S	-	12AU7A/ECC82#, 6189, 6680/12AU7A, ECC802
ECC803	9A	12AX7A, 12AX7A/ECC83, 12AX7A/ECC83/7025, 12AXTWA, 12AX7WA/7025, 6057, 7025, ECC83, ECC803S
ECC803S	-	12AX7A, 12AX7A/ECC83, 12AX7A/ECC83/7025, 12AXTWA, 12AX7WA/7025, 6057, 7025, ECC83, ECC803
ECC804	-	6GA8
ECC807	-	-
ECC808	-	6KX8
ECC813	-	6463
ECC863	-	12DT7
ECC900	-	6HA5, 6HM5/6HA5
ECC903	9A	-
ECC960	-	-
ECC962	-	-
ECC2000	-	-
ECF80	9DC	6BL8, 6BL8/6LN8, 6BL8/ECF80, 6C16, 6LN8, 7643*, E80CF*, ECF80, ECF80/6BL8
ECF80/6BL8	9DC	6BL8, 6BL8/6LN8, 6BL8/ECF80, 6C16, 6LN8, 7643*, E80CF*, ECF80, ECF80/6BL8
ECF82	9AE	6AX8, 6KD8, 6U8, 6U8A, 6U8A/6AX8/6KD8, 6678
ECF83	-	-
ECF86	9MP	6HG8, 6HG8/ECF86
ECF200	10K	6X9, 6X9/ECF200
ECF201	10K	6U9, 6U9/ECF201
ECF202	-	6AJ9
ECF801	9QA	6GJ7, 6GJ7/ECF801
ECF802	9DC	6JW8, 6JW8/ECF802, 6LX8
ECF804	-	-
ECF805	-	6GV7
ECH33	-	-
ECH35	-	6E8G*
ECH36	-	-
ECH42	12BQ	6CU7, 6CU7/ECH42, 6C10
ECH80	-	6AN7
ECH81	9CA	6AJ8, 6AJ8/ECH81, 6C12, 20D4, ECH81/6AJ8, X719
ECH81/6AJ8	9CA	6AJ8, 6AJ8/ECH81, 6C12, 20D4, ECH81, X719
ECH82	-	6E8
ECH83	-	6DS8, 6DS8/ECH83
ECH84	-	6JX8
ECH113	-	6CU7, 6CU7/ECH42, ECH42
ECH200	-	6V9
ECL80	-	6AB8, 6AB8/ECL80
ECL82	9EX	6BM8, 6BM8/ECL82, 6PL12, ECL82/6BM8
ECL82/6BM8	9EX	6BM8, 6BM8/ECL82, 6PL12, ECL82
ECL83	-	6BM8, 6BM8/ECL82, ECL82, ECL82/6BM8
ECL84	9HX	6DX8, 6DX8/ECL84

Tube Type	Basing	Replacement
ECL85	9LY	6GV8, 6GV8/ECL85
ECL86	9LZ	6GW8, 6GW8/ECL86
ECL100	-	6BC8, 6BC8/6BQ7A, 6BQ7A, 6BQ7A/6BZ7, 6BQ7A/6BZ7/6BS8, 6BS8, 6BS8/6BK7B, 6BZ7, 6BZ7/6BQ7A
ECL113	-	-
ECL180	8CK	6BQ7A, 6BQ7A/6BZ7, 6BQ7A/6BZ7/6BS8, 6BS8, 6BS8/6BK7B, 6BZ7, 6BZ7/6BQ7A
ECL821	-	6CH6, 6CH6/EL821
ECLL800	-	6KH8, 6KH8/ECLL800
ED2	-	6AL5, 6AL5/6EB5, 6AL5/5727/6663, 6AL5/EAA91, 6AL5W, 6EB5, 5726, 5726/6AL5W, 5726/6AL5W/6097, 6097, 6663, 6663/6AL5, EAA91, EAA91/6AL5
ED92	-	6AB4, 6AB4/EC92, 6664, 6664/6AB4, EC92, EC92/6AB4
ED500	-	6ED4
EF2	-	-
EF5	-	-
EF9	-	-
EF13	-	6DA6, 6DA6/EF89, EF89, EF89/6DA6
EF22	8V	7A7*
EF36	-	6J7#
EF37	7R	6J7#, 6J7GT*, EF37A
EF37A	-	6J7#, 6J7GT*, EF37
EF39	-	6K7*
EF40	-	-
EF41	-	6CJ5, 6CJ5/EF41
EF42	-	-
EF50	-	-
EF70	-	6487
EF71	-	5899
EF72	-	5840#, 5840/6225#, 5840W, 6225#
EF73	-	6488
EF74	-	6391
EF80	9AQ	6BW7, 6BX6, 6BX6/EF80, 6BY7, 6BY7/EF85, 6EL7, 8D6, 64SPT, EF85, EF85/6BY7, Z152, Z719
EF81	-	6BH5
EF82	-	6CH6, 6CH6/EL821, EL821
EF83	-	6BK8, 6BK8/EF83
EF85	9AQ	6BX6, 6BX6/EF80, 6BY7, 6BY7/EF85, 6BY8, 6F19, 6F26, EF80, EF85/6BY7, W719
EF85/6BY7	9AQ	6BX6, 6BX6/EF80, 6BY7, 6BY7/EF85, 6BY8, 6F19, 6F26, EF80, EF85, W719
EF86	9CQ	6BQ5, 6BQ5/EL84, 6F22, 8D8, 6267, EF86/Z729, EF87, EF806S, EL84, EL84/6BQ5, M8195, Z729
EF86/Z729	9CQ	6BQ5, 6BQ5/EL84, 6F22, 8D8, 6267, EF86, EF87, EF806S, EL84, EL84/6BQ5, M8195, Z729
EF87	-	6267
EF89	9AU	6DA6, 6DA6/EF89, EF89/6DA6
EF89/6DA6	9AU	6DA6, 6DA6/EF89, EF89
EF89F	-	6DG7
EF91	-	6AM6, 6AM6/EF91
EF92	-	6CQ6, 6CQ6/EF92
EF93	7BK	6AU6, 6AU6/6136, 6AU6A, 6AU6A/EF94, 6BA6, 6BA6/5749/6660, 6BA6/EF93, 6BA6W, 6BA6WA, 6BD6, 6BZ6#, 6BZ6/6JH6#, 6CG6, 6F31, 6JH6#, 5749#, 5749/6BA6W, 6136, 6660#, 6660/6BA6#, 7496, 7543, 8425A, 8425A/6AU6A, EF94, M8101, PM04, W727
EF94	7BK	6AU6, 6AU6/6136, 6AU6A, 6AU6A/EF94, 6BA6, 6BA6/5749/6660, 6BA6/EF93, 6BA6W, 5749#, 5749/6BA6W, 6136#, 6660#, 6660/6BA6#, 8425A, 8425A/6AU6A, EF93
EF95	7BD	6AK5, 6AK5/5654, 6AK5/EF95, 6AK5W, 6AK5WA, 6AK5WB, 6F32, 403A, 1220, 5654, 5654/6AK5W, 6096, 6096/6AK5W, 6968, DP61, E95F, EF905, PM05
EF96	7BD	6AG5, 6AG5/EF96
EF97	-	6ES6
EF98	-	6ET6, 6ET6/EF98
EF183	9AQ	6EH7, 6EH7/EF183, 6EJ7, 6F25, 6F29, EF811
EF184	9AQ	6EH7, 6EJ7, 6EJ7/EF184, 6F24, 6F30, EF814
EF190	7CM	6BZ6, 6BZ6/6JH6, 6CB6, 6JH6
EF730	-	5636
EF731	-	5899
EF732	-	5840#, 5840/6225#, 5901, 6225#
EF734	-	6205
EF800	-	-
EF804	-	-
EF806	-	EF806S
EF806S	-	EF806
EF811	9AQ	6EH7, 6EH7/EF183
EF812	-	6EL7
EF814	9AQ	6EJ7, 6EJ7/EF184
EF861	-	-
EF905	7BD	6AK5#, 6AK5/5654#, 6AK5/EF95#, 5654, EF95#
EFL200	10L	6Y9, 6Y9/EFL201, EFL201
EFL201	10L	6Y9, 6Y9/EFL200, EFL200
EH90	7CH	6CS6, 6CS6/EH90
EH900S	-	5915
EK90	7CH	6BE6, 6BE6/5750, 6BE6/EK90, 6BE6W#, 5750#, 5750/6BE6W#, HM04
EL1C	-	3B22, 3B22/EL1C
EL3B	4CQ	6013
EL3C	-	4B24, 6484, EL3C/4B24
EL3C/4B24	-	4B24, 6484, EL3C
EL6B	-	618, 618P, 5892
EL12	-	-
EL22	6AA	-
EL33	-	6AG6G, 6AG6G/EL33, 6M6G*
EL34	8ET	6CA7, 6CA7/EL34, 7D11, 12E13, KT77, KT88

Tube Type	Basing	Replacement
EL34/6CA7	8ET	6CA7, 6CA7/EL34, 7D11, 12E13, EL34, KT77, KT88
EL36	8GT	6CM5, 6CM5/EL36, 6DQ6B, 6DQ6B/6GW6, 6GB7, 6GW6, 6GW6/6DQ6B#, EL36/6CM5, EL360
EL36/6CM5	8GT	6CM5, 6CM5/EL36, 6DQ6B, 6DQ6B/6GW6, 6GB7, 6GW6, 6GW6/6DQ6B#, EL36, EL360
EL37	7AC	6L6#, 6L6GC#, 6L6WGC#, 5932, 5881#, 5881/6L6WGC#
EL38	8EW	6CN6, EL38/6CN6
EL38/6CN6	8EW	6CN6, EL38
EL41	-	6CK5, 6CK5/EL41, 67PT, BF61, N150
EL42	-	-
EL71	-	5902
EL80	-	6M5, 6Q4, 6Q4/EL80
EL81	9AS	6CJ6, 6CJ6/EL81, 6DR6, EL820
EL82	-	6DY5
EL83	-	6CK6, 6CK6/EL83
EL84	9CV	6BQ5, 6BQ5/EL84, 6P15, 6267, 7189, 7189A, 7320, 7320/E84L, EF86, EL84/6BQ5, EL84L, N709, Z729
EL84/6BQ5	9CV	6BQ5, 6BQ5/EL84, 6P15, 7189, 7189A, 7320, 7320/E84L, EL84, EL84L, N709
EL84M	9CV	6BQ5WA, EL84M/6BQ5WA
EL84M/6BQ5WA	9CV	6BQ5WA, EL84M
EL85	-	6BN5, 6BN5/EL85
EL86	9CV	6CN5, 6CW5, 6CW5/EL86, EL86/6CN5
EL86/6CN5	9CV	6CN5, 6CW5, 6CW5/EL86, EL86
EL90	7BZ	6AQ5, 6AQ5/6005/6669, 6AQ5A, 6AQ5A/6HG5, 6HG5, 6005, 6669, 6669/6AQ5A, N727
EL91	-	6AM5, 6AM5/EL91
EL95	7DQ	6DL5, 6DL5/EL95
EL180	9BF	12BV7, 12BV7/12BY7A, 12BY7, 12BY7A/12BV7/12DQ7
EL300	-	6FN5
EL500	9NH	6GB5, 6GB5/EL500, 6GB5/EL504, EL504
EL503	-	8278
EL504	-	6GB5, 6GB5/EL500, 6GB5/EL504, EL500
EL505	-	6KG6, 6KG6A/EL509, 6KG6A/EL519, EL509, EL519
EL508	-	6KW6
EL509	9RJ	6KG6A, 6KG6A/EL509, 6KG6A/EL519, EL519
EL519	-	6KG6A, 6KG6A/EL509, 6KG6A/EL519, EL509
EL802	-	6LD6
EL803	-	6CK6#, 6CK6/EL83#, EL83#
EL820	-	6CK6, 6CK6/EL83, EL83
EL821	-	6CH6, 6CH6/EL821
EL822	-	6CH6, 6CH6/EL821, EL821
EL861	-	-
ELC6C	-	-
ELC5F14	-	6278
ELF86	-	6HG8, 6HG8/ECF86
ELL80	9NJ	6HU8, 6HU8/ELL80
EM34	8EV	6CD7, 6CD7/EM34, 6CD7/EM34-35, 64ME, EM34-35, EM34-35/6CD7, EM34/6CD7, EM34/35
EM34-35	8EV	6CD7, 6CD7/EM34, 6CD7/EM34-35, 64ME, EM34, EM34-35/6CD7, EM34/6CD7, EM34/35
EM34-35/6CD7	8EV	6CD7, 6CD7/EM34, 6CD7/EM34-35, 64ME, EM34, EM34-35, EM34/6CD7, EM34/35
EM34/6CD7	8EV	6CD7, 6CD7/EM34, 6CD7/EM34-35, 64ME, EM34, EM34-35, EM34-35/6CD7, EM34/35
EM34/35	8EV	6CD7, 6CD7/EM34, 6CD7/EM34-35, 64ME, EM34, EM35-35, EM34-35/6CD7, EM34/6CD7
EM35	6R	6U5
EM71	-	-
EM80	9DB	6BR5, 6BR5/EM80, 6DA5, 6DA5/EM81, 65ME, EM80/6BR5, EM81, EM81/6DA5
EM80/6BR5	9DB	6BR5, 6BR5/EM80, 6DA5, 6DA5/EM81, 65ME, EM80, EM81, EM81/6DA5
EM81	9DB	6BR5, 6BR5/EM80, 6DA5, 6DA5/EM81, 65ME, EM80, EM80/6BR5, EM81/6DA5
EM81/6DA5	9DB	6BR5, 6BR5/EM80, 6DA5, 6DA5/EM81, 65ME, EM80, EM80/6BR5, EM81
EM84	9GA	6FG6, 6FG6/EM84, EM84/6FG6, EM840
EM84/6FG6	9GA	6FG6, 6FG6/EM84, EM84, EM840
EM85	-	6DG7, 6DG7/EM85
EM87	9GA	6HU6, 6HU6/EM87
EM840	7S	6FG6, 6FG6/EM84
EMM801	-	-
EN32	-	2050, 2050A
EN70	-	5634, ME1500
EN91	-	2D21, 2D21/5727, 2D21/PL21, 5727, 5727/2D21, PL21
EN92	-	5696, 5696A
EN93	-	6D4
EQ80	-	6BE7, 6BE7/EQ80
EY51	-	6X2, 6X2/EY51
EY80	-	6U3
EY81	9CB	6R3, 6R3/EY81, EY81/6R3
EY81/6R3	9CB	6R3, 6R3/EY81, EY81
EY81F	9BD	6V3
EY82	9BM	6N3, 6U3, EY80
EY82/6N3	9BM	6N3, 6U3, EY82
EY83	-	6AL3*, 6AL3/EY88*, EY88
EY84	-	6374, 6374/EY84
EY86	-	6S2
EY87	-	6S2A, 6S2A/EY87
EY88	9CB	6AL3, 6AL3/EY88
EY500	6EC4	6EC4, 6EC4/EY500A, 6EC4A/EY500, EY500A

Tube Type	Basing	Replacement
EY500A	6EC4	6EC4, 6EC4/EY500A, 6EC4A, 6EC4A/EY500, EY500
EZ3	-	6V4
EZ4	9M	6CA4, 6CA4/EZ81, EZ81, EZ81/6CA4
EZ11	-	6V4
EZ22	-	7Y4
EZ35	6S	6X5G, 6X5GT, 6X5GT/EZ35, U147
EZ40	-	6BT4
EZ41	-	-
EZ80	9M	6V4, 6V4/EZ80
EZ81	9M	6CA4, 6CA4/EZ81, EZ81/6CA4, U709, UU12
EZ81/6CA4	9M	6CA4, 6CA4/EZ81, EZ81, U709, UU12
EZ90	5BS	6X4, 6X4/EZ90, 6202#, U78, V2M70
EZ91	-	6AV4
EZ900	5BS	6X4, 6063
F1EL	-	8278
FA6	-	5677
FF95	-	-
FG17	-	5557, 5557/FG17
FG27A	-	-
FG57	-	5559, 5559/FG57
FG95	-	5560, 5560/FG95
FG104	-	5561, 5561/FG104
FG105	-	-
FG280	-	-
FM1000	-	-
G2	-	2S, 2S/4S, 4S
G75/2D	-	0A3#, 0A3/VR75#, VR75#
G77	-	6C6
G84	4B	2Z2, 2Z2/G84
G105/1D	-	0C3#, OC3/VR105, VR105
G150/3D	-	0D3#, 0D3/VR150, VR150, VR150/0D3
G234	-	-
G/50/4K	-	0A2#, 6073#, 6073/0A2#, 6626, 6626/0A2WA
GL546	-	5696, 5696A
GS10	C	
GY501	-	3BH2
GZ30	5T	5Y3, 5Y3/6087, GT, 5Z4, 5Z4G, 6087
GZ31	-	5AS4A, 5AS4A/5U4GB, 5U4G, 5U4GB, 5U4GB/5AS4A
GZ32	5DA	5AQ4, 5AQ4/GZ32, 5V4, 5V4GA, GZ32, GZ37
GZ33	-	-
GZ34	5DA	5AR4, 5AR4/GZ34, 52KU, 53KU, 54KU, GZ30, GZ32, GZ33, GZ34/5AR4, GZ37, R52, U54, U77
GZ34/5AR4	5DA	5AR4, 5AR4/GZ34, 52KU, 53KU, 54KU, GZ30, GZ32, GZ33, GZ34, GZ37, R52, U54, U77
GZ37	5DA	5AR4, 5AR4/GZ34, CV378, GZ34, GZ34/5AR4, GZ37/CV378
GZ37/CV378	-	CV378, GZ37
GZ	5D	2S, 2S/4S, 4S
GZ-33	-	-
H2-10	-	2X2A
H36	-	-

Tube Type	Basing	Replacement
H52	-	5AS4A, 5AS4A/5U4GB, 5U4GB, 5U4GB/5AS4A
H63	5M	6F5, 6F5GT
H250	6F	6C6
H1112	-	-
H1208	-	-
HAA9	-	-
HAA91	6BT	12AL5
HABC80	-	19T8
HBC80	-	-
HBC90	7BT	12AT6, 12AT6/HBC90
HBC91	7BT	12AV6
HC85	-	-
HCC85	9AJ	17EW8, 17EW8/HCC85
HCH81	-	12AJ7
HD14	5Z	1H5, 1H5/HD14/DAC32, 1H5G, 1H5GT, 1H5GT/G, DAC32
HD30	-	3B4, 3B4/DL98, 3B4WA, DL98
HD51	-	0A2, 0A2WA, 6073, 6073/0A2, 6626, 6626/0A2WA
HD52	-	0B2, 0B2WA, 6074, 6074/0B2, 6627, 6627/0B2WA
HD93	-	1BX2, 1X2A, 1X2A/1X2B, 1X2B, 1X2B/1X2A, 1X2C, 1C2C/1BX2
HD94	6AM	6BQ6GA, 6BQ6GA/6CU6, 6BQ6GTB, 6BQ6GTB/6CU6, 6CU6
HD96	6AM	25BQ6GTB, 25BQ6GTB/25CU6
HF61	-	6CJ5, 6CJ5/EF41, EF41
HF93	7BK	12BA6, 12BA6/HF93
HF94	7BK	12AU6
HF100	-	4C22, HF100/4C22
HF100/4C22	-	4C22, HF100
HF121	-	12AC5
HF125	-	203H
HF300	-	HF300/T300, T300
HF300/T300	-	HF300, T300
HK24	-	25T, 25T/HK24
HK90	7CH	12BE6, 12BE6/HK90
HL86	-	30CW5
HL90	-	19AQ5
HL92	7CV	50C5
HL94	7CV	30A5, 35C5, 35C5A
HL94/30A5	7CV	30A5, 35C5, 35C5A, HL94
HM04	7CH	6BE6, 6BE6/5750, 6BE6/EK90, 6BE6W#, 5750#, 5750/6BE6W#, EK90
HP6	-	6AM6, 6AM6/EF91, EF91
HP45	B	
HY61	5AW	4Y25, 5S1, 807, 807W, 807WA, 5933, 5933WA, 8018, P17A, QE06/50, QE06140, QV05-25, RK39
HY65	-	-
HY69	-	-
HY75	-	-
HY90	5BQ	35W4
HY113	-	HY123
HY115	-	HY145
HY123	-	HY113
HY125	-	HY155
HY145	-	1U4, HY115
HY155	-	HY125
HY1269	-	-

Tube Type	Basing	Replacement
HZ50	6G	12Z3, 14Z3
HZ90	5BS	12X4
KBC32	-	-
KD21	-	0A3#, 0A3/VR75#, VR75#
KD24	-	0C3, 0C3/VR105, 0C3A, VR105
KD25	-	0D3, 0D3/VR150, 0D3A, VR150, VR150/0D3
KF35	-	-
KK24	-	0C3, 0C3/VR105, VR105
KK32	-	-
KL35	-	1F5G*
KR1	4G	1V, 1V/6Z3, 1V/KR1, 6Z3, AD
KR25	-	2A5, 2A5/KR65, KR65
KR65	6B	2A5, 2A5/KR65
KT32	7AC	25L6, 25L6GT/25W6GT, 25W6GT
KT33	-	25A6
KT61	-	6M6G*
KT63	7S	6F6, 6F6G, 6F6GT
KT66	7AC	6L6GC, 6L6WGC#, 5881#, 5881/6L6WGC#, 7027A#, 7581A, 7581A/KT66
KT71	7AC	50L6GT
KT77	8N	6CA7, 6CA7/EL34, EL34, EL34/6CA7
KT81	6AA	-
KT88	8EF	6550#, 6550A, KT88/6550A
KT88/6550A	8EF	6550A, KT88
KT90	-	-
KTW63	7R	6J7GT, 6K7
KTZ63	7R	6J7, 6K7GT
KTZ63M	-	6J7GT
KX642	-	-
KY80	-	5J2
L63	6Q	6J5, 6J5GT
L63B	6Q	6J5
L77	6BS	6C4, 6C4/6135, 6C4/EC90, 6135, EC90
LA	-	6A4
LC97	-	3FQ5, 3FY5, 3GK5, 3GK5/3FQ5
LC900	7GM	3HA5, 3HA5/3HM5, 3HK5, 3HK5/3HA5, 3HM5, 3HM5/3HA5
LCC189	-	5ES8, 5ES8/YCC189, YCC189
LCF80	9DC	6LN8, 6LN8/LCF80
LCF86	9MP	5HG8, 5HG8/LCF86
LCF200	-	5X9
LCF201	10K	5U9, 5U9/LCF201
LCF801	9QA	5GJ7, 5GJ7/LCF801, 5GX7
LCF802	9DC	6LX8, 6LX8/LCF802
LCH200	-	5V9
LCL82	9EX	11BM8
LCL84	9HX	10DX8, 10DX8/LCL84
LCL85	9LY	10GV8, 10GV8/LCL85
LCL200	-	10DX8, 10DX8/LCL84
LF183	9AQ	4EH7, 4EH7/LF183, 4EJ7, 4EJ7/LF184, LF184, YF183
LF184	9AQ	4EH7, 4EH7/LF183, 4EJ7, 4EJ7/LF184, LF183, YF184
LF200	10L	-
LFL200	10L	11Y9, 11Y9/LFL200
LL86	9CV	10CW5, 10CW5/LL86
LL88	-	20AQ3, 20AQ3/LY88
LL500	9NH	18GB5, 18GB5/LL500
LL505	-	27KG6

Tube Type	Basing	Replacement
LL521	-	21KQ6
LN119	9EX	50BM8, 50BM8/UCL82
LN152	-	6AB8, 6AB8/ECL80, ECL80
LN309	-	-
LN319	-	13GC8
LY81	-	11R3
LY88	9CB	20AQ3, 20AQ3/LY88
LY500	-	28EC4
LZ319	9DC	9A8, 9A8/PCF80
LZ329	9DC	9A8, 9A8/PCF80
LZ339	-	9EN7
M7	-	117L7
M709	9CV	-
M2057	-	-
M5039	-	-
M8030	-	-
M8063	-	6AM6, 6AM6/EF91, EF91
M8079	-	6AL5#, 6AL5/6EB5#, 6AL5/5726/6663#, 6AL5/EAA91#, 6AL5W#, 6EB5#, 5726#, 5726/6AL5W#, 5726/6AL5W/6097#, 6097#, 6663#, 6663/6AL5#, EAA91#, EAA91/6AL5#
M8080	6BG	6C4, 6C4/6135, 6C4/EC90, 6135, EC90
M8081	7BF	6J6#, 6J6A#, 6J6WA#, 6101
M8083	-	-
M8096	-	5763
M8098	-	5651A
M8100	-	6AK5#, 6AK5/5654#, 6AK5/EF95#, 5654, EF95#
M8101	-	6BA6#, 6BA6/5749/6660#, 6BA6/EF93#, 5749#, 6660#, 6660/6BA6#, EF93#
M8108	7BK	-
M8121	-	5840, 5840/6225, 6225
M8136	9A	12AU7A/ECC82#, 12AU7WA, 618, 5814#, 6189, 6189/12AU7WA
M8137	9A	12AX7A, 12AX7A/ECC83, 12AX7A/ECC83/7025, 12AX7WA, 12AX7WA/7025, 5751, 7025, ECC83
M8138	-	6202
M8161	-	-
M8162	9A	12AT7/ECC81, 12AT7WA, 6201, 6679/12AT7#
M8180	-	6AK5#, 6AK5/5654#, 6AK5/EF95#, 5654, EF95#
M8190	-	5783#
M8196	-	5725#
M8204	-	5727#
M8212	-	6AL5#, 6AL5/6EB5#, 6AL5/5726/6663#, 6AL5/EAA91#, 6AL5W#, 6EB5#, 5726#, 5726/6AL5W#, 5726/6AL5W/6097#, 6097#, 6663#, 6663/6AL5#, EAA91#, EAA91/6AL5#
M8223	-	0A2#, 0A2WA, 6073#, 6073/0A2#, 6626, 6626/0A2WA
M8224	-	0B2#, 0B2WA, 6074#, 6074/0B2#, 6627, 6627/0B2WA
M8232	-	6J4WA

Tube Type	Basing	Replacement
M8245	7BZ	6AQ5#, 6AQ5/6005/6669#, 6AQ5A#, 6AQ5A/6HG5#, 6HG5#, 6005, 6669, 6669/6AQ5A
ME1501	-	2050
ML381	-	-
ML6442	-	-
ML8533	-	-
ML8906AL	-	-
MU14	-	6BT4
MV6-5	-	6SA7, 6SA7GT
N2ED	-	6HT5
N14	6X	1C5, 1C5/DL35/N14, 1C5GT, 1C5GT/G, DL35
N15	-	3Q5GT, 3Q5GT/G
N16	7AP	3Q5GT, 3Q5GT/G
N17	7BA	3S4, 3S4/DL92, DL92, 1P10
N18	7BA	3Q4, DL95
N19	6BX	3V4, 3V4/DL94, DL94
N22LL	-	19KF6
N25	-	3C4, 3C4/DL96, DL96
N30EL	-	6LF6
N47	-	6AM5, 6AM5/EL91, EL91
N63	7S	6F6, 6F6GT
N66	-	6K6GT
N77	-	6L6GC, 6L6GT
N78	-	6AM5, 6AM5/EL91, EL91
N119	-	6BJ5
N142	-	45B5
N144	-	6AM5, 6AM5/EL91, EL91
N147	-	6M6G*
N148	6AA	7C5
N150	-	6CK5, 6CK5/EL41, 67PT, BF61, EL41
N152	-	21A6
N153	-	15A6
N154	-	16A5
N155	-	6BN5, 6BN5/EL85, EL85
N308	8GT	25E5, 25E5/PL36
N309	-	15A6
N329	-	16A5
N359	-	21A6
N369	9EX	16A8, 16A8/PCL82
N378	-	15CW5, 15CW5/PL84
N379	-	15CW5, 15CW5/PL84
N389	-	25GF6
N707	-	-
N709	-	6BQ5, 6BQ5/EL84, EL84, EL84/6BQ5
N727	7BZ	6AQ5, 6AQ5/6005/6669, 6AQ5A, 6AQ5A/6HG5, 6HG5, 6005, 6669, 6669/6AQ5A, EL90
NE48	-	991
NL604	-	7014, NL604/7014
NL604/7014	-	7014, NL604
NL604L	-	7015, NL604L/7015
NL604L/7015	-	7015, NL604L
NL606	-	7016, NL606/7016
NL606/7016	-	7016, NL606
NL623	-	-
NL653	-	5835, 5835/NL653, NL653/5835
NL653/5835	-	5835, 5835/NL653, NL653
NL660	-	-
NL673	-	-
NL677	-	-
NL714	-	7021, NL714/7021
NL714/7021	-	7021, NL714
NL716	-	6855, NL716/6855
NL716/6855	-	6855, NL716
NL770L	-	-
NL840	-	8754, NL840/8754
NL840/8754	-	8754, NL840
NL842	-	CK1917, NL842/CK1917
NL842/CK1917	-	CK1917, NL842
NL1022A	-	-
NL1051A	-	-
NL5441	-	-
NL5853S	-	-
NL8422	-	5991, NL8522/5991
NL8422/5991	-	5991, NL8522
NU1106	-	6AZ6
OBC3	-	12SQ7
OF1	-	6S7
OF5	-	12K7GT
OH4	-	12A8
OM3	-	6H6*
OM4	-	6R7G
OM5	-	6S7*
OM6	-	6K7
OM10	-	-
OSW2190	-	6AC7, 6AC7/1852, 6AC7/1852/6134, 6AC7/EL34, 1852, 6134, EL34
OSW2192	-	6AG7, 6AG7/6AK7, 6AG7W, 6AK7
OSW2600	-	6AC7, 6AC7/1852, 6AC7/1852/6134, 6AC7/EL34, 1852, 6134, EL34
OSW2601	-	6AG7, 6AG7/6AK7, 6AG7W, 6AK7
OSW3104	8R	6SA7
OSW3105	8Q	6SQ7
OSW3106	7AC	6V6, 6V6GT*, 6V6GTA
OSW3107	-	5AR4/GZ34, 5CG4, 5V4GA, 5Z4, 5Z4G, GZ34, GZ34/5AR4
OSW3109	-	6H6
OSW3110	6R	6E5
OSW3111	8N	6SK7, 6SK7GT
OSW3112	-	6J5, 6J5GT
P17A	-	4Y25, 5S1, 807, 807W, 807WA, 5933, 5933WA, 8018, HY61, QE06/50, QE06140, QV05-25, RK39
PABC80	9E	9AK8, 9AK8/PABC80, PABC80/9AK8
PABC80/9AK8	9E	9AK8, 9AK8/PABC80, PABC80
PC86	-	4CM4
PC88	-	4DL4
PC89	-	7F7
PC93	-	4BS4
PC95	7FP	4GK5
PC97	-	4FY5, 4GK5
PC200	-	-
PC900	7GM	4HA5, 4HA5/4HM5, 4HA5/PC900, 4HM5, 4HM5/4HA5, 4HQ5
PCC18	9A	7AU7
PCC84	9DD	7AN7, 7AN7/PCC84, 7EK7, 30L1, 30L15,

cont'd

Tube Type	Basing	Replacement
PCC84, cont'd		B319, PCC84/7AN7
PCC84/7AN7	9DD	7AN7, 7AN7/PCC84, 7EK7, 30L1, 30L15, B319, PCC84
PCC85	9DE	9AQ8, 9AQ8/PCC85, PCC85/9AQ8
PCC85/9AQ8	9DE	9AQ8, 9AQ8/PCC85, PCC85
PCC88	9DE	7DJ8, 7DJ8/PCC88, 7ES8, PCC189
PCC89	-	7FC7
PCC186	-	7AU7
PCC189	-	7ES8
PCC805	-	7EK7
PCE800	-	9GB8
PCF80	9DC	8A8, 9A8, 9A8/PCF80, 9EA8, 9U8, 9U8A, 30C1, LZ319, LZ329
PCF80/9A8	9DC	8A8, 9A8, 9EA8, 9U8, 9U8A, 30C1, PCF80, LZ319, LZ329
PCF82	9DX	9U8, 9U8A
PCF86	9MP	7HG8, 7HG8/PCF86, 8HG8, 8HG8/PCF86
PCF200	-	8X9
PCF201	-	8U9
PCF800	-	9EN7
PCF801	9QA	8GJ7, 8GJ7/PCF801, 8GX7, PCF806
PCF802	9DC	9JW8, 9JW8/PCF802
PCF805	-	7GV7
PCF806	-	8GJ7, 8GJ7/PCF801
PCH200	-	9V9
PCL82	9EX	16A8, 16A8/PCL82, 30PL12, N369
PCL83	-	-
PCL84	9HX	15DQ8, 15DQ8/PCL84
PCL85	9LY	18GV8, 18GV8/PCL85
PCL86	-	14GW8, 14GW8/PCL86
PCL88	-	16GK8
PCL800	9GK	16GK8
PCL801	-	13GC8
PD500	-	9ED4
PF9	-	6K7
PF86	-	4HR8, 4HR8/PF86
PF88	-	7ED7
PF818	-	-
PFL200	-	16Y9
PH4	8A	6A8
PJ8	-	-
PL21	-	2D21#, 2D21/5727, 2D21/PL21, 5727, 5727/2D21
PL36	8GT	25E5, 25E5/PL36, 30P4, 30P19, N308
PL36/25E5	8GT	25E5, 30P4, 30P19, N308, PL36
PL81	9AS	21A6, 21B6, 213PEN, N152, N339, N359, PL820
PL82	-	16A5, 16A5/PL82
PL83	-	15A6, 15A6/PL83
PL84	9CV	15CW5, 15CW5/PL84, 30P18, N379, PL84
PL86	-	-
PL87	-	-
PL300	-	35FN5
PL302	-	25GF6
PL500	9NH	27GB5, 27GB5/PL500, 27GB5/PL504, 28GB5, PL504
PL504	-	27GB5, 27GB5/PL500, cont'd
PL504, cont'd		27GB5/PL504, 28GB5, PL500
PL505	-	40KG6, 40KG6A/PL509, 49KG6A/PL519, PL509, PL519
PL508	-	17KW6
PL509	9RJ	40KG6, 40KG6A, 40KG6A/PL509, 40KG6A/PL519, 49KG6A, PL519
PL519	-	40KG6, 40KG6A, 40KG6A/PL509, 40KG6A/PL519, 49KG6A, PL509
PL521	9RJ	29KQ6, 29KQ6/PL521, 29LE6
PL801	-	12FB5
PL820	-	21A6
PL1267	4V	0A4#, 0A4/PL1267, 0A4G#, 1267, 1267/0A4, WTT132, Z300T
PLL80	-	12HU8
PM04	7BK	6BA6, 6BA6/5749/6660, 5749#, 6660, 6660/6BA6, EF93
PM05	7BD	6AK5#, 6AK5/5654#, 6AK5/EF95#, 5654, DP61, EF95
PM07	-	6AM6, 6AM6/EF91, EF91
PM95	-	6AK6
PP6BG	-	6M6G#
PY31	-	25Y4, U31
PY32	-	-
PY80	-	19X3
PY81	9CB	17Z3, 17Z3/PY81, A61, PY83, PY800, PY801, U153, U193, U251, U349
PY82	-	19Y3
PY83	4G	17Z3, 17Z3/PY81
PY88	9CB	30AE3, 30AE3/PY88
PY301	-	19CS4, U191, U339
PY500	6EC4	42EC4, 42EC4/PY500A, 42EC4A, 42EC4A/PY500, PY500A
PY500A	6EC4	42EC4, 42EC4/PY500A, 42EC4A, 42EC4A/PY500, PY500A
PY800	4G	17Z3, 17Z3/PY81
PY801	-	17Z3, 17Z3/PY81
QA2400	-	-
QA2401	6BG	6C4, 6C4/6135, 6C4/EC90, 6135, EC90
QA2404	6BG	6AL5, 6AL5/6EB5, 6AL5/5726/6663, 6AL5/EAA91, 6AL5W, 6EB5, 5726#, 5726/6AL5W, 5726/6AL5W/6097, 6097, 6663, 6663/6AL5, EAA91, EAA91/6AL5
QA2406	9A	12AT7/ECC81
QA2407	-	6201
QA2408	-	5692, 6SN7GTB
QB5/1750	-	6079
QB65	-	6SN7GT, 6SN7GTB
QB309	9A	12AT7, 12AT7/ECC81
QB329	-	-
QB339	-	-
QBL5	-	6076, 6076/QBL5
QE03/10	-	5763
QE05/40	-	6146
QE05/40H	-	6159
QE06/50	-	4Y25, 5S1, 807, cont'd

Tube Type	Basing	Replacement
QE06/50, cont'd		807W, 807WA, 5933, 5933WA, 8018, HY61, P17A, QE06140, QV05-25, RK39
QE06140	-	4Y25, 5S1, 807, 807W, 807WA, 5933, 5933WA, 8018, HY61, P17A, QE06/50, QV05-25, RK39
QK209	-	5721, 5721/QK209
QK324	-	7452, 7452/QK324
QL77	6BG	6C4, 6C4/6135, 6C4/EC90, 6135, EC90
QM328	-	-
QM556	-	6X4W#
QM557	-	5654
QM558	-	5725
QM559	-	5726
QN77	-	6AM5*, 6AM5/EL91*, EL91*
QQE02/5	-	6939
QQE03/12	-	6360
QQE03/20	-	6252
QQE04/5	-	7377, 7377/QQE04/5
QQE06/40	-	5894
QQV02-6	-	6939
QQV03-10	-	6360
QQV03-20	-	6252
QQV03-20A	-	-
QQV06-40A	-	-
QS150/40	-	0D3, 0D3/VR150, 0D3A, VR150, VR150/0D3
QS150C1	-	0A2, 0A2WA, 6073, 6073/0A2, 6626, 6626/0A2WA
QS150C2	-	0A2, 0A2WA, 6073, 6073/0A2, 6626, 6626/0A2WA
QS150C3	-	0D3, 0D3/VR150, 0D3A, VR150, VR150/0D3
QS1205	-	0A3, 0A3/VR75, 0A3A, VR75
QS1206	-	0C3, 0C3/VR105, 0C3AVR105
QS1207	-	0A2, 0A2WA, 6073, 6073/0A2, 6626, 6626/0A2WA
QS1208	-	0B2, 0B2WA, 6074, 6074/0B2, 6627, 6627/0B2WA
QS1209	-	5651A
QS1210	-	0A2#, 0A2WA, 6073#, 6073/0A2#, 6626, 6626/0A2WA
QS1211	-	0B2#, 0B2WA, 6074#, 6074/0B2#, 6627, 6627/0B2WA
QS2404	-	6AL5#, 6AL5/6EB5#, 6AL5/5726/6663#, 6AL5/EAA91#, 6AL5W#, 6EB5#, 5726#, 5726/6AL5W#, 5726/6AL5W/6097#, 6097#, 6663#, 6663/6AL5#, EAA91#, EAA91/6AL5#
QS2406	-	6201, 12AT7WA#, 12AT7WB#, 6679/12AT7#
QV03-12	-	5763
QV05-25	-	4Y25, 5S1, 807, 807W, 807WA, 5933, 5933WA, 8018, HY61, P17A, QE06/50, QE06140, RK39
QV06-20	-	6146
QW77	-	6CQ6, 6CQ6/EF92, EF92
QZ77	-	6AM6, 6AM6/EF91, EF91
QZ2406	-	12AT7WB
R3	-	1W4
R10	-	6305
R12	-	6X2
R16	-	1T2
R17	-	-
R18	-	-
R19	9Y	1BX2, 1X2A, 1X2A/1X2B, 1X2B, 1X2B/1X2A, 1X2C, 1X2C/1BX2
R20	-	2J2
R52	5DA	5Y3, 5Y3/6087, 5Y3G, 5Y3G/GT, 5Y3GT, 5Y3GT/G, 5Z4G, 6087
R144	-	6AM6, 6AM6/EF91, EF91
RE1	4C	80
RE2	-	81
RJ2	5T	5Y3, 5Y3/6087, 5Y3G, 5Y3G/GT, 5Y3GT, 5Y3GT/G, 6087
RK28	-	803
RK31	-	830B
RK34	-	2C34, 2C34/RK34
RK39	5AW	4Y25, 5S1, 807, 807W, 807WA, 5933, 5933WA, 8018, HY61, P17A, QE06/50, QE06140, QV05-25
RK57	-	805
RK58	-	838
RK60	-	1641, 1641/RK60
RK61	-	-
RK72	-	-
RK73	-	-
RL21	-	2D21, 2D21/5727, 2D21/PL21, 5727, 5727/2D21, PL21
RL1267	-	0A4G
RS2	-	5Y3, 5Y3/6087, 5Y3G, 5Y3G/GT, 5Y3GT, 5Y3GT/G, 5Z4, 6087
RS1029	-	6360
RZ	-	-
S6F12	-	6AM6, 6AM6/EF91, EF91
S6F33	-	-
S340	3Z	5960
S856	-	0A2, 0A2WA, 6073, 6073/0A2, 6626, 6626/0A2WA
S860	-	0B2, 0B2W, 6074, 6074/0B2
SM150-30	-	0A2, 6073, 6073/0A2, 6626, 6626/0A2WA
SN4	-	1D21, 1D21/SN4, 631P1, 631P1/SN4
SP6	-	6AM6, 6AM6/EF91, EF91
SR2	-	0G3, 0G3/85A2, 85A2
SR3	-	0B2, 6074, 6074/0B2
SR55	-	0B2, 6074, 6074/0B2
SR56	-	0A2, 6073, 6073/0A2, 6626, 6626/0A2
STR85/10	-	0G3, 0G3/85A2, 85A2, 5651A
STV108/30	-	0B2, 0B2WA, 6074, 6074/0B2, 6627, 6627/0B2WA
STV150/30	-	0A2, 0A2WA, 6073, 6073/0A2, 6626, 6626/0A2WA
SU61	-	6X2
T2M05	7BF	6J6, 6J6A, 6101#
T6D	-	2B35, 2B35/EA50, EA50
T60	-	8005
T66G-GT	-	884#, 884W#
T77	-	6C6

Tube Type	Basing	Replacement
T300	-	HF300, HF300/T300
TD24	-	-
TE36	-	6754
TE40	8N	6888
TH62	-	6E8G*
TJ880	-	-
TM12	-	6J4
TS229	-	-
TS383	-	-
TT23	-	-
TT24	-	-
TTZ63	-	6J7
TY66G	-	884#, 884W#
TZ20	-	-
TZ40	-	-
U19	-	-
U26	-	2J2
U31	-	25Z4GT
U37	-	1T2
U41	3C	1B3GT, 1B3GT/1G3GTA, 1G3GT, 1G3GT/1G3GTA, 1G3GTA, 1G3GTA/1B3GT
U43	-	6X2
U45	-	6X2*
U49	-	2J2
U50	5T	5Y3, 5Y3/6087, 5Y3G, 5Y3G/GT, 5Y3GT, 5Y3GT/G, 6087
U51	-	5Y3, 5Y3/6087, 5Y3G, 5Y3G/GT, 5Y3GT, 5Y3GT/G, 5W4GT, 5W4GT/G, 6087
U52	5T	5AS4A, 5AS4A/5U4GB, 5U4G, 5U4GB, 5U4GB/5AS4A
U54	5DA	5AR4, 5AR4/GZ34, GZ34, GZ34/5AR4
U70	5AA	6X5G*
U74	5AA	35Z4GT
U76	5BS	35Z4GT
U77	-	5AR4, 5AR4/GZ34, GZ34, GZ34/5AR4
U78	5BS	6X4*, 6202#, EZ90, V2M70
U82	-	7Y4*
U118	-	31A3
U119	-	38A3
U142	-	31A3
U145	-	31A3
U147	-	6X5G, 6X5GT, EZ35
U149	-	7Y4
U150	-	6BT4
U151	-	6X2
U152	-	19X3
U153	-	17Z3, 17Z3/PY81
U154	-	19Y3
U191	-	19CS4, PY301, U339
U192	-	19Y3
U193	-	17Z3, 17Z3/PY81
U251	-	17Z3, 17Z3/PY81
U291	-	-
U309	-	19X3
U319	-	19Y3
U329	-	25BR3
U339	-	19CS4, PY301, U191
U349	-	17Z3, 17Z3/PY81
U381	-	38A3
U707	5DA	6X4
U709	9M	6CA4, 6CA4/EZ81, EZ81, EZ81/6CA4

Tube Type	Basing	Replacement
U718	-	6BT4*
U4020	-	-
UABC80	-	-
UAF42	-	12S7
UBC41	-	14L7, 14L7/UBC41
UBC80	-	14G6
UBC81	-	14G6
UBF80	-	17C8, 17C8/UBF80
UBF89	-	19FL8
UC92	-	9AB4
UCC85	-	26AQ8
UCF80	-	-
UCH42	-	14K7
UCH80	-	14Y7
UCH81	-	19D8, 19D8/UCH81
UCL82	9EX	10PL12, 48A8, 50BM8, 50BM8/UCL82, LN119
UCL83	9EX	-
UCL86	-	-
UF41	-	12AC5, 12AC5/UF41
UF80	-	19BX6
UF85	-	19BY5
UF86	-	-
UF89	-	12DA6, 12DA6/UF89
UL41	-	45A5, 45A5/UL41
UL84	9CV	10P18, 45B5, 45B5/UL84, N119
UM80	-	19BR5
UM84	-	12FG6
UN954	-	954
UN955	-	955
UQ80	-	12BE7
UR1C	-	-
UU5	-	6BT4
UU9	-	6BT4*
UU12	9M	6CA4, 6CA4/EZ81, EZ81, EZ81/6CA4
UV216	4B	81
UX216B	4B	81
UY41	-	31A3, 31A3/UY41
UY42	-	31A3
UY82	-	55N3
UY85	-	38A3
UY89	-	38A3
V2M70	-	6X4, 6202, EZ90, U78
V61	-	6BT4
V70D	-	-
V99	-	-
V153	4G	17Z3, 17Z3/PY81, PY81
V177	-	6CQ6, 6CQ6/EF92, EF92
V199	-	-
V201	-	-
V311	-	31A3
V312	-	31A3
V741	6BG	6C4, 6C4/6135, 6C4/EC90, 6135, EC90
V884	-	6CQ6, 6CQ6/EF92, EF92
V886	-	6AM5, 6AM5/EL91, EL91
VG18	-	-
VL5631	-	-
VP6	-	6CQ6, 6CQ6/EF92, EF92
VP12D	-	12C8
VR75	4AJ	0A3, 0A3/VR75, 0A3A, KD21, VR75
VR90	4AJ	0B3, 0B3/VR90, 0B3A
VR105	4AJ	0C3, 0C3/VR105, 0C3A, 0C3W, KD24, WT269

Tube Type	Basing	Replacement
VR105-30	4AJ	0C3, 0C3/VR105, 0C3A, 0C3W, KD24, VR105, WT269
VR150	4AJ	0D3, 0D3/VR150, 0D3A, 0D3W, 150C3, KD25, VR150/0D3, VR150W, WT294
VR150/0D3	4AJ	0D3, 0D3A, 0D3W, 150C3, KD25, VR150, VR150W, WT294
VR150W	4AJ	0D3W
VSM70	5BS	6X4
VT83	-	83
VT138	-	6E5#, 1629
VT139	-	0D3, 0D3, 0D3A, VR150, VR150/0D3
VT202	-	9002
VT203	-	9003
W17	6AR	1F3, 1T4, DF91
W25	-	1AF4, 1AF4/1AJ4, 1AJ4
W26	-	-
W61	7R	6K7, 6K7GT
W63	7R	6K7, 6K7GT
W76	-	12K7*
W77	-	6CQ6, 6CQ6/EF92, EF92
W81	8V	7A7
W110	-	13EC7
W118	-	12AC5
W119	-	13EC7
W142	-	12AC5
W143	8V	7A7*
W145	-	12AC5
W147	-	6K7*
W148	8V	7A7
W149	-	7B7
W150	-	6CJ5, 6CJ5/EF41, EF41
W179	-	-
W719	-	6BY7, 6BY7/EF85, EF85, EF85/6BY7
W727	7BK	6BA6, 6BA6/5749/6660, 6BA6/EF93, 5749#, 6660, 6660/6BA6, EF93
W739	-	6EC7
WD11	-	-
WD12	-	-
WD119	-	-
WD142	-	12S7
WD150	-	6CT7, 6CT7/EAF42, EAF42
WD709	-	6N8
WL624	-	-
WL630	6BS	2050
WL630A	6BS	2050
WT6	7AC	6L6#, 6L6GC#, 1614
WT210-0001	7BN	2D21, 2D21/5727, 2D21/PL21, 5727, 5727/2D21, PL21
WT210-0003	6Q	6Q5, 884, 884/6Q5, 884W
WT210-0004	6BS	2050, 2050A#
WT210-0006	7Q	6H6#
WT210-0007	-	6L6#, 6LG6C#
WT210-0009	5D	6Z4
WT210-0011	4AJ	0C3, 0C3/VR105, 0C3A, VR105
WT210-0012	4C	80
WT210-0013	4C	5Z3
WT210-0018	4AJ	0D3, 0D3/VR150, 0D3A, VR150, VR150/0D3
WT210-0019	4C	83
WT210-0021	6S	6X5#, 6X5GT#
WT210-0025	7Q	117Z6GT
WT210-0028	7AP	3Q5GT#, 3Q5GT/G#
WT210-0029	6Q	6C5#
WT210-0037	8AO	117L7/M7GT
WT210-0040	5BS	6X4#, 6X4W
WT210-0042	5T	5Y3#, 5Y3/6087, 5Y3G#, 5Y3G/GT#, 5Y3GT#, 5Y3GT/G#, 6087
WT210-0048	5T	5AS4A, 5AS4A/5U4GB, 5U4G#, 5U4GB#, 5U4GB/5AS4A
WT210-0060	4R	0Z4#, 0Z4A/0Z4#
WT210-0062	-	5557, 5557/FG17, FG17
WT210-0069	-	5557, 5557/FG17, FG17
WT210-0077	-	2D21#2D21/5727, 2D21/PL21#, 5727, 5727/2D21, PL21#
WT210-0081	8N	6SJ7#, 5697#
WT210-0082	7AC	6V6#, 6V6GTA#
WT210-0083	8BF	7K7
WT210-0084	8B	6N7#, 6N7GT#
WT210-0085	7BZ	50B5#
WT210-0087	8K	6K8#
WT210-0088	6Q	6J5#, 6J5GT#
WT210-0089	7S	6G6G
WT210-0090	6F	6C6#
WT210-0091	4V	0A4G
WT210-0108	-	6AS7G, 6AS7GA#, 6080, 6080W, 6080WA
WT210-0148	-	6AX5GT#
WT210-3000	-	2D21, 2D21/5727, 2D21/PL21, 2D21W, 5727, 5727/2D21, 5727/2D21W, PL21
WT245	6Q	6Q5, 884, 884/6Q5, 884W
WT246	6BS	2050, 2050A
WT261	7Q	6H6#
WT261A	-	6H6#
WT263	5D	6Z4
WT269	4AJ	0C3, 0C3/VR105, 0C3A, VR105
WT270	4C	80
WT270X	5T	5Z3
WT294	4AJ	0D3, OD3/VR150, 0D3A, VR150, VR150/0D3
WT301	4C	83
WT301A	-	83
WT308	6S	6X5GT#
WT377	7Q	117Z6GT
WT389	7AP	3Q5GT#, 3Q5GT/G#
WT390	6Q	6C5#
WT606	7BN	2D21, 2D21/5727, 2D21/PL21, 5727, 5727/2D21, PL21
WTT100	5BS	6X4#, 6X4W
WTT102	5T	5Y3#, 5Y3/6087, 5Y3G#, 5Y3G/GT#, 5Y3GT#, 5Y3GT/G#, 6087
WTT103	7Q	6H6#
WTT108C1	-	0B2, 0B2WA, 6074, 6074/0B2, 6627, 6627/0B2WA
WTT114	4R	0Z4, 0Z4A/0Z4
WTT115	8AV	117N7GT
WTT117	-	5557, 5557/FG17, FG17
WTT118	-	105
WTT119	-	172

Tube Type	Basing	Replacement
WTT122	8N	6SJ7#, 5693
WTT123	7AC	6V6#, 6V6GTA#
WTT124	8BF	6AT6#, 6AT6/6BK6#, 6BK6#, 7K7
WTT125	8B	6N7#, 6N7GT#
WTT126	7BZ	50B5#
WTT127	-	833A
WTT128	8K	6K8#
WTT129	6Q	6J5#, 6J5GT#
WTT130	7S	6G6G
WTT131	6F	6C6#
WTT132	4V	0A4G
WTT135	5T	5AS4A#, 5AS4A/5U4GB#, 5U4G#, 5U4GB#, 5U4GB/5AS4A#
WTT149	-	172
WX12	-	-
X14	-	1A7G, 1A7GT, 1A7GT/G
X17	7AT	1C1, 1R5, 1R5/DK91, DK91
X18	-	1AC6, 1AC6/DK92, DK92
X20	-	1AC6, 1AC6/DK92, DK92
X25	-	1AB6, 1AB6/DK96, 1C3, 1C3/DK96/1AB6, DK96
X61M	-	6E8G
X63	8A	6A8
X64	-	6L7
X65	-	6E8
X66	-	-
X71M	-	12K8
X73	-	6D8*
X73M	-	6D8*
X75	-	-
X76M	-	12K8GT
X77	7CH	6BE6, 6BE6/5750, 6BE6/EK90, 6BE6W#, 5750#, 5750/6BE6W#, EK90
X79	-	6AE8
X81	-	7S7
X99	-	-
X107	6CH	18FX6#
X119	-	19D8
X142	-	14K7
X147	-	6E8G
X148	-	7S7
X150	12BQ	6C10
X155	9AJ	6BC8, 6BC8/6BQ7A, 6BC8/6BZ8, 6BK7*, 6BK7A*, 6BK7B*, 6BQ7, 6BQ7A, 6BQ7A/6BZ7, 6BQ7A/6BZ7/6BS8, 6BS8, 6BS8/6BK7B, 6BZ7, 6BZ7/6BQ7A, 6BZ8
X199	-	-
X201A	-	-
X319	-	6351
X719	-	6AJ8, 6AJ8/ECH81, ECH81, ECH81/6AJ8
X727	7CH	6BE6, 6BE6/5750, 6BE6/EK90, 6BE6W#, 5750#, 5750/6BE6W#, EK90
X6030	-	-
XAA91	-	3AL5
XB91	-	3AL5
XC95	7FP	2ER5, 2FQ5, 2FQ5A, 2FQ5A/2GK5, 2GK5, 2GK5/2FQ5, 2GK5/2FQ5A
XC97	7FP	2FY5, 2FQ5, 2FQ5A, cont'd
XC97, cont'd		2FQ5A/2GK5, 2GK5, 2GK5/2FQ5, 2GK5/2FQ5A
XC900	7GM	2HA5, 2HA5/2HM5, 2HM5, 2HM5/2HA5
XCC82	9A	7AU7
XCC189	9DE	4ES8, 4ES8/XCC189, 4KN8, 4KN8/4RHH8, 4RHH8
XCF80	9DC	4BL8, 4BL8/XCF80, 5EA8, 5EA8/5GH8A, 5GH8A, 5RHP1, 5U8
XCF82	-	5U8
XCF86	-	2HR8
XCF801	9QA	4GJ7, 4GJ7/4GX7, 4GJ7/XCF801, 4GX7
XCH81	-	3AJ8
XCL82	-	8B8
XCL84	-	-
XCL85	9LY	9GV8, 9GV8/XCL85
XF80	-	3BX6
XF85	-	3BY7
XF86	-	-
XF94	7BK	3AU6
XF183	9AQ	3EH7, 3EH7/XF183, 3EJ7, 3EJ7/XF184, XF184
XF184	9AQ	3EH7, 3EH7/XF183, 3EJ7, 3EJ7/XF184, XF183
XL36	-	13CM5
XL84	9CV	8BQ5
XL86	9CV	8CW5, 8CW5/XL86, 8CW5A
XL500	9NH	13GB5, 13GB5/XL500
XXA91	6BT	3AL5
XXB	7BW	3C6, 3C6/XXB
XXD	8AC	14AF7, 14AF7/XXD, 14F7
XXFM	8BZ	7X7, 7X7/XXFM
XXL	5AC	7A4, 7A4/XXL
XY88	9CB	16AQ3, 16AQ3/XY88
Y25	-	1M3, 1M3/DM70, 1N3, 1N3/DM70/1M3, 1N3/DM71, DM70, DM70/1N3, DM71
Y61	6R	6U5
Y64	-	6U5
YC95	7FP	3ER5, 3ER5/YC95
YC97	-	3FQ5, 3FY5, 3GK5, 3GK5/3FQ5
YCC189	9DE	5ES8, 5ES8/YCC189, YCC189
YCF86	-	5HG8, 5HG8/LCF86
YCL84	-	10DX8, 10DX8/LCL84
YCL180	9AJ	5BQ7A, 5BQ7A/5BZ7, 5BZ7
YF183	9AQ	4EH7, 4EH7/LF183
YF184	9AQ	4EJ7, 4EJ7/LF184
YL	-	-
YL84	-	10BQ5
YL86	-	10CW5, 10CW5/LL86
YL1370	-	-
YL1371	-	-
YL1372	-	-
Z14	5Y	1N5GT, 1N5GT/1P5GT, 1N5GT/G, 1P5GT
Z63	7R	6J7
Z70U	-	7710, 7710/Z70U
Z77	-	6AM6, 6AM6/EF91, EF91
Z150	-	6CU7, 6CU7/ECH42, ECH42
Z152	-	6BX6, 6BX6/EF80, EF80
Z300T	-	QA4G
Z329	-	7ED7
Z550M	-	8453

Tube Type	Basing	Replacement
Z719	-	6BX6, 6BX6/EF80, EF80
Z729	-	6BQ5, 6BQ5/EL84, 6267, EF86, EL84, EL84/6BQ5
Z749	-	6EL7
Z759	-	-
Z900T	-	5823
Z1494	-	-
Z1751	-	-
Z3000T	-	0A4G
ZA2	-	954*

Tube Type	Basing	Replacement
ZD17	6AU	1FD9, 1S5, 1S5/DAF91, DAF91
ZD25	-	1AF5, 1AF5/1AH5, 1AH5
ZD152	-	6N8
ZD2669	5DE	3DG4
ZM1000	-	-
ZM1001	-	-
ZM1033	-	-
ZZ1000	-	8228, 8228/ZZ1000
ZZ1030	-	-

2

Picture Tube Substitutes

This section lists, in numeric-alphabetic order, all known receiving tubes. To use this listing, look in the first column for the tube type you want to replace, then find a suitable substitute in the third column. The substitutes are only those tubes which have similar or improved characteristics and can be inserted directly into the socket. No rewiring or mechanical changes are needed, although the associated service controls may have to be readjusted. The symbols which may follow a substitute, refer to special characteristics of the tube and are shown in the key below. **When no substitutes are shown, none are recommended.** The second column contains the basing code. To use this number, turn to Section 3 to reference your tube substitution selection against the diagrammed pin connections.

The X-radiation levels emitted from all television receivers produced after January 15, 1970 must meet rigid government standards. These radiations are primarily produced by high-voltage rectifier tubes, shunt-regulator tubes, and picture tubes. The radiation, however, is carefully controlled in tubes of later design. Whenever possible,only tubes produced after 1970 are suggested for substitution. Although other tubes may appear to be acceptable substitutes, they have not been listed because of the possibility of radiation hazard.

Key to Symbols:

- ~ Omit ion trap.
- & Add ion trap.
- + May require longer or shorter HV lead.
- @ Connect pin 6 to pin 2 or pin 10.
- % Parallel-filament circuits only.
- ^ In some cases it may be necessary to reverse the red and green cathode lead to achieve proper balance.
- < Add high-voltage filter capacitor.
- > Ground the aquadag coating.
- | Change the anode connector.
- ? Change the ion trap.

Tube Type	Basing	Replacement
1VABP4	-	-
1VACP4	-	-
2EP4	8JK	-
3VABP4	7GR	-
5VABP4	7GR	-
5VACP4	-	-
5VACRP	7GR	-
5VADP4	7GR	-
7RP4	12D	7RP4A
7RP4A	12D	7RP4
8AP4	12H	8AP4A
8AP4A	12H	8AP4
8DP4	12AB	-
8DP	-	-
8HP4	12L	8MP4
8LP4	7FA	-
9ACP4	7GR	9AGP4, 9WP4
9ADP4	8HR	-
9AEP4	7GR	-
9AGP4	7GR	9ACP4
9AMP4	-	-
9QP4	12AD	9QP4A
9QP4A	12AD	9QP4
9SP4	8HR	-
9TP4	8HR	-
9UP4	7GR	9VP4/230DB4, 9YP4
9VABP4	7GR	-
9VACP22	14BL	-
9VADP4	7GR	-
9VAEP4	7GR	-
9VAGP4	7GR	-
9VAHP4	7GR	-
9VAJP4	7GR	9VAWP4
9VAKP4	7GR	9VALP4, 9VASP4
9VALP4	7GR	9VASP4
9VAMP4	7GR	-
9VANP4	7GR	-
9VARP4	7GR	-
9VASP4	7GR	9VALP4, 9VAUP4
9VATP4	7GR	-
9VAUP4	7GR	9VALP4, 9VASP4
9VAWP4	7GR	9VAJP4
9VAZP4	7GR	-
9VBAP22	-	-
9VP4	7GR	9YP4
9WP4	7GR	-
9XP4	-	-
9YP4	7GR	-
10ABP4	12L	10ABP4A, 10ABP4B, 10ABP4C, 10AEP4%
10ABP4A	12L	10ABP4, 10ABP4B, 10ABP4C, 10AEP4%
10ABP4B	12L	10ABP4, 10ABP4A, 10ABP4C, 10AEP4%
10ABP4C	12L	10ABP4, 10ABP4A, 10ABP4B, 10AEP4%
10ADP4	12L	-
10AEP4	12L	10ABP4%, 10ABP4A%, 10ABP4B%, 10AEP4C%
10AFP4	-	-
10AJP4	12S	-
10ARP4	7GR	-
10ASP4	7GR	9AEP4, 9VABP4
10ATP4	7GR	-
10AVP4	7GR	9VAHP4
10BP4	12N	10BP4A, 10BP4AC?, 10BP4AD?, cont'd
10BP4, cont'd		10FBP4~, 10FP4A~
10BP4A	12N	10BP4, 10BP4C?, 10BP4D?, 10FP4~, 10FP4A~
10BP4C	12N	10BP4D, 10BP4?, 10BP4A?, 10FP4~, 10FP4A~
10BP4D	12N	10BP4C, 10BP4?, 10BP4A?, 10FP4~, 10FP4A~
10CP4	12N	-
10DP4	12M	-
10EP4	12N	10BP4AI, 10CP4~, 10FP4~I, 10FP4A~I
10FP4	12N	10FP4A, 10BP4&, 10BP4A&, 10BP4B&, 10BP4D&
10FP4A	12N	10FP4A, 10BP4&, 10FP4C&, 10FP4D&
10MP4	12G	10MP4A
10MP4A	12G	10MP4
10RP4	12L	-
10VABP22	-	10VADP22
10VACP4	7GR	-
10VADP22	14BM	-
10VAEP4	7GR	-
10VAFP4	7GR	-
10VAGP4	7GR	-
10VAHP22	14BP	-
10VAP22	14BM	-
11AP4	8HR	11HP4
11BP4	8HR	11HP4
11CP4	8HR	11LP4%
11DP4	8HR	-
11EP4	8HR	11FP4%
11FP4	8HR	11EP4%
11GP4	8HR	11MP4%
11HP4	8HR	11HP4A
11HP4A	8HR	11HP4
11JP4	8HR	-
11KP4	8HR	-
11LP4	8HR	11CP4%
11MP4	8HR	7RP4A
11PN4	-	-
11QP4	7GR	280NB4
11RP4	7GR	11UP4
11SP22	14BJ	11WP22
11TP4	8HR	-
11UP4	7GR	11RP4
11WP22	14BJ	-
12AP4	-	-
12AYP4	8HR	12AZP4%, 12BAP4%, 12BSP4%
12AZP4	8HR	12AYP4%, 12BAP4%, 12BSP4%
12BAP4	8HR	12BSP4, 12AYP4%, 12AZP4%
12BEP4	7FA	-
12BFP4	7GR	-
12BGP4	8HR	-
12BJP4	8HR	-
12BKP4	-	-
12BKP4A	8HR	-
12BLP4	8HR	12BKP4, 12BQP4, 12DHP4+
12BMP4	7GR	12CDP4, 12CDP4A
12BNP4	8HR	12BNP4A
12BNP4A	8HR	12BNP4
12BQP4	8HR	12BKP4, 12BLP4, 12DHP4+
12BRP4	9RS	-
12BSP4	8HR	12AYP4, 12AZP4, 12BAP4
12BTP4	8HR	-
12BUP4	8HR	12BUP4A, 12BUP4B, 12DQP4
12BUP4A	8HR	12BUP4, 12BUP4B, 12BUP4C, 12DHP4, 12DQP4

Tube Type	Basing	Replacement
12BUP4B	8HR	12BUP4, 12BUP4A, 12BUP4C, 12DHP4
12BUP4C	8HR	12BKP4
12BVP4	7GR	-
12BWP4	8HR	-
12BZP4	7GR	-
12CBP4	7FA	-
12CDP4	7GR	12BMP4
12CDP4A	7GR	12BMP4, 12CDP4
12CEP4	7GR	-
12CFP4	7GR	12CNP4, 12CNP4A
12CGP4	-	-
12CHP4	7GR	310BYB4B
12CNP4	7GR	12CNP4A, 12CFP4, 12VAGP4%
12CNP4A	7GR	12CNP4, 12CFP4, 12VAGP4%
12CP4	-	-
12CQP4	8HR	12BKP4
12CSP4	7GR	-
12CTP4	7GR	12DFP4
12CTP4A	7GR	-
12CUP4	8HR	-
12CVP4	7GR	-
12CWP4	7GR	12CWP4A
12CWP4A	7GR	12CWP4
12CXP4	7GR	-
12CZP4	7GR	-
12DCP22	14BH	310FGP22
12DEP4	7GR	-
12DFP4	7GR	12DSP4%
12DGP4	7GR	-
12DGP4A	7GR	-
12DHP4	8HR	12BKP4, 12BLP4+, 12BLP4A+, 12BQP4+, 12BUP4+, 12BUP4A+,12VABP4
12DKP4	7GR	-
12DMP4	8HR	-
12DQP4	8HR	12BKP4, 12DHP4, 12VABP4+
12DSP4	7GR	-
12JP4	12D	12RP4&, 12QP4&, 12QP4A&
12KP4	12N	12KP4A, 12ZP4&, 12ZP4A&
12KP4A	12N	12KP4, 12ZP4&, 12ZP4A&
12LP4	12N	12LP4A, 12LP4C, 12KP4~, 12KP4A~, 12ZP4?, 12ZP4A?
12LP4A	12N	12LP4, 12LP4C, 12KP4~, 12KP4A~, 12ZP4?, 12ZP4A?
12LP4C	12N	12LP4, 12LP4A,12KP~, 12KP4A~, 12ZP4?, 12ZP4A?
12QP4	12D	12QP4A, 12RP4, 12JP4~
12QP4A	12D	12QP4, 12RP4, 12JP4~
12RP4	12D	12QP4, 12QP4A, 12JP4~
12TP4	12D	-
12UP4	12D	12UP4A, 12UP4B?
12UP4A	12D	12UP4, 12UP4B?
12UP4B	12D	12UP4?, 12UP4A?
12VABP4	8HR	12BKP4+, 12DHP4, 12DQP4+
12VACP4	7GR	-
12VADP22	-	-
12VAEP4	7GR	-
12VAFP4	7GR	-
12VAGP4	7GR	-
12VAHP22	14BH	12VALP22, 12VASP22, 12VATP22
12VAJP22	14BH	12VADP22
12VALP22	14BH	12VAHP22, 12VASP22, 12VATP22
12VAMP4	7GR	12VAFP4, 12VAWP4
12VANP4	7GR	-
12VAQP4	7GR	-
12VARP22	14BH	-
12VASP22	14BH	12VALP22, 12VAHP22, 12VATP22
12VATP22	14BH	12VAHP22, 12VALP22, 12VASP22
12VAUP4	7GR	12VAWP4%
12VAWP4	7GR	12VAFP4, 12VAMP4, 12VAUP4%
12VAXP4	7GR	-
12VAZP4	7GR	-
12VBCP4	-	-
12VBDP4	7GR	-
12VBEP4	7GR	-
12VBFP4	7GR	-
12VBGP4	7GR	12VBFP4
12VBHP4	7GR	-
12VBJP4	7GR	-
12VBLP4	7GR	-
12VBLY4	-	-
12VBMP4	7GR	-
12VBMP	-	-
12VBNP4	7GR	12VCGP4
12VBQP4	7GR	12VCJP4
12VBQP	-	-
12VBRP4	7GR	12VCBP4, 12VCKP4, 12VCLP4
12VBSP4	-	-
12VBWP4	7GR	-
12VBXP4	7GR	-
12VBYP4	7GR	-
12VBZP4	7GR	-
12VCAP4	7GR	-
12VCBP4	7GR	12VBRP4, 12VCLP4
12VCBP31	-	-
12VCDP4	7GR	-
12VCEP4	7GR	-
12VCFP4	7GR	-
12VCGP4	7GR	12VBNP4
12VCJP4	7GR	12VBQP4
12VCKP4	7GR	12VBRP4, 12VCBP4
12VCLP4	7GR	12VBRP4, 12VCBP4
12VCUP4	-	-
12VP4	12G	12VP4A
12VP4A	12G	12VP4
12WP4	9CH	-
12XP4	12N	-
12YP4	12N	-
12ZP4	12N	12ZP4A, 12JP4<'l, 12KP4~, 12KP4A~
12ZP4A	12N	12ZP4, 12JP4<'l, 12KP4~, 12KP4A~
13AP4	8HR	-
13DP4	8HR	-
13GP22	14BH	12VADP22, 13MP22
13JP22	14BH	12VAHP22, 12VATP22
13LP22	14BH	12VADP22, 13GP22
13MP22	-	12VADP22, 13GP22
13VAEP22	14BH	-
13VAFP22	14BP	13VAHP22, 13VAGP22
13VAGP22	14BP	13VAHP22
13VAHP22	-	-
13VAJP4	7GR	13VANP4, 13VAQP4
13VAKP22	13D	-
13VAKTC02	-	-
13VAMP22	13E	-
13VANP4	7GR	13VAJP4, 13VAQP4

Tube Type	Basing	Replacement
13VAQP4	7GR	13VAJP4, 13VANP4
13VARP4	7GR	13VASP4
13VASP4	7GR	13VARP4
13VATP22	13R	13VBFP22
13VAUP22	13M	-
13VAWP22	14BP	-
13VAYP22	13L	-
13VAZP4	7GR	-
13VBAP22	13K	13VBMP22
13VBEP22	13M	13VBFP22
13VBFP22	-	-
13VBGP22	-	13VAHP22
13VBHP22	-	13VBFP22
13VBJP22	-	-
13VBLP22	-	-
13VBMP22	-	-
13VBRP22	-	-
13VBSP22	-	13VBJP22
13VBTP22	-	13VAHP22
13VBUP22	-	-
13VBWP22	-	13VBMP22
13VCAP22	-	-
13VCBP22	-	13VCAP22
13YBRP22	-	-
14ACP4	12L	14AEP4~
14AEP4	-	-
14AJP4	8HR	14ASP4~, 14AVP4~
14ARP4	12L	14AUP4%
14ASP4	8HR	14AVP4
14ATP4	12L	-
14AUP4	12L	14AWP4, 14ARP4%
14AVP4	8HR	14ASP4
14AWP4	12L	14AUP4
14BCP22	14AU	-
14BDP4	12L	-
14BP4	12N	14BP4A, 14CP4, 14CP4A, 14EP4, 14CP4B~
14BP4A	12N	14BP4, 14CP4, 14CP4A, 14EP4, 14CP4B~
14CP4	12N	14CP4A, 14BP4, 14BP4A, 14EP4, 14CP4B~
14CP4A	12N	14BP4, 14BP4A, 14CP4, 14EP4, 14CP4B~
14CP4B	12N	14CP4&, 14CP4A&, 14BP4&, 14BPAA&, 14EP4&
14DP4	12D	14CP4>'?, 14CP4A>'?, 14EP4>'?, 14FP4?, 14UP4~
14EP4	12N	14BP4, 14BP4A, 14CP4, 14CP4A, 14CP4B~
14GP4	12L	-
14HP4	12L	-
14KP4	12N	14KP4A
14KP4A	12N	-
14NP4	12L	14NP4A, 14RP4, 14RP4A, 14SP4, 14WP4~, 14XP4%, 14XP4A%, 14ZP4~
14NP4A	12L	14NP4, 14RP4, 14RP4A, 14SP4, 14WP4~, 14XP4%, 14XP4A%, 14ZP4~
14QP4	12L	14QP4A, 14QP4B~, 14HP4
14QP4A	12L	14QP4, 14QP4B~, 14HP4
14QP4B	12L	14QP4&, 14QP4A&, 14HP4&
14RP4	12L	14RP4A, 14NP4, 14NP4A, 14SP4, 14WP4~, 14XP4%, 14XP4A%, 14ZP4~
14RP4A	12L	14RP4, 14NP4, 14NP4A, 14SP4, 14WP4~, cont'd

Tube Type	Basing	Replacement
14RP4A, cont'd		14XP4%, 14XP4A%, 14ZP4~
14SP4	12L	14NP4, 14NP4A, 14RP4, 14RP4A, 14WP4~, 14XP4%, 14XP4A%
14UP4	12D	14DP4&
14VABP22	-	-
14VACP22	-	14VADP22
14VADP22	14BH	14VAHP22
14VAEP22	14BH	14VAGP22, 14VALP22
14VAFP22	14BH	14VABP22
14VAGP22	-	-
14VAHP22	14BH	14VADP22
14VAJP22	14BE	-
14VAKP22	14BE	-
14VALP22	14BH	14VAGP22
14VAMP22	14BH	14VAGP22
14VANP22	-	-
14VAQP22	-	-
14VARP4	8HR	-
14WP4	12L	14ZP4
14XP4	12L	14XP4A, 14NP4%, 14NP4A%, 14RP4%, 14RP4A%, 14SP4%, 14WP4~%, 4ZP4~%
14XP4A	12L	14XP4, 14NP4%, 14NP4A%, 14RP4%, 14RP4A%, 14SP4%, 14WP4~%, 14ZP4~&
14ZP4	12L	14WP4
15ABP22	-	-
15ACP22	14BH	14AFP22
15ADP4	12D	-
15AEP22	14BH	14VAGP22, 14VALP22, 15AGP22
15AFP22	14BH	14VAGP22, 14VALP22, 15AGP22
15AGP22	-	14VAGP22
15AHP22	-	14VAGP22
15AJP22	-	14VADP22, 15NP22
15AP4	12D	15DP4&, 15DP4A&
15CP4	12D	-
15DP4	12D	15DP4A, 15AP4~
15DP4A	12D	15DP4, 15AP4~
15GP22	20A	-
15HP22	20A	-
15JP4	8HR	-
15KP22	14BH	15LP22, 15NP22
15LP22	14BH	14VADP22, 14VAHP22, 15NP22
15MP22	14BK	-
15NP22	14BH	14VADP22, 14VAHP22
15QP22	-	14VADP22, 15NP22
15RP22	14BH	14VADP22, 15LP22, 15NP22
15SP22	14BH	14VAGP22, 15AFP22, 15AGP22
15TP22	-	14VADP22, 15NP22
15UP22	14BH	-
15VACP4	8HR	-
15VADP22	13D	-
15VADTC01	13D	15VAETC01
15VADTC02	-	15VAETC02
15VAEP22	13D	-
15VAETC01	13D	-
15VAETC02	-	-
15VAFP22	13D	15VALP22
15VAGP22	-	-
15VAHP22	-	-
15VAJP4	7GR	15VAJP4A
15VAJP4A	7GR	15VAJP4

PICTURE TUBES

Tube Type	Basing	Replacement
15VAKTC02	-	15VAETC02
15VALP22	13D	15VAFP22
15VANP4	7GR	15VANP4A
15VANP4A	7GR	15VANP4
15VANP	-	-
15VANPA	-	-
15VARP22	-	-
15VASP22	-	15VAGP22
15VATP22	-	-
15VAWP22	-	-
15VAYP22	-	-
15VAZP22	-	15VARP22
15WP22	14BH	15LP22
15XP22	14BH	-
15YP22	14BK	-
15ZP22	14BH	-
16ABP4	12P	16AEP4@
16ACP4	12P	-
16AEP4	12L	16ABP4
16AFP4	12L	-
16ANP4	8HR	16AQP4, 16ASP4%, 16AXP4%
16AP4	12D	16AP4A, 16AP4B
16AP4A	12D	16AP4, 16AP4B
16AP4B	12D	16AP4, 16AP4A
16AQP4	8HR	16ANP4, 16ASP4%, 16AXP4%
16ASP4	8HR	16AXP4, 16ANP4%, 16AQP4%
16ATP4	8HR	-
16AUP4	8HR	16BFP4+%, 16BMP4%
16AVP4	7FA	16BCP4
16AWP4	8HR	16AZP4%
16AXP4	8HR	16ASP4, 16ANP4%, 16AQP4%
16AYP4	8HR	16BDP4%, 16CAP4
16AZP4	8HR	16AWP4%
16BAP4	8HR	16BEP4%
16BCP4	7FA	-
16BDP4	8HR	16AYP4%
16BEP4	8HR	16BAP4+%
16BFP4	8HR	16AUP4+%, 16BMP4+
16BGP4	8HR	16BWP4, 16BHP4%, 16BVP4+, 16CAP4, 16CMP4A+
16BJP4	8HR	16BGP4%, 16BWP4%, 16CMP4A%
16BMP4	8HR	16AUP4%, 16BFP4+
16BNP4	8HR	-
16BRP4	8HR	16BVP4%, 16BWP4%+, 16CAP4%+, 16CEP4%+, 16JCP4%, 16CKP4%, 16CMP4%
16BSP4	8HR	16BKP4+, 16BLP4+, 16BTP4+
16BUP4	8HR	-
16BVP4	8HR	168WP4+, 16CJP4
16BWP4	8HR	16BGP4, 16BVP4+, 16CAP4, 16CJP4+
16BXP4	7FA	-
16BYP4	8HR	-
16CAP4	8HR	16BWP4, 16CEP4, 16BRP4+%, 16VP4+, 16CJP4+, 16CKP4%+, 16CMP4+
16CDP22	14BE	-
16CEP4	8HR	16BWP4, 16CAP4, 16BRP4%+, 16BVP4+, 16CJP4+, 16CKP4%+, 16CMP4+
16CFP4	7GR	-
16CHP4	8HR	16CHP4A, 16BSP4+
16CHP4A	8HR	16CHP4, 16BSP4+
16CJP4	8HR	16BWP4+
16CKP4	8HR	16BRP4%, 16BVP4%, cont'd

Tube Type	Basing	Replacement
16CKP4, cont'd		16BWP4%+, 16CAP4%+, 16CEP4%+, 16CJP4%, 16CMP4%
16CMP4	8HR	16BVP4, 16CJP4, 16BRP4%, 16BWP4+, 16CAP4+, 16CEP4+, 16CKP4%
16CMP4A	8HR	16BGP4+, 16BWP4+
16CNP4	7GR	-
16CP4	12D	-
16CQP4	7GR	-
16CSP22	14BE	16DAP22, 400CUB22
16CTP4	8HR	16BWP4, 16CAP4, 16CEP4, 16BRP4%+, 16BVP4+, 16CJP4+, 16CKP4%+, 16CMP4+, 16CUP4+
16CUP4	8HR	16BVP4, 16CJP4, 16CMP4, 16BRP4%, 16BWP4+, 16CAP4+, 16CEP4+, 16CKP4%, 16CTP4+
16CWP4	7GR	16DCP4, 16DCP4A, 16CXP4
16CWP4A	7GR	-
16CXP4	7GR	16CWP4, 16DCP4, 16DCP4A
16CYP22	14BH	400CVB22
16DAP22	14BE	16CSP22, 400CUB22
16DCP4	7GR	16DCP4A, 16CWP4, 16CXP4
16DCP4A	7GR	16DCP4, 16CWP4, 16CXP4
16DP4	12D	16DP4A
16DP4A	12D	16DP4
16EP4	12D	16EP4A, 16EP4B
16EP4A	12D	16EP4, 16EP4B
16EP4B	12D	16EP4, 16EP4A
16FP4	12D	-
16GP4	12D	16GP4A
16GP4A	12D	16GP4
16GP4B	12D	16GP4
16GP4C	12D	16GP4
16HP4	12N	16HP4A
16HP4A	12N	16HP4
16JP4	12N	16JP4A
16JP4A	12N	16JP4
16KP4	12N	16KP4A, 16RP4, 16RP4A, 16RP4B~, 16TP4
16KP4A	12N	16KP4, 16RP4, 16RP4A, 16RP4B~, 16TP4
16LP4	12N	16LP4A, 16ZP4
16LP4A	12N	16LP4, 16ZP4
16MP4	12N	16MP4A, 16JP4, 16JP4A
16MP4A	12N	16MP4, 16JP4, 16JP4A
16QP4	12D	16XP4
16RP4	12N	16RP4A, 16KP4, 16KP4A, 16TP4, 16RP4B~
16RP4A	12N	16RP4, 16KP4, 16KP4A, 16RP4B~
16RP4B	12N	16KP4&, 16KP4A&, 16RP4&, 16RP4A&, 16TP4&
16SP4	12N	16SP4A, 16WP4A, 16WP4A, 16WP4B
16SP4A	12N	16SP4, 16WP4A, 16WP4B
16TP4	12N	16KP4, 16RP4A, 16RP4, 16RP4A, 16RP4B~
16UP4	12D	16KP4>, 16KP4A>, 16QP4?, 16RP4>, 16RP4A>, 16TP4>, 16XP4?
16VABP4	8HR	16VAQP4, 17EMP4, 17EWP4
16VABP22	14BH	16VACP22
16VACP22	14BH	16VABP22
16VADP4	7GR	-

Tube Type	Basing	Replacement
16VADP22	-	-
16VAEP22	-	-
16VAFP22	14BH	-
16VAGP4	8HR	-
16VAHP22	-	16VACP22
16VAJP4	8HR	-
16VAKP22	14BH	17FBP22
16VALP4	8HR	-
16VAMP22	14BE	16VBGP22
16VANP22	14BH	16VAFP22
16VAQP4	8HR	16VABP4, 16VBNP4, 16VBTP4, 17EWP4, 17FDP4
16VARP4	8HR	-
16VASP22	14BE	16VATP22
16VATP22	14BE	16VASP22
16VAUP22	14BH	-
16VAWP22	14BK	16VBDP22%, 16VBMP22%
16VAXP22	14BH	-
16VAYP4	8HR	16VBCP4A
16VAZP4	8HR	-
16VBCP4	8HR	16VBCP4A, 17ESP4
16VBCP4A	8HR	16VAYP4, 16VBCP4
16VBDP22	14BK	16VAWP22%, 16VBMP22
16VBEP22	14BH	-
16VBFP22	14BE	-
16VBGP22	14BE	-
16VBHP22	14BE	16VBFP22
16VBJP4	7GR	-
16VBKP4	7GR	-
16VBLP4	8HR	-
16VBMP22	14BK	16VAWP22%, 16VBDP22
16VBNP4	8HR	16VAQP4, 16VBQP4, 16VBTP4, 16VBUP4, 16VBZP4
16VBQP4	8HR	16VBZP4
16VBRP4	8HR	-
16VBTP4	8HR	16VAQP4, 16VBNP4, 16VBQP4, 16VBUP4, 16VBZP4
16VBUP4	8HR	16VBNP4, 16VBTP4, 16VBZP4
16VBWP4	8HR	16VBBXP4, 16VBYP4, 16VCXP4
16VBXP4	8HR	16VBWP4, 16VBYP4, 16VCXP4
16VBYP4	8HR	16VBWP4, 16VBBXP4, 16VCXP4
16VBZP4	8HR	16VBQP4
16VCAP4	7GR	-
16VCBBP4	-	-
16VCBP4	7GR	-
16VCDP4	8HR	-
16VCEP4	8HR	-
16VCFP4	8HR	-
16VCGP4	8HR	-
16VCHP4	8HR	-
16VCJP4	8HR	-
16VCKP4	8HR	16VCLP4%
16VCLP4	8HR	16VCKP4%
16VCNP4	8HR	-
16VCQP4	8HR	-
16VCRP4	8HR	-
16VCSP4	8HR	-
16VCWP4	8HR	-
16VCXP4	8HR	16VBWP4, 16VBXP4, 16VBYP4
16VCYP4	8HR	-
16VCZP4	8HR	-
16VP4	12D	16WP4?, 16YP4>
16WP4	12D	16WP4A, 16WP4B>, 16SP4>, 16SP4A>, 16VP4?
16WP4A	12N	16WP4B, 16SP4, cont'd
16WP4A, cont'd		16SP4A, 16WP4>, 16YP4?
16WP4B	12N	16WP4A, 16SP4, 16SP4A, 16WP4>, 16YP4?
16XP4	12D	16QP4, 16UP4?
16YP4	12N	16WP4A?, 16WP4B?, 16SP4?, 16SP4?, 16SP4A%, 16VP4<
16ZP4	12N	16LP4, 16LP4A, 16CP4<
17AP4	12N	17BP4B~
17ASP4	12N	-
17ATP4	12L	17TP4A, 17AVP4, 17AVP4A, 17BUP4, 17CBP4, 17CLP4, 17BJP4~, 17BKP4%, 17BKP4A%, 17BSP4~%, 17CUP4~%, 17DCP4~%
17ATP4A	12L	17ATP4, 17AVP4, 17AVP4A, 17BUP4, 17CBP4, 17CLP4, 17BJP4~, 17BKP4%, 17BKP4A%, 17BSP4~%, 17CUP4~%, 17DCP4~%, 17DJP4%
17AVP4	12L	17AVP4A, 17ATP4, 17ATP4A, 17BUP4, 17CBP4, 17CLP4, 17BJP4~, 17BKP4%, 17BKP4A%, 17BSP4~%, 17CUP4~%, 17DCP4~%, 17DJP4%
17AVP4A	12L	17AVP4, 17ATP4, 17ATP4A, 17BUP4, 17CBP4, 17CLP4, 17BJP4~, 17BKP4%, 17BKP4A%, 17BSP4~%, 17CUP4~%, 17DCP4~%, 17DJP4%
17BJP4	12L	17BSP4%, 17CUP4%, 17DCP4%
17BKP4	12L	17BKP4A, 17ATP4%, 17ATP4A%, 17AVP4%, 17AVP4A%, 17BJP4~%, 17BSP4~%, 17BUP4%, 17CBP4%, 17CLP4%, 17CUP4~%, 17DCP4~%, 17DJP4%
17BKP4A	12L	17BKP4, 17ATP4%, 17ATP4A%, 17AVP4%, 17AVP4A%, 17BJP4~%, 17BSP4~, 17BUP4%, 17CBP4%, 17CLP4%, 17CUP4~%, 17DCP4~%, 17DJP4%
17BMP4	12L	17BNP4~
17BMP4~	12L	-
17BNP4	12L	-
17BP4	12D	17BP4A>, 17BP4B>, 17BP4C>, 17BP4D>, 17AP4>, 17JP4>
17BP4A	12N	17BP4<, 17BP4B, 17BP4C, 17BP4D~, 17AP4, 17JP4
17BP4B	12N	17BP4<, 17BP4A, 17BP4C, 17BP4D~, 17AP4, 17JP4
17BP4C	12N	17BP4<, 17BP4A, 17BP4B, 17BP4D~, 17AP4, 17JP4
17BP4D	12N	17BP4~~<, 17BP4A~~, 17BP4B~~, 17BP4C~~, 17AP4~~, 17JP4~~
17BRP4	8HR	17BZP4~%, 17CAP4~, 17CKP4~, 17CTP4~%, 17CVP4~%, 17DTP4~%, 17DZP4~%
17BSP4	12L	17BJP4%, 17CUP4%, 17DCP4%
17BTP4	12AJ	-
17BUP4	12L	17ATP4, 17ATP4A, cont'd

Tube Type	Basing	Replacement
17BUP4, cont'd		17AVP4, 17AVP4A, 17CBP4, 17CLP4, 17BJP4~, 17BKP4%, 17BKP4A%, 17BSP4~, 17CUP4~%, 17DCP4~%, 17DJP4%
17BVP4	7FA	17BWP4~, 17BYP4~%, 17CSP4~
17BWP4	7FA	17CSP4, 17BYP4%
17BYP4	7FA	17BWP4%, 17BYP4%
17BZP4	8HR	17DSP4
17CAP4	8HR	17BZP4, 17CKP4, 17DTP4, 17BRP4~~, 17CTP4%, 17CVP4%, 17DZP4%
17CBP4	12L	17ATP4, 17ATP4A, 17AVP4, 17AVP4A, 17BUP4, 17CLP4, 17BJP4~, 17BKP4A%, 17BSP4~%, 17CUP4~%, 17DCP4~%, 17DJP4%
17CDP4	8HR	-
17CEP4	12L	17CFP4%
17CFP4	12L	17CEP4%
17CGP4	12L	-
17CKP4	8HR	17CAP4, 17BRP4~~, 17BZP4%, 17CTP4%, 17CVP4%, 17DTP4%, 17BZP4%, 17DSP4
17CLP4	12L	17ATP4, 17ATP4A, 17AVP4, 17AVP4A, 17BUP4, 17CBP4, 17BJP4~, 17BKP4%, 17BKP4A%, 17BSP4~%, 17CUP4~%, 17DCP4~%, 17DJP4%
17CMP4	12L	17CNP4, 17CRP4%, 17CXP4%
17CNP4	12L	17CMP4, 17CRP4%, 17CXP4%
17CP4	12D	17CP4A
17CP4A	12D	17CP4
17CRP4	12L	17CXP4, 17CMP4$, 17CRP4%
17CSP4	7FA	17BWP4, 17BYP4%
17CTP4	8HR	17BZP4%, 17CAP4%, 17CKP4%, 17CVP4%, 17DTP4%, 17DZP4%
17CUP4	12L	17DCP4, 17BJP4, 17BSP4, 17DJP4~~
17CVP4	8HR	17BZP4%, 17CAP4%, 17CKP4%, 17CTP4%, 17DTP4%, 17DZP4%
17CWP4	8HR	17DSP4, 17EFP4%
17CXP4	12L	17CRP4, 17CMP4%, 17CNP4%
17CYP4	12L	17BJP4, 17CZP4
17CZP4	12L	17BJP4, 17CYP4
17DAP4	8JK	17DRP4
17DBP4	12L	17DWP4~%, 17HP4%, 17HP4A%, 17HP4B%, 17HP4C~%, 17KP4%, 17KP4A%, 17RP4%, 17RP4C%
17DCP4	12L	17CUP4, 17BJP4%, 17BSP4%
17DEP4	8JN	-
17DHP4	8HR	17EBP4
17DJP4	12L	17ATP4%, 17ATP4A%, 17AVP4%, 17AVP4A%, 17JP4~%, 17BKP4%, 17BKP4A%, 17BSP4~%, 17BUP4%, 17CBP4%, 17CLP4%, 17CUP4~, 17DCP4~
17DKP4	8JR	17DSP4
17DLP4	8JS	17DSP4, 17EBP4%

Tube Type	Basing	Replacement
17DQP4	7FA	-
17DRP4	8JK	17DAP4
17DSP4	8HR	17EBP4%, 17EFP4%
17DTP4	8HR	17DKP4, 17DSP4
17DWP4	12L	17DWP4A
17DWP4A	12L	-
17DXP4	8JR	17DHP4, 17EBP4, 17DKP4%
17DZP4	8HR	17DTP4%
17EAP4	12AT	-
17EBP4	8HR	-
17EFP4	12L	17DXP4
17EHP4	8HR	-
17EJP22	14AU	17ENP22
17EKP4	12L	-
17ELP4	8HR	-
17ELP4A	8HR	-
17EMP4	8HR	16VABP4, 16VAQP4, 17EWP4
17ENP22	14AU	17EJP22
17EQP4	8HR	17FCP4
17ESP4	8HR	16VBCP4, 16VBPCP4A
17ETP22	14BE	16VACP22, 17FAP22
17EVP22	14BH	16VAFP22
17EWP4	8HR	16VABP4, 17EMP4
17EXP22	14BK	16VBMP22, 17FJP22
17EZP22	14BH	16VABP22, 16VACP22, 17FAP22
17FAP22	14BE	16VACP22
17FBP22	14BH	16VAKP22
17FCP4	8HR	17EQP4
17FDP4	8HR	16VAQP4
17FEP22	14BH	16VACP22, 17FAP22
17FGP22	14BH	16VACP22, 17FAP22
17FHP22	14BH	-
17FJP22	14BK	16VBMP22, 17EXP22
17FKP22	-	16VACP22, 17FAP22
17FP4	12L	17FP4A, 17HP4C~
17FP4A	12L	17FP4, 17HP4C~
17GP4	12M	-
17HP4	12L	17HP4A, 17HP4B, 17HP4C~, 17KP4, 17KP4A, 17RP4, 17RP4C, 17DBP4%, 17DWP4~
17HP4A	12L	17HP4, 17HP4B, 17HP4C~, 17KP4, 17KP4A, 17RP4, 17RP4C, 17DBP4%, 17DWP4~
17HP4B	12L	17HP4, 17HP4A, 17HP4C~, 17KP4, 17KP4A, 17RP4, 17RP4C, 17DBP4%, 17DWP4~
17HP4C	12L	17HP4~~, 17HP4A~~, 17HP4B~~, 17DWP4, 17KP4~~, 17KP4A~~
17JP4	12N	17AP4, 17BP4<, 17BP4A<, 17BP4B<, 17AP4C<, 17AP4D~
17KP4	12P	17KP4A
17KP4A	12P	17KP4
17LP4	12L	17LP4A, 17VP4, 17VP4B
17LP4A	12L	17LP4, 17LP4B~, 17VP4, 17VP4B
17LP4B	12L	17LP4~~, 17LP4A~~, 17VP4~~, 17VP4B~~
17QCP4	12N	-
17QP4	12N	17QP4A, 17QP4B~, 17YP4
17QP4A	12N	17QP4, 17QP4B~, 17YP4
17QP4B	12N	17QP4~~, 17QP4A~~, cont'd

Tube Type	Basing	Replacement
17QP4B, cont'd		17YP4~~
17RP4	12L	17RP4C, 17HP4, 17HP4A, 17HP4B, 17HP4C~, 17KP4, 17KP4A, 17DBP4%, 17DWP4~
17RP4C	12L	17RP4, 17HP4, 17HP4A, 17HP4B, 17HP4C~, 17KP4, 17KP4A, 17DBP4%, 17DWP4~
17SP4	12N	-
17TP4	12M	-
17UP4	12N	17QP4, 17QP4A, 17QP4B~, 17YP4
17VABP22	-	-
17VACP22	14BH	17VADP22
17VADP22	14BE	-
17VAEP22	-	-
17VAFP22	-	-
17VAGP22	-	17VAEP22
17VAHP22	-	-
17VAKP22	14BE	-
17VAMTC01	-	-
17VAMTC02	-	-
17VANP22	13D	17VAYP22
17VANTC01	-	17VAYTC01
17VANTC02	-	17VAYTC02
17VAQP22	14BP	17VAXP22, 17VBAP22
17VARP22	14BE	17VAUP22
17VASP22	14BE	-
17VATP22	13D	17VAZP22
17VATTC02	-	17VAZTC02
17VAUP22	14BE	17VARP22
17VAWP22	14BH	17VASP22
17VAXP22	14BP	17VBAP22
17VAYP22	13D	17VANP22
17VAYTC01	-	-
17VAYTC02	-	-
17VAZP22	13D	17VATP22
17VAZTC02	-	-
17VBAP22	-	-
17VBCP22	-	17VBAP22
17VBDP22	-	-
17VBGP22	-	17VBAP22
17VBHP22	-	17VBAP22
17VBLP22	13M	-
17VBMP22	13L	-
17VBNP22	13K	-
17VBQP22	-	17VBAP22
17VBRP22	-	17VBAP22
17VBSP22	-	17VBAP22
17VBTP22	13L	-
17VBUP22	-	17VBAP22
17VBWP22	-	17VBAP22
17VBXP22	-	17VBAP22
17VBYP22	-	17VBMP22
17VBZP22	-	17VBMP22
17VCDP22	-	-
17VDP22	-	-
17VP4	12L	17VP4B, 17LP4, 17LP4A, 17LP4B~
17VP4B	12L	17VP4, 17LP4, 17LP4A, 17LP4B~
17YP4	12N	17QP4, 17QP4A, 17QP4B~
18VABP22	14BE	18VAJP22, 18VASP22, 18VATP22
18VACP22	-	18VASP22, 18VAHP22
18VADP22	-	18VASP22
18VAEP22	-	18VBMP22
18VAFP22	-	18VBJP22
18VAGP22	-	18VBDP22
18VAHP22	14BE	18VASP22
18VAJP22	14BE	18VAHP22, 18VASP22, 18VATP22, 18VBHP22, 18VBKP22, 18VBTP22
18VAKP22	14BE	-
18VALP22	14BH	18VAHP22, 18VATP22, 18VBDP22, 18VBGP22, 18VBKP22, 19JSP22
18VAMP22	14BH	18VBDP22, 18VBGP22
18VANP22	13C	-
18VAQP22	-	18VASP22
18VARP22	14BE	18VAHP22, 18VASP22
18VASP22	14BE	18VAJP22, 18VBKP22
18VATP22	14BE	18VAJP22, 18VBTP22, 18VASP22
18VAZP22	14BH	18VAZP22
18VBAP22	-	18VASP22
18VBCP22	14BE	18VAHP22, 18VASP22, 19HFP22
18VBDP22	14BH	18VAMP22, 18VBGP22
18VBEP22	14BE	-
18VBFP22	14BE	18VAKP22
18VBGP22	14BH	18VALP22, 18VBDP22, 19JSP22
18VBHP22	14BE	18VAJP22, 18VASP22, 18VBKP22%, 19HXP22
18VBJP22	14BE	-
18VBKP22	14BE	18VAJP22, 18VASP22
18VBMP22	14BE	-
18VBQP4	8HR	-
18VBTP22	14BE	18VAJP22, 18VASP22, 18VATP22, 19HCP22, 19JZP22
18VBUP22	14BH	18VAZP22
18VBWP22	-	18VAKP22
18VBXP22	14BE	-
19ABP4	8JK	-
19ACP4	8HR	19CHP4, 19CKP4, 19DLP4, 19FVP4+%
19AEP4	8HR	-
19AFP4	8HR	19AUP4, 19BAP4%, 19BCP4%
19AHP4	8HR	19ALP4%, 19AYP4
19AJP4	7FA	19CDP4%, 19CQP4%, 19CXP4%
19ALP4	8HR	19AHP4%
19ANP4	8JR	19AYP4, 19YP4%
19AP4	12D	19AP4A, 19AP4B, 19AP4C, 19AP4D
19AP4A	12D	19AP4, 19AP4B, 19AP4C, 19AP4D
19AP4B	12D	19AP4, 19AP4A, 19AP4C, 19AP4D
19AP4C	12D	19AP4, 19AP4A, 19AP4B, 19AP4D
19AP4D	12D	19AP4, 19AP4A, 19AP4B, 19AP4C
19AQP4	8HR	19AXP4%, 19BLP4%, 19XP4%
19ARP4	8HR	19ASP4%, 19CEP4%
19ASP4	8HR	19CEP4, 19ARP4%
19ATP4	8JR	19AFP4, 19AUP4, 19BAP4%, 19BCP4%, 19BMP4%
19AUP4	8HR	19AFP4, 19BAP4%, 19BCP4%
19AVP4	8HR	19AYP4%
19AXP4	8HR	19AYP4, 19AVP4%, 19XP4%
19AYP4	8HR	19AVP4%

Tube Type	Basing	Replacement
19BAP4	8HR	19BCP4, 19AFP4%, 19AUP4%
19BDP4	12L	-
19BEP4	8HR	19CTP4, 19DJP4, 19BSP4%
19BFP4	12L	-
19BHP4	8HR	19AVP4
19BLP4	8HR	19AVP4, 19CYP4
19BMP4	8HR	19AFP4, 19AUP4, 19BAP4%, 19BCP4%
19BNP4	8HR	19BQP4, 19CVP4%, 19DYP4%, 19GMP4%
19BQP4	8HR	19BNP4, 19CVP4%, 19DYP4%, 19GMP4%
19BRP4	8HR	19DKP4, 19CZP4%, 19DAP4%, 19DEP4
19BSP4	8HR	19AVP4, 19BEP4%, 19CTP4%, 19DJP4%
19BTP4	8JR	19AVP4, 19CYP4
19BUP4	-	-
19BVP4	8HR	19AVP4, 19BWP4%
19BWP4	8HR	19AYP4, 19BVP4%
19CAP4	8JR	19AVP4, 19BTP4
19CDP4	7FA	19CQP4, 19CXP4, 19AJP4%
19CEP4	8HR	19BAP4, 19BCP4, 19AUP4%, 19BMP4%
19CFP4	8HR	19CHP4, 19CKP4, 19DLP4, 19FVP4+%
19CGP4	12L	-
19CHP4	8HR	19CKP4, 19DLP4, 19FVP4+%
19CJP4	8HR	19AVP4
19CKP4	8HR	19CHP4, 19FVP4+%
19CLP4	12L	19BDP4, 19CRP4
19CMP4	8HR	19CMP4A, 19CHP4%
19CMP4A	8HR	19CUP4
19CP22	14BE	18VAJP22
19CQP4	7FA	19CDP4, 19CXP4, 19AJP4%
19CUP4	8HR	19DFP4%
19CVP4	8HR	19DYP4, 19GMP4
19CXP4	7FA	19CDP4, 19CQP4, 19AJP4%
19CYP4	8HR	19AVP4
19CZP4	8HR	19DAP4, 19DKP4%
19DAP4	8HR	19CZP4, 19DKP4%
19DBP4	7FA	-
19DCP4	8HR	19EDP4, 19EUP4, 19DWP4%, 19FCP4%, 19FTP4%,19GJP4%, 19GJP4%, 19GJP4A%
19DEP4	8HR	19AVP4
19DFP4	8HR	19CHP4, 19CUP4%
19DHP4	8HR	19DSP4, 19EFP4, 19ESP4, 19DUP4%, 19EAP4%, 19ENP4%, 19ENP4A%, 19FUP4+%
19DJP4	8HR	19BEP4, 19BSP4
19DKP4	8HR	19CZP4%, 19DAP4%
19DLP4	8HR	19CHP4, 19CKP4, 19EDP4
19DNP4	8HR	19DKP4, 19CZP4%, 19DAP4%, 19EDP4
19DP4	12N	19DP4A, 19FP4<?, 19GP4<
19DP4A	12N	19DP4, 19FP4<?, 19GP4<
19DQP4	8HR	19DRP4%, 19FLP4+, 19FRP4%
19DRP4	8HR	19DQP4%, 19FLP4+%, 19FRP4%
19DSP4	8HR	19DHP4, 19EFP4, 19ESP4, 19DUP4%, 19EAP4%, 19ENP4%, 19ENP4A%, 19FUP4+%
19DUP4	8HR	19ENP4, 19ENP4A, 19GHP4
19DVP4	8HR	19DZP4
19DWP4	8HR	19EVP4, 19FCP4, 19GJP4, 19GJP4A, 19EDP4%
19DYP4	8HR	19CVP4, 19GMP4
19DZP4	8HR	19BVP4, 19DVP4
19EAP4	8HR	19DUP4, 19ENP4, 19ENP4A, 19DHP4%, 19DSP4%, 19EFP4%, 19ESP4%
19EBP4	8HR	19GFP4+%
19ECP4	8HR	19FTP4, 19HGP4+
19EDP4	8HR	19DWP4% 19FCP4%, 19GJP4%, 19GJP4A%
19EFP4	8HR	19DHP4, 19DSP4, 19ESP4, 19DUP4%, 19EAP4%, 19ENP4%, 19ENP4A%, 19GHP4%
19EGP4	8HR	19HAP4%
19EHP4	8HR	19EHP4A, 19DCP4, 19FJP4%
19EHP4A	8HR	19EHP4, 19DCP4, 19FJP4%
19EJP4	8HR	19FEP4, 19FEP4A, 19FEP4B
19EKP4	8HR	19EZP4
19ELP4	8HR	19AVP4
19ENP4	8HR	19ENP4A, 19DUP4, 19EAP4, 19DHP4%, 19DSP4%, 19EFP4%, 19ESP4%, 19GHP4
19ENP4A	8HR	19ENP4, 19DUP4, 19EAP4, 19DHP4%, 19DSP4%, 19EFP4%, 19ESP4%, 19GHP4
19EP4	12D	19JP4
19ESP4	8HR	19DHP4, 19DSP4, 19EFP4, 19DUP4%, 19EAP4%, 19ENP4%, 19ENP4A%, 19GHP4
19ETP4	8HR	-
19EUP4	8HR	19DWP4%, 19EDP4, 19FCP4%, 19FTP4%, 19GJP4%, 19GJP4A%
19EXP22	14BE	18VAHP22, 19EYP22, 19GVP22, 19GYP22
19EYP22	14BE	18VAHP22, 18VASP22, 19HCP22, 19FMP22
19EZP4	7FA	19EKP4
19FAPA	8HR	-
19FBP4	8HR	19EGP4, 19HAP4%
19FCP4	8HR	19DWP4, 19GJP4, 19GJP4A, 19EDP4%
19FDP4	-	130V
19FEP4	8HR	19FEP4A, 19FEP4B, 19EJP4
19FEP4A	8HR	19FEP4B
19FEP4B	8HR	19FEP4A
19FGP4	8JR	19EBP4
19FHP4	8HR	-
19FJP4	8HR	19DCP4%, 19EHP4%, 19EHP4A%
19FJP4A	8HR	19DWP4, 19EDP4%, 19FCP4, 19GJP4, 19GJP4A, 19GRP4
19FLP4	8HR	19DQP4+, 19DRP4+, 19FRP4+%, 19FTP4
19FMP22	-	19EYP22
19FNP4	8HR	19GEP4%, 19GKP4%
19FP4	12D	19GP4
19FQP4	8HR	-
19FRP4	8HR	19DQP4%, 19DRP4%, 19FLP4+%
19FSP4	8HR	-
19FTP4	8HR	18VBQP4, 19DWP4, 19FCP4, 19GJP4, 19GJP4A%, cont'd

Tube Type	Basing	Replacement
19FTP4, cont'd		19EDP4%, 19EUP4%, 19FTP4A
19FTP4A	8HR	18VBQP4, 19DCP4%, 19DWP4, 19EDP4, 19EUP4, 19FCP4, 19GJP4, 19GJP4A
19FWP4	8HR	-
19FXP22	14BE	18VAHP22, 19HCP22, 19GWP22
19FYP4	-	-
19FZP4	8HR	-
19GAP4	8HR	-
19GBP4	8HR	19CZP4, 19DAP4, 19DKP4%
19GDP22	14BE	19GZP22
19GEP4	8HR	19FNP4%, 19GEP4A
19GEP4A	8HR	19FNP4%, 19GEP4
19GFP4	8HR	19EBP4+%
19GHP4	8HR	19DHP4%, 19DSP4%, 19DUP4, 19EAP4, 19EFP4%, 19ENP4, 19ENP4A, 19ESP4%
19GJP4	8HR	19GJP4A, 19DWP4, 19EDP4%, 19FCP4
19GJP4A	8HR	19GJP4, 19DWP4, 19EDP4%, 19FCP4
19GKP4	8HR	19FNP4%, 19GEP4%
19GLP22	14BE	18VAHP22%
19GMP4	8HR	19CVP4, 19DYP4
19GP4	12D	19FP4?
19GQP4	8HR	-
19GSP22	14BE	18VAHP2%, 18VASP22, 19EYP22
19GTP4	8HR	-
19GVP22	14BE	18VAHP22^, 19EXP22, 19EYP22, 19GYP22
19GWP22	14BE	18VAHP22, 18VASP22, 19HBP22, 19HCP22, 19EYP22, 19HRP22, 19JYP22
19GXP22	-	18VAHP22, 19EXP22, 19EYP22
19GYP22	14BE	18VAHP22, 19EXP22, 19EYP22, 19GVP22, 19HQP22
19GZP22	14BE	18VAHP2, 18VASP22, 19GDP22, 19EYP22, 19HCP22
19HAP4	8HR	19EGP4%
19HBP22	14BE	18VAHP22, 18VASP22, 19EYP22, 19HCP22, 19GWP22, 19HRP22, 19JYP22
19HCP22	14BE	18VAJP22, 18VASP22, 18VATP22, 18VBTP22, 19HCP22, 19JZP22
19HDP22	-	-
19HFP22	14BE	18VASP22, 18VBCP22, 18VAHP22, 19EYP22, 19HCP22
19HGP4	8HR	19ECP4+
19HJP22	14BE	18VABP22, 18VASP22, 19HCP22
19HJP22A	-	-
19HKP22	-	18VASP22, 19HCP22
19HLP22	-	18VBMP22, 19HMP22
19HMP22	14BE	18VBMP22
19HNP22	14BH	18VAZP22
19HQP22	14BE	18VAHP22, 19GYP22, 19EXP22, 19EYP22
19HRP22	14BE	18VAHP22, 19EYP22, 19HCP22, 19HBP22, 19GWP22, 19JYP22
19HSP22	14BH	-
19HTP22	14BH	18VBJP22
19HXP22	14BE	18VAJP22, 18VASP22, 18VBHP22, 19HCP22
19HYP22	14BH	18VAMP22, 18VBDP22
19JAP22	14BH	18VBDP22
19JBP22	14BE	18VAHP22, 19EXP22, 19EYP22
19JCP22	14BH	18VBJP22
19JDP22	-	18VAHP22, 19EXP22, 19EYP22
19JEP22	-	-
19JFP22	-	-
19JGP22	-	18VAZP22, 19HNP22, 19JWP22
19JHP22	-	18VAHP22, 18VASP22, 19EYP22, 19HCP22
19JKP22	-	18VAHP22, 18VASP22, 19EYP22, 19HCP22
19JLP22	14BE	-
19JNP22	14BE	18VABP22, 18VAJP22, 18VASP22, 18VAMP22, 19HCP22
19JP4	12D	19EP4
19JQP22	-	18VAHP22, 19EXP22, 19EYP22
19JSP22	14BH	18VALP22, 18VBGP22, 18VBDP22
19JVP22	14BH	-
19JWP22	14BH	18VAZP22, 19HNP22
19JXP22	-	18VANP22
19JYP22	14BE	18VAHP22, 18VASP22, 19EYP22, 19HCP22
19JZP22	14BE	18VAJP22, 18VASP22, 18VATP22, 18VBTP22, 19HCP22
19KAP22	14BE	18VBMP22, 19HMP22
19KBP22	-	18VBGP22, 18VBDP22
19KCP22	-	18VBDP22, 18VBGP22
19KDP22	-	18VBJP22, 19HTP22
19KEP22	-	18VAHP22, 18VASP22, 19EYP22, 19HCP22
19KFP22	-	18VASP22, 19HCP22
19KGP22	-	18VBMP22, 19HMP22
19KHP22	-	18VBJP22, 19HTP22
19KJP22	-	18VBDP22, 18VBGP22
19KLP22	14BE	18VAKP22, 18VBMP22, 19HMP22
19QP4	12L	-
19TP22	20A	-
19VABP22	-	RE21FJP22
19VACP22	-	RE21FJP22
19VADP22	13C	-
19VAFP22	14BE	19VEDP22
19VAGP22	14BE	19VAFP22
19VAHP4	8HR	19VBXP4, 19VCJP4, 19VCLP4, 19VCMP4, 19VGXP4, 19HGP4
19VAJP4	8HR	-
19VAKP22	14BE	19VATP22, 19VBRP22, 19VEKP22
19VALP4	8HR	-
19VAMP22	14BE	19VATP22, 19VBRP22, 19VEKP22
19VANP22	14BH	19VBQP22
19VAQP22	14BE	19VBSP22, 19VCBP22, 19VCSP22, 19VDUP22, 19VEFP22, 19VEGP22
19VARP4	8HR	19VDHP4, 19VENP4
19VASP4	8HR	-

Tube Type	Basing	Replacement
19VATP22	14BE	19VBRP22, 19VBWP22, 19VCFP22, 19VCTP22, 19VCYP22, 19VDQP22, 19VDTP22, 19VDXP22, 19VEAP22, 19VEDP22, 19VEKP22, 19VEUP22, 19VHJP22
19VAUP22	14BE	19VATP22, 19VBRP22, 19VCFP22, 19VEKP22
19VAWP22	14BE	19VATP22, 19VBRP22, 19VEKP22
19VAXP22	14BE	19VATP22, 19VBRP22, 19VEKP22, 19VBJP22
19VAYP22	14BH	19VECP22
19VAZP22	14BH	-
19VBAP22	14BE	-
19VBCP22	14BH	19VECP22
19VBDP22	14BE	19VATP22, 19VBRP22, 19VEKP22
19VBFP22	14BE	19VATP22, 19VBRP22, 19VEKP22
19VBGP22	14BE	-
19VBHP22	14BH	19VAZP22
19VBJP22	14BE	19VATP22, 19VAXP22, 19VBRP22, 19VEKP22
19VBKP22	-	-
19VBLP22	13C	-
19VBNP4	8HR	19VEMP4, 20AHP4
19VBNP4A	8HR	19VEMP4, 20AHP4
19VBQP22	14BH	19VCFP22, 19VCNP22, 19VCTP22, 19VCXP22, 19VDFP22, 19VDQP22, 19VDYP22, 19VEAP22, 19VEDP22, 19VHJP22
19VBRP22	-	-
19VBSP22	14BE	19VAQP22, 19VCBP22, 19VCSP22, 19VDMP22, 19VDUP22, 19VEFP22, 19VEGP22
19VBUP22	-	-
19VBWP22	14BE	19VATP22%, 19VBRP22%, 19VCRP22, 19VCTP22%, 19VDFP22%, 19VDQP22%, 19VEAP22%, 19VEBP22%, 19VEKP22
19VBXP4	8HR	19VAHP4, 19VCGP4, 19VCJP4, 19VCLP4, 19VCMP4
19VBYP4	8HR	-
19VBZP4	8HR	19VCMP4, 19VDGP4
19VCAP22	14BH	19VANP22
19VCBP22	14BE	19VCSP22, 19VDMP22, 19VEFP22, 19VEGP22
19VCFP22	14BE	19VATP22, 19VAUP22, 19VBRP22, 19VCNP22, 19VCTP22, 19VCXP22, 19VCYP22, 19VDSP22, 19VDWP22, 19VDYP22, 19VEKP22
19VCGP4	8HR	19VAHP4, 19VBXP4, 19VCJP4, 19VCLP4, 19VCMP4
19VCHP4	8HR	-
19VCJP4	8HR	19VAHP4, 19VBXP4, 19VCJP4, 19VCLP4, 19VCMP4
19VCKP4	8HR	19VEDP4, 20XP4
19VCLP4	8HR	19VBXP4, 19VCGP4, 19VCJP4, 19VCMP4, 20YP4, 20ZP4
19VCMP4	8HR	19VBXP4, 19VCGP4, cont'd
19VCMP4, cont'd		19VCJP4, 19VCMP4, 20YP4, 20ZP4
19VCNP22	14BE	19VBRP22, 19VCFP22, 19VDYP22, 19VEKP22
19VCQP22	-	19VDZP22
19VCQP22A	14BE	19VAFP22, 19VDZP22
19VCRP22	14BE	19VBRP22, 19VEKP22, 19VDQP22%, 19VEAP22%, 19VEDP22%
19VCSP22	14BE	19VAQP22, 19VCBP22, 19VDMP22, 19VDUP22, 19VEFP22, 19VEGP22, 19VEWP22
19VCTP22	14BE	19VBRP22, 19VCFP22, 19VD0% 19VDQP22, 19VDSP22, 19VEAP22, 19VEDP22, 19VHBP22, 19VHJP22, 19VEKP22
19VCUP4	8HR	19VARP4
19VCWP22	13C	-
19VCXP22	14BE	19VBRP22, 19VCFP22, 19VEKP22
19VCYP22	14BE	19VATP22, 19VBRP22, 19VBRP22, 19VCTP22, 19VDTP22, 19VDXP22
19VDAP4	8HR	-
19VDBP22	-	-
19VDCP22	14BP	19VFBP22, 19VFXP22
19VDEP4	8HR	19VCKP4
19VDFP22	14BE	19VBRP22, 19VCTP22, 19VDQP22, 19VEAP22, 19VEDP22, 19VEHP22, 19VEKP22
19VDGP4	8HR	19VBZP4
19VDHP4	8HR	19VARP4
19VDKP22	13D	-
19VDKTC02	-	19VEJTC02
19VDLTC02	-	19VEJTC02
19VDMP22	14BE	19VAQP22, 19VCBP22, 19VCSP22, 19VDUP22, 19VEFP22, 19VEGP22, 19VEWP22
19VDNP22	14BE	19VAQP22, 19VCBP22, 19VEGP22
19VDQP22	14BE	19VBRP22, 19VBWP22%, 19VEKP22, 19VEAP22, 19VEBP22
19VDRP22	13D	19VEMP22, 19VGMP22
19VDSP22	14BE	19VCFP22, 19VCTP22, 19VEDP22, 19VEHP22, 19VHJP22
19VDTP22	14BE	19VBRP22, 19VCTP22, 19VDFP22, 19VDQP22, 19VDSP22, 19VEDP22, 19VHJP22, 19VSP22, 19VATP22, 19VEKP22
19VDUP22	14BE	19VCSP22, 19VDMP22, 19VEFP22, 19VEGP22, 19VEWP22, 19VBRP22, 19VEKP22
19VDWP22	14BE	19VATP22, 19VCFP22, 19VCTP22, 19VDFP22, 19VEAP22, 19VEDP22, 19VHBP22, 19VHJP22, 19VBRP22, 19VEKP22
19VDXP22	14BE	19VATP22, 19VBRP22, 19VCRP22, cont'd

Tube Type	Basing	Replacement
19VDXP22, cont'd		19VDFP22, 19VDQP22, 19VEKP22, 19VSP22
19VDYP22	14BE	19VBRP22, 19VCFP22, 19VCNP22, 19VEKP22
19VDZP22	14BE	-
19VEAP22	14BE	19VATP22, 19VBRP22, 19VBWP22%, 19VCRP22%, 19VCTP22, 19VDFP22, 19VEBP22
19VEB22	-	19VATP22
19VEBP22	14BE	19VBWP22%, 19VDQP22, 19VEAP22
19VECP22	-	-
19VEDP22	14BE	19VEHP22, 19VEKP22, 19VHBP22, 19VHJP22
19VEFP22	14BE	19VCBP22, 19VCSP22, 19VDMP22, 19VDUP22, 19VEGP22, 19VEWP22
19VEGP22	14BE	19VEWP22
19VEHP22	14BE	19VETP22, 19VHBP22, 19VEKP22
19VEJP22	13D	-
19VEJTC01	-	-
19VEJTC02	-	-
19VEJTC05	-	-
19VEKP22	14BE	19VGDP22, 19VELP22, 19VFAP22, 19VHBP22, 19VHJP22
19VELP22	14BE	19VEDP22, 19VEKP22, 19VHBP22, 19VHJP22
19VEMP4	8HR	19VBNP4, 19VBNP4A
19VENP4	8HR	19VARP4, 19VDHP4, 19VERP4
19VEQP22	13D	19VDRP22, 19VGMP22
19VERP4	8HR	19VARP4, 19VDHP4, 19VENP4
19VESP22	-	19VEKP22
19VETP22	14BE	19VEHP22, 19VEKP22, 19VHBP22
19VEUP22	14BE	19VBRP22, 19VEDP22, 19VEHP22, 19VEKP22, 19VHBP22, 19VJP22
19VEWP22	14BE	19VEGP22
19VEXP22	14BE	19VEWP22
19VEZP22	14BE	19VEDP22, 19VEKP22, 19VELP22, 19VHBP22, 19VHJP22
19VFAP22	14BE	19VEDP22, 19VEKP22, 19VELP22, 19VFAP22, 19VHBP22, 19VHP22
19VFBP22	14BP	19VFJP22
19VFDP4	8HR	-
19VFEP4	8HR	-
19VFGP4	8HR	-
19VFHP22	-	19VHTP22
19VFJP22	14BP	19VFBP22
19VFKP22	13L	19VBRP22, 19VEKP22
19VFLP22	-	-
19VFMP22	-	-
19VFQP22	14BE	19VEGP22
19VFSP22	-	19VEKP22
19VFTP4	8HR	-
19VFWP4	8HR	-
19VFXP22	14BP	19VDCP22, 19VFBP22
19VFYP4	8HR	-
19VFZP4	8HR	-
19VGAP4	8HR	-
19VGCP4	8HR	19VGFP4%
19VGDP4	8HR	-
19VGEP4	8HR	-
19VGFP4	8HR	19VGCP4%
19VGKP22	13L	-
19VGKTC02	-	19VEJTC02
19VGLP22	14BP	19VFBP22
19VGMP22	13D	-
19VGNP22	-	19VFBP22
19VGP22	-	19VBAP22
19VGQP22	13G	19VHFP22
19VGQTC01	-	-
19VGRP22	-	19VFBP22
19VGSP4	8HR	-
19VGTP22	-	19VFBP22
19VGUP4	8HR	-
19VGXP4	8HR	19VAHP4
19VGYP4	8HR	19VEHP4
19VGZP22	13K	-
19VHBP22	14BE	19VEHP22, 19VEKP22, 19VETP22, 19VEUP22, 19VHJP22
19VHCP22	14BE	-
19VHDP22	-	-
19VHEP4	8HR	19VGYP4
19VHFP22	13G	19VGQP22
19VHFTC01	-	19VEJTC01
19VHFTC02	-	19VEJTC02
19VHGP4	8HR	19VAHP4, 19VGXP4
19VHJP22	14BE	19VEDP22, 19VEKP22, 19VEUP22, 19VHBP22
19VHKP22	13G	-
19VHNP22	-	19VFBP22
19VHSP22	13K	19VJNP22
19VHTP22	13K	-
19VHUP4	8HR	-
19VHWP22	13B	19VLBP22
19VHXP22	14BP	19VFBP22
19VHYP22	13L	-
19VHZP22	-	19VJCP22
19VJAP4	8HR	-
19VJBP22	13M	19VJCP22, 19VJQP22
19VJCP22	-	-
19VJDP22	13P	-
19VJFP22	13N	19VJTP22
19VJGP4	8HR	-
19VJKP22	-	19VJMP22
19VJLP22	-	-
19VJMP22	13N	19VJTP22
19VJNP22	13K	-
19VJQP22	13M	19VJBP22
19VJRP22	-	19VGZP22
19VJSP22	14BE	-
19VJTP22	13N	19VJMP22
19VJUP22	13N	19VJMP22
19VJWP22	13L	-
19VJXP22	-	-
19VJYP22	13M	19VJXP22
19VKDP22	13N	19VJMP22
19VKEP22	13N	-
19VKFP22	-	-
19VKGP22	-	19VLNP22
19VKHP22	-	19VFBP22
19VKJP22	-	19VFBP22
19VKLP22	-	19VFBP22
19VKMP22	-	19VMFP22
19VKNP22	14BE	19VEGP22
19VKRP22	14BE	19VKTP22
19VKSP22	13N	19VJTP22

Tube Type	Basing	Replacement
19VKTP22	14BE	19VKRP22
19VKUP22	-	19VJTP22
19VKWP22	13B	-
19VKXP22	-	19VJDP22
19VKYP22	-	-
19VLAP22	14BE	19VEKP22
19VLBP22	13B	-
19VLCP22	-	-
19VLDP22	-	19VJTP22
19VLEP22	-	-
19VLGP22	-	19VJTP22
19VLHP22	-	19VLEP22
19VLKP22	-	-
19VLMP22	-	19VKXP22, 19VJDP22
19VLNP22	-	-
19VLP22	-	19VJMP22
19VLQP22	-	19VMEP22
19VLRP22	-	-
19VLSP22	-	-
19VLTP22	-	19VMFP22
19VLUP22	-	-
19VLWP22	-	19VJTP22
19VLXP22	-	-
19VLYP22	-	19VLSP22
19VLZP22	-	19VMFP22
19VMAP22	-	19VJTP22
19VMBP22	-	19VJTP22
19VMCP22	-	19VMFP22
19VMDP22	-	-
19VMEP22	-	-
19VMFP22	-	-
19VMGP22	-	-
19VMHP22	-	19VLXP22
19VMJP22	-	19VLXP22
19VMKP22	-	19VMFP22
19VMLP22	-	19VMFP22
19VMNP22	-	19VLXP22
19VMRP22	-	19VMFP22
19VMSP22	-	19VMFP22
19VMTP22	-	19VLBP22
19VMWP22	-	-
19VMYCP22	-	-
19VMYP22	-	19VMFP22
19VMZP22	-	-
19VNAP22	-	19VLXP22
19VNBP22	-	19VMFP22
19VNCP22	-	19VMFP22
19VNDP22	-	19VMFP22
19VNEP22	-	19VMFP22
19VNFP22	-	-
19VNHP22	-	-
19VNJP22	-	19VMFP22
19VP22	14W	-
19VSP22	14BE	-
19VUP4	8HR	-
19XP4	8HR	19AVP4, 19AXP4%, 19AYP4%
19YP4	8JR	19ANP4%, 19AVP4
19ZP4	8JS	19AQP4%, 19AVP4, 19BHP4
20ABP4	8HR	19VCJP4, 19VCLP4, 19VCMP4, 20RP4, 20YP4, 20ZP4
20ADP4	8HR	19VERP4
20AEP4	8HR	19VCMP4, 20SP4+
20AFP4	8HR	20AFP4A
20AFP4A	8HR	-
20AGP4	8HR	-
20AHP4	8HR	19VBUP4, 19VBNP4A, 19VEMP4
20AKP4	8HR	-
20AZP4	-	-
20BP4	12D	-
20CP4	12D	20CP4B, 20CP4C, 20DP4, 20DP4B, 20CP4A>, 20CP4D>, 20DP4A> 20DP4C>, 20DP4D~>
20CP4A	12N	20CP4D, 20DP4A, 20DP4C, 20CP4<, 20CP4B<, 20CP4C<, 20DP4<, 20DP4B<, 20DP4D~
20CP4B	12D	20CP4, 20CP4C, 20DP4, 20DP4B, 20CPA>, 20CPAD>, 20DP4A>, 20DP4C>, 20DP4D~>
20CP4C	12D	20CP4, 20CP4B, 20DP4, 20DP4B, 20CP4A>, 20CP4D>, 20DP4A>, 20DP4C>, 20DP4D~>
20CP4D	12N	20CP4A, 20DP4A, 20DP4C, 20CP4<, 20CP4B<, 20CP4C<, 20DP4<, 20DP4B<, 20DP4D~
20DP4	12D	20DP4B, 20CP4B, 20CP4C, 20DP4A>, 20DP4C>, 20DP4D~>, 20CP4A>, 20CP4D>
20DP4A	12N	20DP4C, 20CP4A, 20CP4D, 20DP4<, 20DP4B<, 20DP4D~, 20CP4<, 20CP4B<, 20CP4C<
20DP4B	12D	20DP4, 20CP4, 20CP4B, 20CP4C, 20DP4A>, 20DP4C>, 20DP4D~>, 20CP4A>, 20CP4D>
20DP4C	12N	20DP4A, 20CP4A, 20CP4D, 20DP4<, 20DP4B<, 20DP4D~, 20CP4<, 20CP4B<, 20CP4C<
20DP4D	12N	20DP4A~~, 20DP4C~~, 20CP4A~~, 20CP4A~~, 20CP4D~~
20FP4	12M	20GP4>
20GP4	12L	20FP4<
20HP4	12M	20HP4B, 20HP4C, 20HP4A>, 20HP4D>, 20HP4E~>, 20JP4>, 20LP4>, 20MP4>
20HP4A	12L	20HP4D, 20JP4, 20LP4, 20MP4, 20HP4<, 20HP4B<, 20HP4C<, 20HP4E~
20HP4B	12M	20HP4, 20HP4C, 20HP4A>, 20HP4D>, 20HP4E~>, 20JP4>, 20LP4>, 20MP4>
20HP4C	12M	20HP4, 20HPB, 20HPA>, 20HPD>, 20HP4E~>, 20JP4>, 20LP4>, 20MP4>
20HP4D	12L	20HP4A, 20JP4, 20LP4, 20MP4, 20HP4>, 20HP4B>, 20HP4C>, 20HP4E~
20HP4E	12L	20HP4A~~, 20JP4~~, 20LP4~~, 20MP4~~
20JP4	12P	-
20LP4	12L	20HP4A, 20HP4D, 20JP4, 20MP4, 20HP4<, 20HP4B<, 20HP4C<, 20HP4E~
20MP4	12L	20HP4A, 20HP4C<, 20HP4E~ 20HP4D, 20JP4, cont'd

Tube Type	Basing	Replacement
20MP4, cont'd		20LP4, 20HP4<, 20HP4B<,
20RP4	8HR	19VCJP4, 19VCLP4, 19VCMP4, 19VEMP4, 20AB4, 20YP4, 20ZP4
20SP4	8HR	19VCJP4, 19VCLP4, 19VCMP4, 20AEP4+
20TP4	8HR	20XP4
20UP4	8HR	19VALP4
20VABP22	-	-
20VACP22	-	-
20VADP22	14BE	20VABP22, 20VAGP22
20VAEP22	14BE	20VABP22
20VAFP22	14BE	20VACP22
20VAGP22	14BE	20VADP22, 20VABP22
20VAHP22	14BH	22ALP22
20VAJP22	14BE	20VABP22, 20VASP22, 22ALP22
20VAKP22	14BE	20VABP22, 20VALP22
20VALP22	14BE	20VABP22, 20VAKP22
20VAMP22	14BE	20VANP22, 20VABP22, 20VAGP22
20VANP22	14BE	20VABP22, 20VAGP22, 20VAMP22
20VAQP4	8HR	-
20VARP22	-	20VABP22, 20VAGP22
20VASP22	14BE	20VABP22, 20VAFP22
20VATP4	8HR	-
20WP4	8HR	-
20WP4A	8HR	20WP4
20XP4	8HR	19VCKP4
20YP4	8HR	19VCLP4, 19VCMP4, 20ABP4, 20RP4, 20ZP4
20ZP4	8HR	19VCMP4, 20ABP4, 20RP4, 20YP4
21ACP4	12N	21ACP4A, 21AMP4, 21AMP4A, 21BSP4, 21CUP4, 21AMP4B~
21ACP4A	12N	21ACP4, 21AMP4, 21AMP4A, 21BSP4, 21CUP4, 21AMP4B~
21AFP4	12M	21BCP4~>, 21YP4>, 21YP4A>, 21YP4B~>
21ALP4	12L	21ALP4A, 21ALP4B, 21ATP4, 21ATP4A, 21ATP4B, 21BTP4, 21CMP4, 21CWP4, 21DNP4
21ALP4A	12L	21ALP4, 21ALP4B, 21ATP4, 21ATP4A, 21ATP4B, 21BTP4, 21CMP4, 21CWP4, 21DNP4
21ALP4B	12L	21ALP4, 21ALP4A, 21ATP4, 21ATP4A, 21ATP4B, 21BTP4, 21CMP4, 21CWP4, 21DNP4
21AMP4	12N	21AMP4A, 21ACP4, 21ACP4A, 21BSP4, 21CUP4, 21AMP4B~
21AMP4A	12N	21AMP4, 21ACP4, 21ACP4A, 21BSP4, 21CUP4, 21AMP4B~
21AMP4B	12N	21AMP4~~, 21AMP4A~~, 21ACP4~~, 21ACP4A~~, 21BSP4~~, 21CUP4~~
21ANP4	12M	21ANP4A
21ANP4A	12M	21ANP4
21AP4	12D	-
21AQP4	12D	21AQP4A, 21ACP4>, 21ACP4A>, 21AMP4>, 21AMP4A>, 21AMP4B~>, 21BSP4>, 21CUP4>
21AQP4A	12D	21AQP4, 21ACP4>, 21ACP4A>, 21AMP4>, 21AMP4A>, 21AMP4B~>, 21BSP4>, 21CUP4>
21ARP4	12N	21ARP4A
21ARP4A	12N	21ARP4
21ASP4	12M	21AYP4>, 21XP4>, 21XP4A>, 21XP4B~>
21ATP4	12L	21ATP4A, 21ATP4B, 21ALP4, 21ALP4A, 21ALP4B, 21BTP4, 21CMP4, 21CWP4, 21DNP4, 21BAP4~, 21BNP4~, 21CVP4~
21ATP4A	12L	21ATP4, 21ATP4B, 21ALP4B, 21BTP4, 21CMP4, 21CWP4, 21DNP4, 21BAP4~,
21ATP4B	12L	21ATP4, 21ATP4A, 21ALP4, 21ALP4A, 21ALP4B, 21BTP4, 21CMP4, 21CWP4, 21DNP4, 21BAP4~, 21BNP4~, 21CVP4~
21AUP4	12L	21AUP4A, 21AUP4B, 21AUP4C~, 21AVP4, 21AVP4A, 21AVP4B, 21AVP4C~, 21BDP4~
21AUP4A	12L	21AUP4, 21AUP4B, 21AUP4C~, 21AVP4, 21AVP4A, 21AVP4B, 21AVP4C~, 21BDP4~
21AUP4B	12L	21AUP4, 21AUP4A, 21AUP4C~, 21AVP4, 21AVP4A, 21AVP4B, 21AVP4C~, 21BDP4~
21AUP4C	12L	21AVP4C, 21BDP4, 21AUP4~~, 21AUP4A~~, 21AUP4B~~, 21AVP4~~, 21AVP4A~~, 21AVP4B~~
21AVP4	12L	21AVP4A, 21AVP4B, 21AVP4C~, 21AUP4, 21AUP4A, 21AUP4B, 21AUP4C~, 21BDP4~
21AVP4A	12L	21AVP4, 21AVP4B, 21AVP4C~, 21AUP4, 21AUP4A, 21AUP4B, 21AUP4C~, 21BDP4~
21AVP4B	12L	21AVP4, 21AVP4A, 21AVP4C~, 21AUP4, 21AUP4A, 21AUP4B, 21AUP4C~, 21BDP4~
21AVP4C	12L	21AUP4C, 21BDP4, 21AVP4~~, 21AVP4A~~, 21AVP4B~~, 21AUP4~~, 21AUP4A~~, 21AUP4B~~
21AWP4	12N	21AWP4A~
21AWP4A	12N	21AWP4~~
21AXP22	14W	21FBP22
21AXP22A	14W	-
21AYP4	12L	21XP4, 21XP4A, 21XP4B~, 21ASP4<
21BAP4	12L	21BNP4, 21CBP4A, 21CBP4B, 21CVP4, 21DRP4, 21FLP4
21BCP4	12L	21YP4B, 21YP4~~, 21YP4A~~
21BDP4	12L	21AUP4C, 21AVP4C, 21AUP4B~~, 21AVP4B~~
21BNP4	12L	21BAP4, 21CBP4A, 21CVP4, 21DRP4, 21FLP4
21BRP4	8GM	-
21BSP4	12N	21ACP4, 21ACP4A, 21AMP4, 21AMP4A, 21CUP4, 21AMP4B~
21BTP4	12L	21ALP4B%, 21ATP4A%, cont'd

Tube Type	Basing	Replacement
		21BTP4, cont'd21BAP4%, 21BNP4%, 21CMP4%, 21CVP4~%, 21CWP4%, 21DNP4%
21CBP4	12L	21CBP4A, 21CBP4B, 21DLP4, 21DQP4, 21DRP4, 21FLP4, 21CKP4%
21CBP4A	12L	21CBP4B, 21DLP4, 21DQP4, 21DRP4, 21FLP4, 21CKP4%
21CBP4B	12L	12CBP4A, 21DLP4, 21DQP4, 21DRP4, 21FLP4, 21CKP4%
21CDP4	12L	21CDP4A, 21CKP4~%
21CDP4A	12L	21CDP4, 21CKP4~%
21CEP4	8HR	21EMP4
21CEP4A	8HR	21DWP4%
21CGP4	12L	21CHP4~
21CHP4	12L	21CGP4~~
21CKP4	12L	21CBP4A%, 21DLP4%, 21DRP4%, 21FLP4%
21CLP4	12AJ	-
21CMP4	12L	21DNP4, 21CBP4A~, 21CKP4%, 21DRP4~, 21FLP4~
21CQP4	7FA	21CSP4%
21CSP4	7FA	21CQP4%
21CUP4	12N	21ACP4, 21ACP4A, 21AMP4, 21AMP4A, 21BSP4, 21AMP4B~
21CVP4	12L	21BAP4, 21BNP4, 21CMP4, 21DNP4
21CWP4	12L	21ALP4B, 21ATP4A, 21BTP4, 21CMP4, 21DNP4
21CXP4	12L	21DSP4%
21CYP22	14AL	21FBP22
21CYP22A	14AL	-
21CZP4	8HR	21DEP4~, 21DEP4A~, 21DHP4%~, 21DKP4%~, 21DKP4A%~
21DAP4	8HR	21DMP4%
21DEP4	8HR	21DEP4A, 21CZP4**, 21DHP4%, 21DKP4%, 21DKP4A%
21DEP4A	8HR	21DEP4
21DFP4	8HR	21CZP4~%, 21DEP4%, 21DEP4A%, 21DWP4%, 21EMP4
21DHP4	8HR	21CEP4, 21CEP4A, 21DMP4, 21DKP4%, 21DKP4A%
21DJP4	12L	21CBP4A%, 21CKP4%, 21DLP4%, 21DRP4%, 21FLP4%
21DKP4	8HR	21DKP4A, 21DEP4%, 21DEP4%, 21DEP4A%, 21DHP4%
21DKP4A	8HR	21DKP4, 21DEP4%, 21DEP4%, 21DEP4A%, 21DHP4%
21DLP4	12L	21DQP4
21DMP4	8HR	-
21DNP4	12L	21CBP4A~%, 21CKP4~%, 21CMP4%, 21DRP4~%, 21FLP4~%
21DP4	12M	-
21DQP4	12L	21CBP4%, 21CBP4A%, 21CBP4B%, 21CKP4%, 21DLP4%, 21FLP4%
21DRP4	12L	21CBP4A, 21DLP4, 21FLP4, 21CKP4%
21DSP4	12L	21CXP4
21DVP4	12L	21ENP4, 21CKP4~%, 21DJP4~%
21DWP4	8HR	21CEP4A%, 21DFP4%
21EAP4	8JK	-
21ELP4	12L	-
21EMP4	8HR	-
21ENP4	12L	21CBP4A~%, 21CKP4~%, 21CMP4%, 21DJP4~%, 21DNP4%, 21DRP4%, 21FLP4~%
21EP4	12D	21EP4A%, 21EP4A>, 21EP4B>, 21EP4C>~, 21ZP4C
21EP4A	12N	21EP4B, 21EP4C~, 21ZP4C
21EP4B	12N	21EP4A, 21EP4C~, 21ZP4C
21EP4C	12N	21EP4A~~, 21EP4B~~, 21ZP4C
21EQP4	8JR	-
21ERP4	8JR	-
21ESP4	8JS	21FAP4
21EVP4	8JK	-
21EXP4	8JR	21EQP4%
21EZP4	8JR	-
21FAP4	8JR	-
21FBP22	14AU	-
21FBP22A	14AU	21GUP22
21FCP4	8HR	-
21FDP4	8KW	-
21FJP22	14AU	21FKP22
21FJP22A	14AU	-
21FKP22	14AU	21FJP22
21FLP4	12L	21CBP4A, 21DLP4, 21DRP4, 21CKP4%
21FMP4	8HR	-
21FNP22	-	22JP22
21FP4	12M	21FP4A>, 21FP4D~>
21FP4A	12L	21FP4<, 21FP4D~
21FP4C	12L	21FP4<, 21FP4A, 21FP4D~
21FP4D	12L	21FP4A~~, 21FP4C~~
21FQP22	-	22UP22
21FUP4	8HR	21GTP4%
21FVP4	8HR	21FWP4
21FWP4	8HR	21FVP4
21FXP4	8HR	-
21FYP4	8HR	-
21FZP4	8HR	-
21GAP4	8HR	21GAP4A
21GAP4A	8HR	21GAP4
21GBP4	8HR	-
21GCP4	8HR	-
21GEP4	8HR	-
21GFP22	14BE	20VAGP22
21GHP4	8HR	-
21GJP4	8HR	-
21GKP4	8HR	-
21GLP22	-	22JP22, 22UP22
21GNP22	-	22JP22, 22UP22
21GRP22	-	22UP22
21GSP4	8HR	-
21GTP4	8HR	21FUP4%
21GUP22	14AU	21FBP22
21GVP22	14AU	21GYP22, 21FJP22
21GWP22	14BE	20VAGP22, 21HBP22, 22JP22, 22UP22
21GXP22	-	21FJP22
21GYP22	14AU	21FJP22, 21GVP22
21GZP22	-	21FJP22
21HAP22	-	21FJP22
21HBP22	14BE	21GWP22, 22JP22

Tube Type	Basing	Replacement
21JP4	12N	21JP4A
21JP4A	12N	21JP4
21KP4	12D	21KP4A>
21KP4A	12P	21FP4D, 21KP4<
21MP4	12M	-
21VABP22	14BE	21VAKP22, 21VACP22, 22AMP22
21VACP22	14BE	21VACP22, 21VAJP22, 21VAKP22, 21VAQP22, 21VAUP22, 21VAZP22, 21VBEP22
21VADP22	14BE	21VARP22, 21VBEP22
21VAFP22	-	21VACP22
21VAGP4	8HR	-
21VAJP22	-	21VACP22
21VAKP22	14BE	21VACP22
21VALP22	14BE	21VALP22, 21VAXP22, 21VBCP22
21VAMP22	-	21VACP22
21VAQP22	14BE	21VACP22, 21VAUP22, 21VAZP22, 21VBEP22, 21VBHP22
21VARP22	14BE	21VBEP22
21VASP4	8HR	22ZP4
21VATP4	8HR	22AFP4
21VAUP22	14BE	21VACP22, 21VAKP22, 21VAQP22, 21VAZP22, 21VBEP22, 21VBHP22
21VAWP22	-	21VACP22
21VAXP22	14BE	21VBCP22
21VAYP22	-	21VACP22
21VAZP22	14BE	21VACP22, 21VAQP22, 21VAUP22, 21VBEP22, 21VBHP22
21VBAP22	-	21VARP22
21VBCP22	14BE	21VBCP22, 21VAXP22
21VBDP22	-	-
21VBEP22	14BE	21VBHP22
21VBFP22	-	21VBHP22
21VBGP22	-	21VBHP22
21VBHP22	14BE	21VBEP22
21VBJP22	14BE	-
21VBLP22	14BP	-
21VBMP22	-	21VBLP22
21VBNP22	-	-
21VBQP22	-	-
21WP4	12N	21WP4A, 21WP4B~
21WP4A	12N	21WP4, 21WP4B~
21WP4B	12N	21WP4~~, 21WP4A~~
21XP4	12L	21XP4A, 21XP4B~
21XP4A	12L	21XP4, 21XP4B~
21XP4B	12L	21XP4~~, 21XP4A~~
21YP4	12L	21YP4A, 21YP4B~, 21BCP4~
21YP4A	12L	21YP4, 21YP4B~, 21BCP4~
21YP4B	12L	21BCP4, 21YP4~~, 21YP4A~~
21ZP4	12D	21ZP4A>, 21ZP4B>, 21ZP4C~>
21ZP4A	12N	21ZP4B, 21ZP4C~
21ZP4B	12N	21ZP4A, 21ZP4C~
21ZP4C	12N	21ZP4A~~, 21ZP4A~~
22ABP22	-	22JP22
22ACP22	-	-
22ADP22	14BE	22UP22
22AEP22	14BE	-
22AFP4	12D	21VATP4
22AGP22	-	22UP22
22AHP22	14BE	20VAKP22, 20VALP22, 22ASP22, 22UP22
22ALP22	14BE	20VAJP22, 20VASP22, 22UP22
22AMP22	14BE	21VABP22, 21VAKP22, 22JP22
22ANP22	14BH	20VAHP22
22AP4	12D	22AP4A
22AP4A	12D	22AP4
22AQP22	14BE	-
22ARP22	14BE	20VAGP22, 22JP22, 22UP22
22ASP22	14BE	22UP22
22ATP22	14BE	-
22AVP22	-	22VABP22
22AWP22	-	20VACP22
22EP22	-	22JP22
22JP22	14BE	20VAGP22, 22ARP22, 22UP22
22KP22	14BE	22JP22
22LP22	-	22UP22
22QP22	14BE	20VAGP22, 22ARP22, 22JP22
22RP22	14BE	22KP22, 22UP22, 22JP22
22SP22	14BE	20VAGP22, 20VAMP22, 20VANP22, 22JP22, 22UP22
22TP4	8HR	-
22UP22	14BE	22ASP22
22VABP4	8HR	22VAHP4%
22VACP4	8HR	-
22VADP4	12L	-
22VAEP4	8HR	-
22VAGP4	8HR	-
22VAHP4	8HR	-
22VAMP4	8HR	22VANP4, 22VANP4A, 23GBP4, 23HFP4
22VANP4	8HR	22VAMP4
22VANP4A	8HR	22VAMP4
22VARP4	8HR	23GJP4, 23GWP4
22VASP4	8HR	23FCP4, 23FDP4, 23FHP4
22VATP4	8HR	23HGP4, 23JEP4
22VAUP4	8HR	-
22VAXP4	8HR	22VABP4, 22VAZP4, 22VBDP4
22VAYP4	8HR	23JQP4
22VAZP4	8HR	22VABP4, 22VAXP4, 22VBDP4
22VBAP4	8HR	22VAXP4, 22VAZP4, 22VBDP4
22VBCP4	8HR	-
22VBDP4	8HR	22VAXP4, 22VAZP4
22VP22	-	-
22WP22	14BE	-
22XP22	-	22UP22
22YP22	14BE	22JP22, 22UP22
22ZP4	8HR	21VASP4
23ACP4	12L	23TP4
23AFP4	12L	23BMP4, 23BTP4, 23YP4, 23CDP4%
23AHP4	12L	23ASP4, 23AUP4, 23CZP4, 23AZP4%
23AKP4	8JR	23CVP4%
23ALP4	8HR	23AMP4%, 23CQP4%, 23MP4%, 23MP4A%, 23VP4%
23AMP4	8HR	23ALP4%, 23CQP4%, 23MP4%, 23MP4A%, 23VP4%
23ANP4	12L	23ATP4, 23BKP4, 23BLP4
23AQP4	8HR	23AMP4, 23VP4, 23ALP4%, 23AMP4%, 23CQP4%, 23FP4%, 23FP4A%, 23MP4%, 23MP4A%
23ARP4	8HR	23AXP4%
23ASP4	12L	-
23ATP4	12L	23ANP4, 23BKP4, 23BLP4
23AUP4	12L	23AHP4, 23ASP4, 23CZP4, 23AZP4%

Tube Type	Basing	Replacement
23AVP4	8HR	23CP4, 23CP4A, 23DXP4, 23AYP4%, 23BSP4%, 23HP4%, 23SP4%
23AWP4	12L	23BJP4
23AXP4	8HR	23EJP4
23AYP4	8HR	23BSP4, 23SP4, 23AVP4%, 23CP4%, 23CP4A%, 23DXP4%, 23HP4%
23AZP4	12L	23ASP4%
23BAP4	8HR	23BP4
23BCP4	8HR	23CEP4%, 23CMP4%
23BDP4	12L	23BTP4
23BEP4	8HR	23BEP4A
23BEP4A	8HR	23BEP4
23BGP4	8HR	23BHP4
23BHP4	8HR	23BGP4
23BJP4	12L	23AWP4
23BKP4	12L	23BLP4, 23ANP4%, 23ATP4%
23BLP4	12L	23BKP4, 23ANP4%, 23ATP4%
23BMP4	12L	23AFP4, 23BTP4, 23XP4, 23YP4, 23CDP4%
23BNP4	8HR	23ADP4, 23AVP4, 23CP4, 23CP4A, 23DXP4, 23GP4, 23AYP4%, 23BEP4%, 23BEP4A%, 23BQP4%, 23BSP4%, 23CBP4%, 23SP4%, 23UP4%
23BP4	8HR	23BAP4, 23CP4A
23BQP4	8HR	23CBP4, 23UP4, 23DGP4%
23BRP4	8JR	23BYP4, 23CSP4, 23RP4, 23CUP4%, 23DP4%
23BSP4	8HR	23AVP4%, 23AYP4%, 23CP4%, 23CP4A%, 23DXP4%, 23GP4%, 23HP4%, 23SP4%
23BTP4	12L	23AFP4, 23BMP4, 23XP4, 23YP4, 23CDP4%
23BVP4	12L	23CTP4%, 23EAP4%
23BXP4	12L	23EDP4
23BYP4	8JR	23BRP4, 23CSP4, 23RP4, 23CUP4%, 23DP4%
23BZP4	12L	-
23CAP4	12L	-
23CBP4	8HR	23BQP4, 23UP4
23CDP4	12L	-
23CEP4	8HR	23ARP4%, 23CMP4%
23CGP4	12L	23ASP4%
23CMP4	8HR	23BCP4%
23CP4	8HR	23CP4A, 23AVP4, 23DXP4, 23AYP4%, 23BSP4%, 23HP4%, 23SP4%
23CP4A	8HR	23CP4, 23AVP4, 23DXP4, 23AYP4%, 23BSP4%, 23HP4%, 23SP4%
23CQP4	8HR	23FP4%, 23FP4A%, 23VP4%
23CSP4	8JR	23BRP4, 23BYP4,23RP4, 23CUP4%, 23DP4%
23CTP4	12L	23BVP4%, 23EAP4%
23CUP4	8JR	23DP4, 23BRP4%, 23BYP4%, 23CSP4%, 23RP4%
23CVP4	8JR	23AKP4%
23CWP4	8JR	23CXP4%
23CXP4	8JR	23CWP4%
23CZP4	12L	23ASP4
23DAP4	8HR	23DCP4%
23DBP4	8HR	23NP4
23DCP4	8HR	23DAP4%
23DEP4	8HR	-
23DFP4	8HR	-
23DHP4	8HR	23DJP4, 23UP4%
23DJP4	8HR	23DHP4, 23UP4%
23DKP4	12L	23DTP4, 23FLP4%
23DLP4	12L	23DLP4A, 23ENP4, 23FBP4, 23FWP4%, 23FWP4A%
23DLP4A	12L	23DLP4, 23ENP4, 23FBP4, 23FWP4%, 23FWP4A%
23DNP4	12L	-
23DP4	8JR	23CUP4, 23RBP4%, 23BYP4%, 23CSP4%, 23RP4%
23DQP4	8HR	-
23DSP4	8HR	23DSP4A
23DSP4A	-	23DSP4
23DVP4	8HR	23DVP4A, 23DZP4%, 23EWP4%, 23EWP4A%
23DVP4A	8HR	23DVP4, 23DZP4%, 23EWP4% 23EWP4A%
23DWP4	-	-
23DYP4	8HR	22VANP4, 22VANP4A, 23ESP4%, 23FVP4%
23DZP4	8HR	23EWP4, 23EWP4A, 23DVP4%, 23DVP4A%
23EAP4	12L	23BVP4%, 23CTP4%
23ECP4	12L	-
23EDP4	12L	-
23EFP4	8HR	23FCP4%, 23FDP4%, 23FHP4%, 23GJP4%, 23GJP4A%, 23GWP4%
23EGP22	14BE	-
23EHP4	-	-
23EKP4	12L	23FLP4
23ELP4	12L	-
23ENP4	12L	23DLP4, 23DLP4A, 23FBP4, 23FWP4%, 23FWP4A%
23EP4	8KP	-
23EQP4	8HR	23ERP4%, 23EWP4A
23ERP4	8HR	23EQP4%
23ESP4	8HR	22VANP4, 22VANP4A, 23DYP4%, 23ETP4%, 23FVP4, 23FVP4A, 23GTP4
23ETP4	8HR	22VANP4%, 22VANP4%, 22VANP4A%, 23FVP4B%, 23GTP4, 23FMP4%, 23HMP4%, 23HXP4%
23EWP4	8HR	23EWP4A, 23DZP4, 23DVP4%, 23DVP4A%
23EWP4A	8HR	23EWP4, 23DZP4, 23DVP4%, 23DVP4A%
23EYP4	12L	-
23EZP4	8HR	-
23FAP4	8HR	23HLP4
23FBP4	12L	23DLP4, 23DLP4A, 23ENP4, 23FWP4%, 23FWP4A%
23FCP4	8HR	22VASP4, 23FDP4, 23FHP4, 23GJP4, 23GJP4A, 23GWP4, 23EFP4%
23FDP4	8HR	22VASP4, 23FCP4, 23FHP4, 23GJP4, 23GJP4A, 23GWP4, 23EFP4%
23FEP4	12L	-
23FHP4	8HR	22VASP4, 23FCP4, 23FDP4, 23GJP4, 23GJP4A, 23GWP4, 23EFP4%
23FKP4	8HR	-
23FLP4	12L	23DKP4%, 23GTP4%
23FMP4	8HR	22VANP4, 22VANP4A, cont'd

Tube Type	Basing	Replacement
23FMP4, cont'd		23ETP4%, 23FVP4B, 23GTP4%, 23HMP4%, 23HXP4
23FNP4	12L	-
23FP4	8HR	23FP4A, 23VP4%
23FP4A	8HR	23FP4, 23VP4%
23FRP4	8HR	22VAHP4%, 23JFP4%
23FSP4	8HR	23JBP4
23FVP4	8HR	22VANP4, 22VANP4A, 23FVP4A, 23DYP4%, 23ESP4, 23TP4%, 23FMP4, 23HMP4
23FVP4A	8HR	22VANP4, 22VANP4A, 23FVP4, 23DYP4%, 23ESP4, 23ETP4%, 23FMP4, 23HMP4
23FVP4B	8HR	22VAMP4
23FWP4	12L	23FWP4A, 23DLP4%, 23DLP4A%, 23ENP4, 23FBP4%
23FWP4A	12L	23FWP4, 23DLP4%, 23DLP4A%, 23ENP4%, 23FBP4%
23GBP4	8HR	22VAMP4, 23HFP4, 23HFP4A
23GDP4	8HR	23DVP4, 23DVP4A, 23DDZP4%, 23EWP4%, 23EWP4A%
23GFP4	8HR	-
23GGXP4	8HR	23HQP4%
23GHP4	8HR	23DWP4%
23GJP4	8HR	22VAHP4%, 22VARP4, 23GJP4A, 23GWP4
23GKP4	12L	23GRP4%
23GP4	8HR	23AVP4, 23CP4, 23CP4A, 23HP4, 23ADP4%, 23AYP4%, 23BSP4%, 23SP4%
23GRP4	12L	23GKP4%
23GSP4	8HR	23HZP4%
23GTP4	8HR	23ETP4, 23FMP4%, 23HMP4%, 23HXP4%
23GUP4	8HR	-
23GVP4	8HR	22VABP4
23GWP4	8HR	22VAHP4, 22VARP4, 23GJP4, 23GJP4A
23GYP4	12L	-
23HAP4	12L	-
23HBP4	8HR	-
23HCP4	8HR	-
23HFP4	8HR	22VAMP4, 23HFP4A, 23GBP4
23HFP4A	8HR	22VAMP4, 23HFP4, 23GBP4
23HGP4	8HR	22VATP4, 23JEP4
23HKP4	8HR	-
23HLP4	8HR	-
23HMP4	8HR	23DYP4%, 23ESP4%, 23ETP4%, 23FMP4%, 23FVP4%, 23FVP4A%, 23GTP4%, 23HXP4%
23HP4	8HR	22VANP4%, 22VANP4A%, 23AVP4, 23CP4, 23CP4A, 23DXP4, 23AYP4%, 23BSP4%, 23FVP4B%, 23SP4%
23HQP4	8HR	23GXP4%
23HQP4A	8HR	23GXP4%
23HRP4	8HR	23HWP4
23HTP4	8HR	-
23HUP4	8HR	23HUP4A, 23JLP4
23HUP4A	8HR	23HUP4, 23JLP4
23HWP4	8HR	23HWP4A
23HWP4A	8HR	23HWP4
23HXP4	8HR	23ETP4%, 23FMP4, cont'd

Tube Type	Basing	Replacement
23HXP4, cont'd		23GTP4%, 23HMP4%
23HZP4	8HR	23GSP4%, 23HGP4%
23JBP4	8HR	-
23JEP4	8HR	23HZP4%
23JEP4A	8HR	22VATP4, 23HGP4, 23HZP4%
23JFP4	8HR	22VAHP4%, 23FRP4%
23JGP4	8HR	23FRP4
23JHP4	8HR	-
23JLP4	8HR	22VAHP4%, 23HUP4, 23HUP4A,
23JP4	7FA	-
23JQP4	8HR	22VAYP4
23JRP4	8HR	23FP4A
23KP4	8JS	23KP4A
23KP4A	8JS	23KP4
23MP4	8HR	23MP4A, 23ALP4%, 23AMP4%, 23CQP4%, 23VP4%
23MP4A	8HR	23MP4, 23ALP4%, 23AMP4%, 23CQP4%, 23VP4%
23NP4	8HR	-
23RP4	8JR	22VAHP4%, 23BRP4, 23BYP4, 23CSP4, 23CUP4%, 23DP4%
23SP4	8HR	23AYP4, 23BSP4, 23AVP4%, 23CP4%, 23CP4A%, 23DXP4%, 23UP4%
23TP4	12L	23ACP4, 23BTP4
23UP4	8HR	23QP4, 23CBP4
23VABP22	14BE	23VALP22, 23VANP22, 23VAQP22, 23VAXP22, 23VBSP22
23VACP22	14BE	23VADP22, 23VALP22, 23VANP22, 23VAQP22, 23VARP22, 23VBRP22, 23VBTP22
23VADP22	14BE	23VACP22, 23VAMP22, 23VAQP22, 23VAYP22, 23VBYP22, 23VCFP22
23VAEP22	14BE	23VAEP22, 23VAJP22
23VAGP22	-	-
23VAHP22	14BE	23VANP22, 23VATP22, 23VAQP22, 23VALP22, 23VBRP22, 23VBYP22
23VAJP22	14BE	23VAEP22, 23VAJP22
23VALP22	14BE	23VAHP22, 23VALP22, 23VANP22, 23VARP22
23VAMP22	14BE	23VADP22, 23VAYP22, 23VAQP22, 23VBWP22, 23VBYP22, 23VCFP22
23VANP22	14BE	23VAHP22, 23VANP22, 23VAQP22, 23VALP22, 23VARP22, 23VATP22
23VAQP22	14BE	23VAMP22, 23VAQP22, 23VANP22, 23VDBP22, 23VBRP22, 23VBWP22, 23CFP22
23VARP22	14BE	23VAHP22, 23VALP22, 23VAQP22, 23VANP22
23VASP22	14BE	23VALP22, 23VANP22, 23VAQP22, 23VAYP22
23VATP22	14BE	23VANP22, 23VALP22, 23VANP22, 23VAQP22, 23VAYP22
23VAUP22	-	23VALP22, 23VANP22, 23VAQP22, 23VAYP22
23VAWP22	-	23VAQP22, 23VAYP22
23VAXP22	14BE	23VABP22, 23VALP22, 23VANP22, cont'd

Tube Type	Basing	Replacement
23VAXP22, cont'd		23VATP22, 23VBSP22, 23VAQP22, 23VAYP22
23VAYP22	-	23VACP22, 23VADP22, 23VAQP22, 23VBAP22, 23VCFP22
23VAZP22	14BE	23VATP22, 23VALP22, 23VANP22
23VBAP22	14BE	23VACP22, 23VALP22, 23VAMP22, 23VAPQ22, 23VAYP22, 23VBWP22, 23VBYP22, 23VCFP22
23VBCP22	14BE	23VAQP22, 23VBP22, 23VAYP22
23VBDP22	14BE	23VAQP22, 23VBAP22, 23VAYP22, 23VBWP22, 23VBYP22, 23VCFP22
23VBEP22	-	23VANP22, 23VAQP22, 23VAYP22
23VBFP22	14BE	23VBUP22
23VBGP22	14BE	23VALP22, 23VANP22, 23VAQP22, 23VAYP22
23VBHP22	14BE	23VANP22, 23VAQP22, 23VALP22, 25SP22
23VBJP22	14BE	23VAQP22, 23VAYP22
23VBKP22	14BE	23VBMP22
23VBMP22	14BE	-
23VBNP22	14BE	23VAQP22, 23VAYP22, 23VBCP22, 23VBKP22, 25BKP22
23VBPP22	14BE	23VAHP22, 23VAMP22, 23VAQP22, 23BCFP22
23VBRP22	-	23VAQP22, 23VAYP22
23VBSP22	14BE	23VABP22, 23VALP22, 23VAQP22, 23VANP22, 23VAXP22
23VBTP22	14BE	23VACP22, 23VANP22, 23VAQP22, 23VAYP22
23VBUP22	14BE	-
23VBWP22	14BE	23VALP22, 23VAMP22, 23VAQP22, 23VAYP22, 23VBAP22, 23VCEP22
23VBXP22	-	-
23VBYP22	14BE	23VADP22, 23VAHP22, 23VAQP22, 23VAYP22, 23VAMP22, 23VBAP22, 23VBDP22, 23VCFP22
23VCEP22	14BE	23VCGP22
23VCFP22	14BE	23VCGP22
23VCGP22	-	-
23VCHP22	-	23VCGP22
23VCJP22	-	23VCGP22
23VCKP22	-	23VCGP22
23VCMP22	14BE	23VCGP22
23VCNP22	-	23VCGP22
23VCRP22	-	-
23VP4	8HR	23ALP4%, 23AMP4%, 23CQP4%, 23MP4%, 23MP4A%
23WP4	8HR	23FP4A
23XP4	12L	23AFP4, 23BMP4, 23BTP4, 23YP4, 23CDP4%
23YP4	12L	23AFP4, 23BMP4, 23BTP4, 23XP4, 23CDP4%
23ZP4	12L	23BKP4
24ADP4	12N	24CP4%, 24CP4A%, 24CP4B~%, 24TP4%, 24VP4%, 24VP4A%
24AEP4	12L	-
24AHP4	8HR	24ALP4, 24AQP2%, 24AWP4%, 24AXP4%
24AJP4	12L	24ATP4
24ALP4	8HR	24AHP4, 24AQP4%, 24AWP4%, 24AXP4%
24AMP4	7FA	-
24ANP4	12L	24DP4, 24DP4A, 24EP4~, 24ASP4~%, 24AUP4~, 24YP4%, 24ZP4~%
24AP4	12D	24AP4A
24AP4A	12D	24AP4
24AP4B	12D	24AP4
24AQP4	8HR	24HP4%, 24ALP4%, 24AWP4%, 24AXP4%
24ASP4	12L	24EP4%, 24AUP4%
24ATP4	12L	24AJP4%
24AUP4	12L	24AEP4
24AVP4	8JK	-
24AWP4	8HR	-
24AXP4	8HR	24AHP4%, 24ALP4%, 24AQP4%, 24AWP4%
24BAP4	8HR	-
24BCP4	12L	-
24BEP4	8KW	-
24BP4	12M	-
24CP4	12N	24CP4A, 24ADP4, 24QP4, 24TP4, 24VP4, 24YP4A, 24CP4B~
24CP4A	12N	24CP4, 24ADP4, 24QP4, 24TP4, 24VP4, 24VP4A, 24CP4B~
24CP4B	12N	24CP4~~, 24CP4A~~, 24DP4~~, 24QP4~~, 24TP4~~, 24VP4~~, 24VP4A~~
24DP4	12L	24DP4A, 24ANP4, 24AEP4~, 24ASP4~, 24AUP4~, 24YP4%, 24ZP4~%
24DP4A	12L	24DP4, 24ANP4, 24AEP4~, 24ASP4~%, 24AUP4~, 24YP4%, 24ZP4~%
24QP4	12N	24ADP4, 24CP4, 24CP4A, 24TP4, 24VP4, 24VP4A, 24CP4B~
24TP4	12N	24ADP4, 24CP4, 24CP4A, 24QP4, 24VP4, 24VP4A, 24CP4B~
24VABP22	14BQ	A66-501X
24VP4	12N	24VP4A, 24ADP4, 24CP4, 24CP4A, 24QP4, 24TP4, 24CP4B~
24VP4A	12N	24VP4, 24ADP4, 24CP4, 24CP4A, 24QP4, 24TP4, 24CP4B~
24XP4	12D	-
24YP4	12L	24ANP4, 24AUP4, 24DP4, 24DP4A, 24ZP4, 24AEP4~, 24ASP4%
24ZP4	12L	24AEP4, 24ANP4, 24AUP4, 24DP4, 24DP4A, 24YP4, 24ASP4%
25ABP22	14BE	23VALP22, 23VANP22, 23VARP22, 23VATP22, 23VAQP22, 25AJP22, 25XP22
25ACP22	-	-

Tube Type	Basing	Replacement
25ADP22	14BE	23VAQP22, 25AJP22
25AEP22	14BE	23VANP22, 23VATP22, 25AP22, 25YP22, 25XP22
25AFP22	14BE	23VALP22, 23VANP22, 23VAQP22, 23VARP22, 23VATP22, 25AJP22, 25BMP22, 25AP22, 25XP22
25AGP22	-	23VAQP22, 25AJP22
25AHP22	-	23VBXP22, 25AHP22
25AJP22	14BE	23VADP22, 23VALP22, 23VANP22, 23VAQP22, 23VARP22, 23VBJP22, 23VBRP22, 23VBYP22, 25BHP22
25AKP22	14BE	25AMP22
25ALP22	14BE	23VBMP22, 25ALP22A
25ALP22A	14BE	23VBMP22, 25ALP22
25AMP22	14BE	23VAEP22, 23VAJP22
25ANP22	14BE	23VABP22, 23VALP22, 23VANP22, 23VAQP22, 23VARP22, 25AP22, 25XP22, 25BAP22, 25AJP22, 25AJP22
25AP22	14BE	23VALP22, 23VANP22, 23VARP22, 23VAQP22, 25AJP22, 25ABP22, 25XP22
25AP22A	14BE	23VALP22, 23VANP22, 23VARP22, 25ABP22, 25XP22
25AQP22	14BE	23VABP22, 23VAXP22, 23VANP22, 23VAQP22, 25AJP22, 25AAP22, 25XP22
25ASP22	14BE	23VAYP22, 23VAQP22, 25AJP22
25ATP22	14BE	-
25AVP22	-	-
25AWP22	14BE	23VACP22, 23VALP22, 23VANP22, 23VAQP22, 23VARP22, 23VAYP22, 23VBNP22, 23VBTP22, 25BKP22, 25AJP22
25AXP22	-	23VAQP22, 23VAYP22, 25AJP22
25AYP22	14BE	25ALP22, 25ACP22, 25AMP22
25AZP22	14BE	23VACP22, 23VALP22, 23VANP22, 23VAQP22, 23VARP22, 23VBPR22, 23VBT22, 25AJP22
25BAP22	14BE	23VABP22, 23VCGP22, 23VALP22, 23VAXP22, 25BAP22, 25ANP22, 25GP22A
25BCP22	14BE	23VALP22, 23VANP22, 23VCGP22, 23VAYP22, 25BAP22
25BDP22	14BE	23BAP22, 23VAQP22, 23VBYP22, 23VAYP22,
25BFP22	14BE	23VACP22, 23VALP22, 23VANP22, 23VAQP22, 23VARP22, 23VBTP22
25BGP22	14BE	23VALP22, 23VANP22, 23VANP22, 23VAQP22, 25AP22, 25XP22, 25AJP22
25BHP22	14BE	23VAP22, 23VANP22, 23VAQP22, 23VARP22, 23VBRP22, 23VBYP22, 25AJP22
25BJP22	-	23VANP22, cont'd

Tube Type	Basing	Replacement
25BJP22, cont'd		23VAQP22, 25AP22, 25AJP22, 25XP22
25BKP22	14BE	23VAQP22, 23VBAP22, 23VAYP22, 25AJP22
25BLP22	14BE	-
25BMP22	14BE	23VALP22, 23VANP22, 23VARP22, 23VAQP22, 25AP22, 25AJP22, 25XP22
25BNP22	-	-
25BP22	14BE	23VANP22, 23VATP22, 25AP22, 25XP22, 25YP22
25BP22A	14BE	23VANP22, 23VATP22, 25YP22
25BRP22	-	23VANP22, 23VAQP22, 25AP22, 25AJP22, 25XP22
25BSP22	-	23VAQP22, 25AJP22
25BTP22	-	-
25BVP22	-	23VANP22, 23VAQP22, 25AP22, 25AJP22, 25XP22
25BWP22	-	23VAQP22, 25AJP22
25BXP22	-	23VANP22, 23VAQP22, 25AP22, 25XP22, 25AJP22
25BYP22	-	-
25BZP22	-	23VANP22, 23VAQP22, 25AP22, 25XP22, 25AJP22
25CAP22	14BE	25ALP22, 25AYP22
25CBP22	14BE	23VANP22, 23VAQP22, 25AP22, 25XP22, 25AJP22
25CP22	14BE	23VAQP22, 23VANP22, 22VALP22, 23VARP22, 25AJP22, 25ABP22, 25AP22, 25XP22
25CP22A	-	-
25CQP22	14BE	25VAZP22%, 25VBGP22%, 25VBNP22, 25VBRP22%, 25VCBP22, 25VCFP22, 25VCUP22, 25VEBP22
25DP4	8HR	-
25EGP22	-	-
25FP22	14BE	23VANP22, 23VAQP22, 25FP22A, 25AP22, 25XP22, 25AJP22
25FP22A	14BE	23VANP22, 25FP22
25GP22	-	23VANP22, 23VAQP22, 25AP22, 25XP22, 25AJP22
25GP22A	14BE	23VABP22, 23VALP22, 23VANP22, 23VARP22, 25ABP22, 25ANP22, 25BAP22, 25GP22
25HP4	8HR	-
25JP4	8HR	-
25KP4	8HR	25LP4%
25LP4	8HR	25KP4%
25RP22	-	23VANP22, 23VAQP22, 25AP22, 25XP22, 25AJP22
25SP22	14BE	23VBP22, 23VANP22, 23VAQP22, 25AP22, 25XP22
25TP4	8HR	-
25UP22	-	-
25VABP22	14BE	25VSDP22, 25VAJP22, 25VALP22%, 25VAMP22, 25VATP22, 25VBMP22, 25VBXP22, 25VCGP22, 25VDAP22, 25VDEP22, 25VDJP22, 25VDXP22
25VACP22	14BE	25VAFP22, 25VAQP22, 25VBQP22, 25VBWP22, 25VBZP22, cont'd

Tube Type	Basing	Replacement
25VACP22, cont'd		25VCP22, 25VCEP22, 25VCKP22, 25VCUP22, 25VCWP22, 25VCZP22, 25VDAP22, 25VDRP22, 25VDSP22, 25VDXP22, 25VEHP22, 25VEXP22
25VADP22	14BE	25VABP22, 25VALP22%, 25VATP22, 25VBMP22, 25VBXP22, 25VCGP22, 25VCKP22, 25VDAP22, 25DEP22, 25VDJP22,
25VAEP22	14BE	25VABP22, 25VADP22, 25VALP22%, 25VBXP22, 25VCGP22, 25VCKP22, 25VDAP22, 25VDEP22, 25VDJP22, 25VDXP22
25VAFP22	14BE	25VABP22, 25VACP22, 25VAEP22, 25VAWP2, 25VBUP22, 25VBWP22, 25VCBP22, 25VCEP22, 25VCKP22, 25VCUP22, 25VCZP22, 25VDAP22, 25VDRP22, 25VDXP22, 25VEHP22, 25VEXP22
25VAGP22	14BE	25VBEP22, 25VBYP22, 25VCBP22, 25VCXP22, 25DFP22, 26HP22
25VAJP22	14BE	25VABP22, 25VALP22, 25VCGP22, 25VCKP22
25VAKP22	14BE	25VALP22, 25VBLP22, 25VCGP22, 25VCKP22%
25VALP22	14BE	25VABP22%, 25VADP22%, 25VBMP22%, 25VCGP22%, 25VCKP22%
25VAMP22	14BE	25VABP22, 25VALP22%, 25VATP22, 25VBXP22, 25VCKP22, 25VDAP22, 25VDEP22, 25VDJP22, 25VDXP22
25VAQP22	14BE	25VABP22, 25VACP22, 25VAEP22, 25VAWP22, 25VBUP22, 25VCBP22, 25VCEP22, 25VCKP22, 25VCUP22, 25VCZP22, 25VDAP22, 25VDRP22, 25VEDP22, 25VEHP22, 25VEXP22
25VARP22	14BE	25VASP22
25VASP22	14BE	25VARP22
25VATP22	14BE	25VADP22, 25VAJP22, 25VAMP22, 25VBXP22, 25VCGP22, 25VCKP22, 25VDAP22, 25VDEP22, 25VDJP22, 25VDXP22
25VAUP22	14BE	25VBAP22, 25VABP22
25VAWP22	14BE	25VACP22, 25VAFP22, 25VAQP22, 25VBQP22, 25VBZP22, 25VCBP 25VCUP22, 25VCWP22, 25VDSP22, 25VEDP22, 25VEHP22
25VAXP22	14BE	25VBCP22, 25VCSP22, 25VDGP22, 25VDHP22, 26EP22
25VAYP22	-	-
25VAZP22	14BE	25VBGP22, 25VBRP22, 25VCBP22%, cont'd
25VAZP22, cont'd		25VCHP22%, 25VCRP22%,
25VBAP22	14BE	25VAUP22
25VBCP22	14BE	25VAXP22, 25VCSP22, 25VDGP22, 25VDHP22
25VBDP22	14BE	25VAXP22
25VBEP22	14BE	25VAGP22, 25VBTP22, 25VBYP22, 25VCDP22, 25VCTP22, 25VCXP22, 25VDFP22
25VBGP22	14BE	25VAZP22, 25VBRP22, 25VCBP22%, 25VCFP22, 25VCHP22%, 25VCRP22%, 25VCWP22%,
25VBHP22	-	-
25VBJP22	14BE	25VAXP22, 25VBEP22, 25VDXP22
25VBKP22	14BE	25VAGP22, 25VBEP22, 25VBYP22%, 25VCDP22, 25VCTP22%, 25VCXP22, 26HP22
25VBLP22	14BE	25VAEP22, 25VAKP22, 25VALP22, 25VCKP22%, 25VEXP22
25VBMP22	14BE	25VABP22, 25VADP22, 25VALP22%, 25VCKP22, 25VDAP22, 25VDEP22, 25VDXP22
25VBNP22	14BE	25VCNP22, 25VEBP22
25VBQP22	14BE	25VAWP22, 25VBWP22, 25VCFP22, 25VCNP22, 25VCUP22, 25VCZP22, 25VEBP22, 25VEDP22
25VBRP22	14BE	25CBP22%, 25VCFP22%, 25VCHP22%, 25VCRP22%, 25VCYP22%
25VBSP22	14BE	25VBEP22
25VBTP22	14BE	25VBEP22, 25VBYP22, 25VCDP22, 25VDFP22
25VBUP22	14BE	25VEXP22
25VBWP22	14BE	25VACP22, 25VAFP22, 25VAQP22, 25VBQP22, 25VBZP22, 25VCBP22, 25VCEP22, 25VCUP22, 25VCWP22, 25VCZP22, 25VDRP22, 25VEAP22, 25VEDP22, 25VEXP22
25VBXP22	14BE	25VATP22, 25VABP22, 25VDJP22
25VBYP22	14BE	25VBEP22, 25VCDP22, 25VCMP22, 25VCTP22
25VBZP22	14BE	25VAWP22, 25VBQP22, 25VBWP22, 25VCBP22, 25VCEP22, 25VCNP22, 25VCUP22, 25VCWP22, 25VDCP22, 25VDRP22, 25VEBP22, 25VEDP22, 25VEHP22, 25VEXP22, 26DP22
25VCAP22	-	-
25VCBP22	14BE	25VAWP22, 25VAZP22%, 25VBGP22, 25VBNP22, 25VBRP22, 25VBWP22, 25VCEP22, 25VCFP22, 25VCRP22, 225VCUP22, 25VCYP22, 25VEXP22
25VCDP22	14BE	25BYP22, 25VBEP22, 25CMP22, 25VCTP22

Tube Type	Basing	Replacement
25VCEP22	14BE	25VAFP22, 25VAQP22, 25VBQP22, 25VBZP22, 25VCBP22, 25VCUP22, 25VCWP22, 25VCZP22, 25VDRP22, 25VDSP22, 25VEDP22, 25VEHP22, 25VEXP22
25VCFP22	14BE	25VBNP22, 25VCBP22, 25VCRP22, 25VCUP22, 25VCYP22, 25VEBP22
25VCGP22	14BE	25VALP22%, 25VCKP22, 25VAXP22
25VCHP22	14BE	25VAZP22%, 25VBGP22%, 25VBGP22%, 25VBNP22, 25VBRP22%, 25VCBP22, 25VCRP22, 25VCYP22, 25EBP22
25VCJP22	14BE	25VBEP22
25VCKP22	14BE	25VATP22, 25VDAP22, 25VDEP22, 25VDKP22, 25VDXP22, 25VCZP22, 25VCKP22, 25DXP22
25VCLP22	-	25VAXP22
25VCMP22	14BE	25VBYP22, 25VCDP22, 25VCTP22, 25VDFP22
25VCNP22	14BE	25VACP22, 25VCZP22, 25VDAP22, 25VDXP22, 25VEBP22
25VCQP22	-	-
25VCRP22	14BE	25VAZP22%, 25VBNP22, 25VBRP22%, 25VCBP22, 25VCFP22%, 25VCHP22, 25VCYP22, 25VEBP22
25VCSP22	14BE	25VAXP22, 25VBCP22, 25VDHP22, 26EP22
25VCTP22	14BE	25VCXP22
25VCUP22	14BE	25VAWP22, 25VBWP22, 25VCEP22, 25VCNP22, 25VCZP22, 25VDCP22, 25VDRP22, 25VEBP22, 25VEDP22, 25VEHP22, 25VEXP22, 25VFAP22
25VCWP22	14BE	25VAWP22, 25VBZP22, 25VCEP22, 25VCZP22, 25VDCP22, 25VDRP22, 25VEBP22, 25VEDP22, 25VEHP22, 25VEXP22, 26DP22, 26KP22
25VCXP22	14BE	25VCTP22, 25VCZP22, 25VDAP22, 25VDXP22
25VCYP22	14BE	25VBNP22, 25VBRP22%, 25VCB22, 25VCFP22, 25VCHP22, 25VCRP22, 25VEBP22
25VCZP22	14BE	25VACP22, 25VBPWP22, 25VCNP22, 25VDCP22, 25VDMP22, 25VDRP22, 25VDXP22, 25VEBP22, 25VEDP22, 25VEHP22, 25VEXP22, 25VFAP22, 25VFGP22
25VDAP22	14BE	25VCKP22, 25VDEP22, 25VDKP22, 25VDNP22, 25VDXP22, 25VCZP22
25VDBP22	14BE	25VDMP22, 25VDRP22, 25VEHP22, 25VEXP22, 25VFAP22, 25VCZP22
25VDCP22	14BE	25CZP22, 25VCUP22, 25VEBP22, 25VCZP22
25VDEP22	14BE	25VCKP22, 25VDAP22, 25VDXP22, 25VCZP22
25VDFP22	14BE	25VBEP22, 25VCMP22, 25VCTP22, 25VCXP22
25VDGP22	14BE	25VBGP22, 26EP22
25VDHP22	14BE	25VAXP22
25VDJP22	14BE	25VABP22, 25VATP22, 25VBXP22
25VDKP22	14BE	25VCKP22, 25VDAP22, 25VDXP22, 25VCZP22
25VDMP22	14BE	25VCZP22, 25VDBP22, 25VDNP22, 25VDRP22, 25VDSP22, 25VDWP22, 25VEDP22, 25VEHP22, 25VEXP22, 25VFAP22
25VDNP22	14BE	25VDMP22, 25VDRP22, 25VEXP22, 25VFAP22, 25VCZP22
25VDRP22	14BE	25VCZP22, 25VDBP22, 25VDMP22, 25VDNP22, 25VDSP22, 25VDP22, 25VEDP22, 25VEHP22, 25VEXP22, 25VFAP22
25VDSP22	14BE	25VAWP222, 25VBWP22, 25VCEP22, 25VCNP22, 25VCUP22, 25VCZP22, 25VDCP22, 25VDRP22, 25VRBP22, 25VEDP22, 25VEXP22
25VDSP22-1	-	-
25VDWP22	14BE	25VDMP22, 25VDRP22, 25VEXP22, 25VFAP22
25VDXP22	14BE	25VATP22, 25VCKP22, 25VDAP22, 25VDBP22, 25VDEP22, 25VDKP22, 25VDMP22, 25VDNP22, 25VEDP22
25VEBP22	14BE	25VCZP22, 25VDMP22
25VECP22	14BE	25VCXP22
25VEDP22	14BE	25VATP22, 25VAWP22, 25VBWP22, 25VCEP22, 25VCUP22, 25VCZP22, 25VDMP22, 25VDRP22, 25VDSP22, 25VEHP22, 25VEXP22, 25VFAP22
25VEFP22	-	25VDRP22
25VEGP22	-	25VDRP22
25VEHP22	14BE	25VAWP22, 25VBWP22, 25VCEP22, 25VDBP22, 25VDMP22, 25VDRP22, 25VEDP22, 25VEXP22, 25VFAP22, 25VCZP22
25VEJP22	-	25VCZP22
25VEKP22	14BP	-
25VEMP22	13K	25VEQP22, 25VESP22, 25VETPP22, 25VEWP22, 25VFBP22
25VENP22	-	25VCZP22
25VEQP22	13K	25VEMP22, 25VESP22, 25VEWP22, 25VFBP22
25VERP22	14BE	25VDWP22
25VESP22	13K	25VEMP22, 25VEQP22, 25VETP22, 25VEWP22, 25VFB22
25VETP22	13K	25VEMP22, 25VEQP22, cont'd

Tube Type	Basing	Replacement
25VETP22 cont'd		25VESP22, 25VEWP22, 25VFBP22
25VEUP22	13K	-
25VEWP22	13K	25VEMP22, 25VEQP22, 25VESP22, 25VETP22, 25VFBP22, 25VFKP22
25VEXP22	14BE	25VCUP22, 25VCZP22, 25VDBP22, 25VDMP22, 25VDNP22, 25VDRP22, 25VDSP22, 25VDWP22, 25VEDP22, 25VEHP22, 25VFAP22
25VEYP22	13L	-
25VEZP22	14BE	-
25VFAP22	14BE	25VCUP22, 25VCZP22, 25VDBP22, 25VDMP22, 25VDNP22, 25VDRP22, 25VDSP22, 25VDWP22, 25VEDP22, 25VEHP22, 25VEXP22
25VFBP22	13K	25VEMP22, 25VEQP22, 25VESP22, 25VETP22, 25VEWP22
25VFCP22	-	25VETP22
25VFEP22	13L	-
25VFGP22	14BE	25VBWP22, 25VCNP22, 25VCZP22, 25VDCP22, 25VDRP22, 25VEBP22, 25VEDP22, 25VEHP22, 25VEXP22
25VFHP22	-	25VFBP22
25VFJP22	13L	25VFJP22
25VFKP22	13K	25VEMP22, 25VESP22, 25VETP22, 25VEWP22, 25VFBP22
25VFLP22	-	25VEZP22
25VFMP22	13K	-
25VFNP22	13L	-
25VFQP22	-	-
25VFRP22	-	-
25VFSP22	13L	-
25VFTP22	13K	-
25VFXP22	-	25VFTP22
25VFYP22	-	25VFMP22
25VFZP22	13L	-
25VGAP22	13L	25VGDP22
25VGBP22	-	25VEZP22
25VGDP22	-	-
25VGEP22	-	-
25VGHP22	-	25VGEP22
25VGJP22	-	25VEUP22
25VGKP22	-	25VFTP22
25VGLP22	-	25VFMP22
25VGMP22	-	25VGEP22
25VGNP22	-	-
25VGQP22	-	25VHAP22
25VGRP22	-	-
25VGSP22	-	25VHGP22
25VGTP22	-	-
25VGUP22	-	-
25VGWP22	-	-
25VGXP22	-	-
25VGYP22	-	-
25VGZP22	-	-
25VHAP22	-	-
25VHBP22	-	-
25VHCP22	-	25VGZP22

Tube Type	Basing	Replacement
25VHDP22	-	-
25VHEP22	-	25VHGP22
25VHFP22	-	-
25VHGP22	-	-
25VHJP22	-	25VHAP22
25VHKP22	-	-
25VHLP22	-	25VHBP22
25VHMP22	-	25VHJP22, 25VHAP22
25VHNP22	-	25VHAP22
25VHQP22	-	25VHJP22, 25VHAP22
25VHRP22	-	25VGNP22
25VHSP22	-	-
25VHTP22	-	25VGEP22
25VHUP22	-	-
25VP22	-	-
25WP22	14BE	23VAQP22, 23VABP22, 23VALP22, 23VANP22, 25AP22, 25XP22, 23VARP22, 25AJP22, 25ABP22, 25ANP22
25XP22	14BE	23VAHP22, 23VALP22, 23VAQP22, 23VANP22, 23VARP22, 25AP22, 25XP22, 25ABP22, 25BGP22, 25GP22A
25YP22	14BE	23VANP22, 23VAQP22, 25AP22, 25XP22, 25AJP22, 25BP22, 25BP22A
25ZP22	14BE	23VALP22, 23VANP22, 23VAQP22, 23VARP22, 25AP22, 25XP22, 25AJP22
26AP22	14BE	25VABP22, 25VAQP22, 25VCKP22, 25VDXP22
26DP22	14BE	25VAQP22, 25VBZP22, 25VCUP22, 25VCWP22, 25VCZP22, 25VDCP22
26EP22	14BE	25VAXP22, 25VBEP22, 25VCSP22, 25VDGP22, 25VDHP22
26FP22	14BE	25VABP22, 25VAEP22, 25VBZP22, 25VCDP22, 25VCUP22, 25VCZP22, 25VDXP22, 26DP22, 25VAQP22
26GP22	14BE	25VAQP22, 25VBZP22, 25VCUP22, 25VCWP22, 25VDCP22, 26DP22
26HP22	14BE	25VAGP22, 25VBEP22, 25VBKP22, 25VBYP22, 25VCTP22, 25VCXP22, 25VDFP22
26KP22	14BE	25VABP22, 25VAQP22, 25VBZP22, 25VCUP22A, 25VCWP22, 25VCZP22, 26DP22
27ABP4	8HR	27ADP4, 27AGP4, 27AFP4%
27ACP4	12L	27YP4
27ADP4	8HR	27AFP4%
27AEP4	8HR	27ZP4%
27AFP4	8HR	27ADP4%
27AGP4	8HR	27ABP4, 27ADP4, 27AFP4%
27AP4	12M	-
27EP4	12D	27GP4
27GP4	12D	27EP4
27LP4	12N	-
27MP4	12D	-
27NP4	12N	27RP4

Tube Type	Basing	Replacement
27RP4	12N	27NP4, 27RP4A~
27RP4A	12N	27NP4, 27RP4~
27SP4	12L	27UP4~, 27VP4~, 27XP4~
27UP4	12L	27SP4, 27VP4~, 27XP4~
27VP4	12L	27XP4
27WP4	12AJ	-
27XP4	12L	-
27YP4	12L	27ACP4
27ZP4	8HR	27AEP4%
30BP4	12D	-
37GDA85X	-	-
37GDA86X	-	-
37JGG68	-	-
51GGB91X	-	-
51GGB95Y	-	-
140BXB4	-	-
140CYB4	-	-
150ARB22	-	-
150AYB22	-	-
150CAB22	-	-
160CB22	-	-
160DB22	-	-
160DB22-TC02	-	-
160DB22-TC04	-	-
180RB22B	-	-
180RB22G	-	-
180RB22R	-	-
200LB22	-	-
200XB4	-	-
230ACB4	7GR	9AGP4, 9YP4
230AUB4	-	-
230RB4	7GR	9WP4
230TB4	7GR	9WP4
230ZB4	7GR	9YP4
250RB22	-	-
250TB22	-	-
270AEB22	-	-
270AKB22	-	-
270ALB22	-	-
270JB22	-	-
270KB22	-	-
270KCB22	-	-
270LB22	-	270KB22
270MB22	-	-
270NB22	-	-
270RB22	-	-
270SB22	-	-
270TB22	-	-
270UB22	-	-
270VB22	-	-
310AVB4	7GR	12CNP4
310EUB4	-	-
310FGB22	-	-
310GNB4	-	-
310HEB4	-	-
320ADB22	-	-
320AEB22	-	13JP22
320AGB22	-	-
320AHB22	-	-
320AKB22	-	-
320ARB22	-	-
320AWB22	-	320BUB22Q
320AYB22	-	-
320BJB22	-	-
320BKB22	-	-
320BMB22	-	-
320BNB22	-	-
320BTB22	-	-
320BUB22Q	-	-
320CB22	-	-
320GB22	14BH	12DCP22, 13GP22
320HB22	-	-
320KB22	-	13GP22
320LB22	-	13JP22
320NB22	-	-
320RB22	-	13GP22
320WB22	-	-
330ADB22	-	-
330GB22	-	-
370AB22	14BH	14VADP22, 15NP22
370ACB22	-	-
370AFB22	-	-
370AGB22	-	-
370ARB22	-	-
370AUB22	-	-
370AWB22	-	-
370BKB22	-	-
370BVBK1U	-	-
370BYB22-TC01	-	-
370CAB22	-	-
370CB22	14BH	14VAMP22, 15AEP22, 15NP22
370CFB22	-	-
370CGB22	-	-
370CHB22	-	370CUB22
370CNB22	-	-
370CR22	14BH	15NP22
370CRB22-TC03	-	370EHB22-TC01
370CSB22-TC01	-	-
370CUB22	-	-
370DEB22	-	-
370DFB22	-	370FMB22
370DHB22	-	-
370DJB22	-	-
370DJB22(LY)	-	-
370DJB22-TC09	-	-
370DJB22-TC12(Y)	-	-
370DKB22	-	-
370DRB22-TC01	-	370EHB22-TC01
370DYB22	-	-
370DZB22	-	370FNB22
370EB22	-	-
370ECB22	-	-
370EGB22	-	-
370EHB22-TC01	-	-
370EKB22	-	-
370ESB22	-	-
370ETB22	-	-
370FCB22	-	-
370FCB22A	-	-
370FJB22	-	-
370FLB22	-	370FLB22Q
370FMB22	-	-
370FNB22	-	-
370FTB22	-	-
370FVB22	-	-
370FXB22	-	-
370GHB22	-	-
370GHP22	-	-
370GKB22	-	-
370HAB22	-	-
370HEB22	-	370HRB22
370HRB22	-	-
370HYB22	-	-
370IB22	-	13VBFP22

Tube Type	Basing	Replacement
370JDB22	-	13VBLP22
370JFB22	-	13VBFP22
370JKB22	-	370FCB22
370JLB22	-	-
370JTB22	-	-
370JXB22	-	-
370KAB22	-	-
370KLB22	-	370KUB22
370KNB22	-	-
370KSB22	-	-
370KSB22(VY)	-	-
370KSB22(Y)	-	-
370KSB22(YB)	-	-
370KSB22-TC01	-	-
370KSB22-TC02	-	-
370KSB22-TC12	-	-
370KSB22-TC13	-	-
370KSB22-TC15	-	-
370KTB22	-	-
370KUB22	-	-
370KZB22	-	-
370LFB22	-	-
370LGB22	-	-
370LGB22-TC01	-	-
370LGB22-TC09	-	-
370LHB22	-	-
370LJB22	-	-
370LMB22	-	-
370LUB22	-	370MVB22, 370KUB22
370MEB22	-	-
370MLB22	-	-
370MVB22	-	-
370NB22	-	13VAEP22
370NVB22	-	-
370NWB22	-	-
370PJB22	-	-
370RKB22	-	-
370RKB22-TC14	-	-
370SKB22Y	-	-
370WB22	-	370AUB22
400AFB4A	8HR	16BGP4
400BDB22	-	-
400BNB22	-	-
400BRB22	-	-
400CLB22	-	-
400CUB22	-	-
400CVB22	14BH	16CYP22
400DFB22	-	-
400DKB22	-	-
400DMB22	-	-
400EHB22	-	-
410AYB22	-	-
410BRB22	-	-
420AB22	-	-
420ACB22	-	-
420ADB22	-	-
420AFB22	-	-
420AGB22	-	-
420AHB22	-	-
420AKB22	-	-
420ALB22	-	-
420AMB22	-	-
420ASB22	-	-
420ATB22	-	-
420AUB22	-	420AZB22
420AVB22	-	420BAB22
420AWB22	-	-
420AYB22	-	-
420AZB22	-	-
420BAB22	-	-
420BCB22	-	-
420BFB22	-	-
420BGB22	-	-
420BHB22	-	-
420BJB22	-	-
420BKB22	-	-
420BLB22	-	-
420BMB22	-	-
420BNB22	-	-
420BRB22	-	-
420BSB22	-	-
420BTB22	-	-
420BUB22	-	-
420BVB22	-	-
420BZB22	-	420BAB22
420CB22	-	-
420CDB22	-	-
420CEB22	-	-
420CFB22	-	-
420CUB22	-	-
420CYB22	-	-
420CZB22	-	-
420DB22	-	-
420DCB22	-	420AZB22
420DHB22	-	-
420DMB22	-	-
420EB22	-	-
420ESB22	-	-
420ETB22	-	-
420FB22	-	-
420FTB22	-	-
420FTB22(Y)	-	-
420GB22	-	-
420HB22	-	-
420JB22	-	-
420KB22	-	-
420LB22	-	-
420MB22	-	-
420NB22	-	-
420RB22	-	-
420SB22	-	-
420UB22	-	-
420WB22	-	-
420XB22	-	-
420YB22	-	-
430DB22	-	-
440ACB4	8HR	16VAJP4, 17ESP4
440AJB22	-	-
440AKB22	-	-
440ALB22	-	-
440ASB22	-	-
440AXB22	-	-
440AZB22	-	-
440BCB22	-	-
440BDB22	-	-
440BEB22	-	16VAUP22
440BFB22	-	-
440BGB4	-	-
440BHB22	-	-
440BKB22	-	17FAP22
440BNB22	-	-
440CLB22	-	-
440HB4	8HR	17EMP4, 17ESP4

Tube Type	Basing	Replacement
440VB4	8HR	17EWP4
470AB4	8HR	19AVP4
470ACB4	8HR	19AYP4
470BFB22	-	-
470BGB22	-	-
470BHB22	-	-
470BIB22	-	-
470BJB22	-	-
470BKB22	-	-
470BLB22	-	-
470BMB22	-	-
470BNB22	-	-
470BRB22	-	-
470BTB22	-	-
470BUB22	-	-
470BVB22	-	-
470BWB22	-	-
470BXB22	-	-
470BYB22	-	-
470CAB22	-	-
470CB4	8HR	19AVP4
470CDB22	-	-
470CEB22	-	-
470CFB22	-	-
470CJB22	-	-
470CKB22	-	-
470CLB22	-	-
470CNB22	-	-
470CTB22	-	-
470CVB22	-	-
470CWB22	-	-
470CYB22	-	-
470CZB22	-	-
470DEB22	-	-
470DFB22	-	-
470DHB22	-	-
470DKB22	-	-
470DMB22	14BH	-
470DTB22	-	-
470DUB22	-	-
470DXB22	-	-
470ECB22	-	-
470EHB22	-	-
470EJB22	-	-
470EMB22	-	-
470EVB22	-	-
470EYB22	-	-
470EZB22	-	-
470FCB22	-	-
470FGB22	-	470GNB22
470FJB22	-	-
470FRB22	-	-
470FTB22	-	-
470FXB22	-	-
470GEB22	-	-
470GEB22-TC05	-	-
470GMB22	-	-
470GNB22	-	-
470GVB22	-	-
470GZB22	-	-
470HAB22	-	470GEB22
470HAB22-TC01	-	470GEB22-TC01
470HDB22	-	-
470HEB22	-	-
470HFB22	-	-
470HJB22	-	-
470HRB22	-	-
470HTB22	-	-
470HUB22	-	-
470HWB22	-	-
470HXB22	-	-
470JAB22	-	470GEB22
470JWB22	-	-
470JZB22	-	-
470KCB22	-	-
470KDB22	-	-
470KRB22	-	-
470KUB22	-	-
470KZB22	-	-
470LXB22	-	-
470RDB22	-	-
470RDB22-TC03	-	-
490AB22	14BE	18VAHP22^, 19EXP22^, 19GYP22^
490ACB22	14BE	18VAHP22^, 18VASP22, 19EXP22^, 19GYP22^
490ADB22	14BE	18VAHP22^, 18VASP22, 19EXP22^, 19GYP22^
490AEB22	14BE	18VASP22, 18VANP22^, 19EYP22^, 19GWP22^
490AFB22	14BE	18VASP22, 18VAHP22^, 19EYP22^, 19GWP22^
490AGB22	14BE	18VAHP22^, 18VASP22, 19HCP22, 19EYP22^, 19GWP22^
490AHB22	14BE	18VASP22, 18VAHP22, 19EXP22, 19GYP22^
490AHB22A	14BE	18VAHP22, 19EXP22^, 19GVP22^
490AHB22B	-	-
490AJB22	14BE	18VAHP22, 18VASP22, 19HCYP22, 19EYP22, 19GWP22
490AJB22A	14BE	18VAHP22, 19EYP22, 19GWP22
490AJB22B	-	-
490AKB22	14BE	18VAHP22^, 18VASP22, 19EYP22^, 19EXP22, 19GVP22^
490ALB22	14BE	18VANP22^, 18VASP22, 19EXP22, 19EYP22^, 19GVP22^
490AMB22	14BE	18VAHP22^, 18VASP22, 18VAMP22, 19EXP22^, 19GVP22^
490AMB22A	-	-
490ANB22	14BE	18VASP22, 18VAHP22^, 19EXP22^, 19GVP22^
490ARB22	-	18VAHP22, 18VASP22, 19HCYP22, 19EYP22
490ARBP22	14BE	18VAHP22^, 19EYP22^, 19GWP22^
490ASB22	14BE	18VASP22, 18VAHP22, 19HCP22, 19EYP22,EYP22^, 19GWP22
490ASB22A	-	-
490ASB22B	-	-
490AXB22	-	18VAKP22
490BAB22	14BE	18VAHP22, 18VASP22, 19EXP22, 19GVP22, 19GXP22
490BCB22	14BE	18VAHP22, 18VASP22, 19HCP22, 19GWP22, 19EYP22

PICTURE TUBES

Tube Type	Basing	Replacement
490BDB22	14BE	18VASP22, 18VAHP22, 19HCP22, 19EYP22, 19GWP22^, 19GWP22^
490BDB22C	-	-
490BDB22D	-	-
490BGB22	14BE	18VASP22, 18VAHP22, 19GVP22, 19EP22, 19EXP22
490BHB22	14BE	18VAHP22, 18VASP22, 19HCP22, 19EXP22, 19GVP22
490BKB22	14BE	18VAKP22
490BLB22	-	-
490BNB22	14BH	19HNP22, 19JWP22
490BRB22	14BE	18VAHP22, 18VASP22, 19EYP22, 19GWP22
490BSB22	-	19HNP22, 19JWP22
490BUB22	14BE	18VAKP22
490BVB22	14BH	19HNP22, 19JWP22
490BXB22	14BH	19HNP22, 19JWP22
490BYB22	-	490BLB22
490BZB22	-	-
490BZB22A	-	-
490CAB22	-	18VAHP22, 18VASP22, 19HCP22, 19EYP22
490CAB22A	-	-
490CB22	14BE	18VAHP22, 18VASP22, 19EXP22^, 19GVP22^
490CDB22	-	19HNP22
490CHB22	14BE	18VAHP22, 18VASP22, 19EXP22
490CHB22A	-	-
490CJB22	-	-
490CMB22	-	490BLB22
490CSB22	-	18VAHP22, 18VASP22, 19HCP22, 19EYP22
490CSB22A	-	-
490CUB22	14BE	18VAHP22, 18VASP22, 19HCP22, 19EYP22
490DB22	14BE	18VAHP22, 18VASP22, 19HCP22, 19EYP22, 19EXP22^, 19GVP22^
490DB22A	-	-
490DCB22	-	18VAKP22
490DEB22	-	19HNP22, 19JWP22
490DHB22	-	18VBDP22, 18VBEP22,
490DJB22	-	18VBDP22, 18VBEP22
490DKB22	-	18VBDP22, 18VBEP22
490EB22	14BE	18VAHP22^, 18VASP22, 19EXP22^, 19GVP22^
490EB22A	14BE	18VAHP22^, 19EXP22^, 19GVP22^
490FB22	14BE	18VAHP22^, 18VASP22, 19EXP22^, 19GVP22^
490GB22	14BE	18VAHP22^, 18VASP22, 19EXP22^, 19GVP22^
490HB22	14BE	18VAHP22^, 18VASP22, 19EXP22, 19GVP22
490JB22	14BE	18VAHP22^, 18VASP22, 19EXP22, 19GVP22
490JB22A	14BE	18VAHP22^, 19EXP22, 19GVP22
490KB22	14BE	18VAHP22^, 18VASP22, 19EXP22^, 19GVP22
490KB22A	14BE	18VAHP22^, 19EXP22^, 19GVP22^
490LB22	14BE	18VAHP22^, 18VASP22, 19EXP22^, 19GVP22^
490MB22	-	-
490MBP22	14BE	18VAHP22^, 19EXP22^, 19GVP22^
490NB22	14BE	18VAHP22, 18VASP22, 19HCP22, 19EYP22^, 19GWP22^
490RB22	14BE	18VANP22, 18VASP22, 19HCP22, 19EYP22^, 19GWP22^
490SB22	14BE	18VANP22, 18VASP22, 19HCP22, 19EYP22^, 19GWP22
490TB22	14BE	18VAHP22, 18VASP22, 19HCP22, 19EYP22^, 19GWP22^
490UB22	14BE	18VAHP22^, 18VASP22, 19EXP22^, 19CVP22^
490VB22	-	18VAHP22, 18VASP22, 19HCP22, 19EYP22
490VBB22	14BE	18VAHP22^, 19EXP22^, 19GVP22^
490WB22	14BE	18VAHP22^, 18VASP22, 19EYP22^, 19GWP22^
490XB22	14BE	18VAHP22^, 19EYP22^, 19GWP22^
490XP22	-	18VAHP22, 18VASP22, 19HCP22, 19EYP22
490YB22	14BE	18VAHP22^, 18VASP22, 19HCP22, 19EYP22^, 19GWP22^
490ZB22	14BE	18VAHP22^, 18VASP22, 19HCP22, 19EYP22^, 19GWP22^
500KB4	8HR	20TP4
510AB22	-	19VBRP22, 19VMFP22, 510ABEB22, 510ABFB22, 510ACFB22
510ABEB22	-	19VBRP22, 19VMFB22, 510ABEB22, 510ABFB22, 510ABYB22
510ABFB22	-	19VBRP22, 19VMFP22, 510ABEB22, 510ABYB22
510ABMB22	-	19VBRP22, 19VMFP22, 510ABEB22, 510ABFB22, 510ABYB22, 510ACFB22
510ABYB22	-	19VBRP22, 19VMFP22, 510ABEB22, 510ABFB22, 510ACFB22
510ACB22	-	19VBRP22, 19VMFP22, 510ABEB22, 510ABFB22, 510ABYB22, 510ACFB22
510ACFB22	-	19VBRP22, 19VMFP22, 510ABEB22, 510ABFB22, 510ABYB22, 510ACFB22
510ADB22	-	19VBRP22, 19VMFP22, 510ABEB22, 510ABFB22, 510ABYB22, 510ACFB22
510AEB22	-	-
510AFB22	-	19VBRP22
510AGB22	-	19VBRP22
510AHB22	-	-
510AJB22	-	19VBRP22
510AKB22	-	-
510ALB22	-	510CZB22
510AMB22	-	19VBRP22
510ANB22	-	510DZB22
510ARB22	-	19VCBP22

Tube Type	Basing	Replacement
510ASB22	-	19VBRP22
510ATB22	-	19VBRP22
510AUB22	-	510AUB22
510AVB22	-	510CZB22
510AWB22	-	19VBRP22
510AXB22	-	19VBRP22
510AYB22	-	19VBRP22
510AZB22	-	19VBRP22
510BAB22	-	510DZB22
510BCB22	-	510DZB22
510BDB22	-	510DZB22
510BEB22	-	510DZB22
510BFB22	-	510DZB22
510BGB22	-	19VBRP22
510BHB22	-	19VATP22, 19VBRP22
510BJB22	-	510DZB22
510BLB22	-	-
510BNB22	-	19VBRP22
510BRB22	-	19VBRP22
510BSB22	-	19VBRP22
510BTB22	-	19VBRP22
510BUB22	-	19VBRP22
510BVB22	-	19VBRP22
510BWB22	-	510DZB22
510BXB22	-	510DZB22
510BYB22	-	510DZB22
510BZB22	-	-
510CAB22	-	510DZB22
510CB22	-	19VBRP22
510CDB22	-	-
510CEB22	-	-
510CFB22	-	-
510CGB22	-	19VBRP22
510CHB22	-	-
510CJB22	-	-
510CKB22	-	-
510CLB22	-	-
510CMB22	-	19VCBP22
510CNB22	-	19VBRP22
510CRB22	-	19VBRP22
510CSB22	-	19VBRP22
510CTB22	-	510CZB22
510CUB22	-	19VBRP22
510CVB22	-	510DZB22
510CWB22	-	19VBRP22
510CXB22	-	19VBRP22
510CYB22	-	510CZB22
510CZB22	-	-
510DAB22	-	-
510DB22	-	-
510DCB22	-	510DZB22
510DEB22	-	-
510DFB22	-	19VBQP22
510DJB22	-	-
510DKB22	-	-
510DLB22	-	-
510DRB22	-	510AEB22
510DSB22	-	510AEB22
510DTB22	-	-
510DVB22	-	-
510DXB22	-	-
510DZB22	-	-
510EB22	-	510DZB22
510EFB22	-	510BLB22
510ELB22	14BE	19VBRP22, 19VEDP22
510ERB22	14BE	19VBRP22
510ETB22	-	19VEDP22

Tube Type	Basing	Replacement
510FB22	-	510DZB22
510FLB22	-	510GLB22
510FVB22-TC01	-	-
510FWB22	-	-
510FXB22	-	-
510FYB22	-	19VJCP22Y
510FZB22	-	-
510GB22	-	-
510GCB22-TC01	-	510LNB22-TC01
510GJB22	-	-
510GLB22	-	-
510GSB22	-	19VJCP22
510HCB22	-	-
510HFB22-TC02	-	19VEJTC02
510HFB22-TC05	-	19VEJTC05
510HJB22	-	-
510HKB22	-	-
510HMB22TC01	-	510LNB22TC01
510HNB22	-	-
510HRB22	-	19VJXP22Y
510HSB22	-	-
510HWB22	-	-
510HZB22	-	-
510JB22	-	510CZB22
510JGB22	-	-
510JHB22	-	-
510JUB22-TC03	-	510LRB22-TC01
510JVB22	-	19VJXP22
510JYB22	-	19VJXP22
510JZB22-TC01	-	510LNB22-TC01
510KB22	-	510DZB22
510KDB22	-	-
510KEB22	-	-
510KHB22	-	-
510KLB22	-	510MVB22
510KTB22	-	19VJXP22
510KWB22	-	19VJCP22
510LAB22	-	-
510LB22	-	-
510LCB22	-	19VJXP22
510LCB22(Y)	-	-
510LCB22-TC11	-	-
510LDB22	-	19VJXP22
510LEB22	-	19VJXP22
510LGB22	-	-
510LNB22-TC01	-	-
510LRB22-TC01	-	-
510LTB22	-	-
510LUB22	-	-
510LZB22	-	19VJCP22
510MAB22	-	-
510MB22	-	19VBRP22
510MCB22	-	-
510MDB22	-	-
510MHB22	-	19VJCP22
510MJB22	-	-
510MKB22	-	-
510MRB22	-	19VJCP22
510MVB22	-	-
510MWB22	-	A48ABW20X, A48ABW25X
510NB22	-	19VBRP22
510NEB22	-	19VJXP22
510NFB22-TC01	-	510LNB22-TC01
510NHB22	-	19VJXP22
510NJB22	-	A48ABW20X, A48ABW25X
510NLB22	-	19VJXP22
510NVB22	-	-

Tube Type	Basing	Replacement
510NYB22	-	-
510RB22	-	19VBRP22
510RCB22	-	19VJXP22
510RDB22	-	19VJXP22
510REB22	-	-
510RJB22	-	-
510RKB22	-	-
510RSB22	-	A48ABW20X, A48ABW25X
510RUB22	-	19VJXP22
510RWB22	-	-
510RWB22-TC01	-	510LRB22-TC01
510SB22	-	19VBRP22
510SFB22	-	-
510SGB22	-	-
510SJB22	-	-
510SKB22	-	19VLXP22
510SLB22	-	19VLXP22
510TB22	-	510DZB22
510TDB22	-	510KHB22
510TFB22	-	-
510TFB22AW	-	-
510TLB22	-	19VJDP22
510TSB22	-	19VLXP22
510UB22	-	510CZB22
510UEB22	-	19VMFP22
510UEB22-TC10	-	-
510UEB22-TC11	-	-
510UEB22-TC13(Y)	-	-
510UEB22TC-03	-	-
510UFB22	-	19VJTP22
510UJB22	-	19VJTP22
510UKB22	-	A48ABW20X, A48ABW25X
510UTB22	-	19VMFP22
510UXB22	-	19VJXP22
510UZB22	-	19VLXP22
510VB22	-	510DZB22
510WB22	-	510DZB22
510WEB22	-	19VMFP22
510WNB22	-	19VMFP22
510WNB22-TC01	-	-
510XB22	-	510DZB22
510XDB22	-	-
510XTB22	-	-
510YB22	-	510DZB22
510YTB22	-	19VLXP22
510YTB22-TC03	-	-
510YWB22	-	A48ABK01X, A48ABK05X
510YWB22-TC05	-	-
510YWB22A	-	-
510YWB22A-TC02(Y)	-	-
510ZB22	-	510AKB22
510ZDB22	-	-
510ZEB22	-	-
520XB22	-	-
550AB22A	-	22JP22
670BDB22	-	670BTB22, 670CTB22, 670CUB22, 670DRB22, 670EAB22, 670EKB22, 670ETB22
670BTB22	-	670BDB22, 670CTB22, 670CUB22, 670DRB22, 670EAB22, 670EKB22, 670ETB22
670CTB22	-	670BDB22, 670BTB22, 670CUB22, 670DRB22, 670EAB22, 670EKB22, 670ETB22
670CUB22	-	670BDB22, 670BTB22, 670CTB22, 670DRB22, 670EAB22, 670EKB22, 670ETB22
670DRB22	-	670BDB22, 670BTB22, 670CTB22, 670CUB22, 670EAB22, 670EKB22, 670ETB22
670EAB22	-	670BDB22, 670BTB22, 670CTB22, 670CUB22, 670DRB22, 670EKB22, 670ETB22
670EKB22	-	670BDB22, 670BTB22, 670CTB22, 670CUB22, 670DRB22, 670EAB22, 670ETB22
670ETB22	-	670BDB22, 670BTB22, 670CTB22, 670CUB22, 670DRB22, 670EAB22, 670EKB22
710AB22	-	-
3701	-	370FCB22
3701B22	-	-
3702	-	-
3702B22-TC	-	-
3707	-	-
3708	-	370HRB22
3710	-	370FCB22
3720	-	370FXB22
3720B22	-	-
3901B22	-	-
4201B22	-	-
5108B22	-	19VMFP22
5108B22-TC20(C)	-	-
5108B22TC	-	-
5110B22	-	A48ABW20X, A48ABW25X
5120B22	-	A48ABW20X, A48ABW25X
(VB)E2280	-	-
A20JKU10X	-	-
A22JCM00X	-	-
A22JCN00X	-	-
A26JAS30X	-	-
A26JAS31X	-	-
A26JGZ01X02	-	-
A26JGZ31X	-	-
A26JGZ91X	-	-
A29JBP00X	-	-
A31-265W	7GR	12VAMP4
A33AAB10X	-	-
A33AAB15X	-	-
A33AAB16X	-	A33AAB10X
A33AAB20X	-	A33AAB30X, A33AAB35X
A33AAB30X	-	A33AAB30X, A33AAB35X
A33AAB35X	-	A33AAB30X, A33AAB35X
A33AAB36X	-	-
A33AAF00X	-	-
A33AAL01X	-	-
A33AAL02X	-	-
A33AAL03X	-	-
A33AAS40X	-	-
A33AAS50X	-	-
A33ABE60X	-	-
A33ABE61X	-	-
A33ABE62X	-	-
A33ABE67X	-	-
A33ABL20X	-	-
A33ABL30X	-	-

Tube Type	Basing	Replacement
A33ABM26X	-	-
A33ABM40X	-	-
A33ABW60X	-	-
A33ABW62X	-	-
A33ACB00X	-	-
A33ACB01X	-	A33AAB30X, A33AAB35X
A33ACB02X	-	A33ACB00X
A33ACB21X	-	A33ACB30X, A33ACB35X
A33ACB51X	-	A33ACB00X
A33ACD02X	-	A33ACD30X
A33ACD25X	-	-
A33ACD30X	-	-
A33ACD42X	-	A33ACD25X
A33ACF60X	-	-
A33ACS01X	-	-
A33ADU00X	-	-
A33ADU96X	-	-
A33JKL00X	-	A33AAB10X
A33JKL60X	-	A33AAB10X
A34EAC50X	-	-
A34EAE00X	-	A34EAE01X
A34EAE00X-AT1620	-	-
A34EAE01X	-	-
A34EAEAT1620	-	-
A34JAJ01X	-	A34JAN90X
A34JAN00X	-	-
A34JAN30X	-	A34JAN90X
A34JAN31X	-	A34JAN90X
A34JAN32X	-	A34JAN90X
A34JAN40X	-	A34JAN90X
A34JAN51X	-	A34JAN00X
A34JAN80X	-	A34JAN90X
A34JAN90X	-	-
A34JAN92X	-	-
A34JAN99X	-	-
A34JBB00X	-	-
A34JBB02X01	-	-
A34JBB02X06	-	-
A34JBB02X	-	-
A34JBB60X	-	-
A34JBH53X	-	-
A34JBH80X	-	-
A34JBH99X	-	-
A34JBU00X	-	-
A34JBU10X	-	-
A34JCD00X	-	A34JCD90X
A34JCD01X	-	A34JCD91X
A34JCD05X	-	-
A34JCD30X	-	A34JCD90X
A34JCD31X	-	A34JCD91X
A34JCD90X	-	-
A34JCD91X	-	-
A34JEB60X	-	-
A34JER00X01	-	-
A34JFN02X	-	-
A34JFQ40X	-	-
A34JFQ41X	-	A34JFQ40X
A34JFQ42X	-	A34JFQ40X
A34JFQ90X	-	-
A34JFY00X	-	A34JAN90X
A34JFY50X	-	A34JAN90X
A34JFY80X	-	A34JAN90X
A34JFY83X	-	A34JAN90X
A34JFY83X-MW	-	-
A34JFY90X	-	A34JAN90X
A34JHS10X	-	-
A34JJB92X	-	-

Tube Type	Basing	Replacement
A34JLL00X	-	A34JAN90X
A34JLL20X	-	-
A34JLN60X	-	A34JAN90X
A34JMW20X	-	-
A34JRY00X	-	-
A34JRY01X	-	-
A34JRY04X	-	-
A34JRY24X	-	-
A34JTJ00X	-	-
A34JTY21X	-	-
A36ACG10X	-	-
A36EBH01X	-	-
A36JAR01X	-	A63JAR50X
A36JAR10X	-	A36JAR50X
A36JAR32X	-	A34JAR50X
A36JAR39X	-	A36JAR50X
A36JAR50X01	-	-
A36JAR50X03	-	-
A36JAR50X07	-	-
A36JAR50X	-	-
A36JAR53X	-	A36JAR83X
A36JAR80X	-	A36JAR90X
A36JAR83X	-	-
A36JAR90X	-	-
A36JAR93X	-	-
A36JFT00X	-	-
A36JHP30X	-	-
A36JJR03X	-	-
A37-570X	-	13VBFP22
A38AAB10X	-	-
A38ABE60X	-	-
A38ABE61X	-	-
A38ABE62X	-	-
A38ABE63X	-	-
A38ABE67X	-	-
A38ABL30X	-	-
A38ABW60X	-	-
A38ABW61X	-	-
A38ABW62X	-	-
A38ACF60X	-	-
A38EAC50X	-	-
A38JAK70X	-	-
A38JBN90X	-	-
A38JDB36X	-	-
A38JFH00X	-	-
A38JFH30X	-	-
A38JFH90X	-	-
A38JJB90X	-	-
A38JRL00X	-	-
A38JSM00X	-	-
A41JHP90X	-	-
A41JHR00X	-	-
A43AAZ00X	-	-
A43AAZ21X	-	-
A43AAZ25X	-	-
A43ACR24X	-	-
A43ADU615X	-	-
A43ADU616X	-	-
A43JAY00X	-	-
A43JAY02X	-	-
A43JBE50X	-	-
A43JBT01X	-	-
A43JCT02X	-	A43JAY02X
A43JDS00X	-	-
A43JEH05X01	-	-
A43JEH05X	-	-
A43JFF00X	-	-

PICTURE TUBES

Tube Type	Basing	Replacement
A43JGW01X	-	-
A43JLP06X	-	-
A44-18W	8HR	19VCUP4
A44JFZ00X	-	-
A46JDL00X	-	-
A46JDL60X	-	-
A46JHL00X	-	-
A46JHR00X	-	-
A46JHR30X	-	-
A46JJL00X	-	-
A48AAB00X	-	-
A48AAB10X	-	A48AAB11X
A48AAB11X	-	A48AAB10X
A48AAB20X	-	A48AAB30X, A48AAB35X
A48AAB21X	-	A48AAB26X
A48AAB22X	-	A48AAB21X, A48AAB26X
A48AAB26X	-	A48AAB21X, A48AAB26X
A48AAB27X	-	A48AAB21X, A48AAB26X
A48AAB30X	-	A48AAB35X
A48AAB35X	-	A48AAB30X
A48AAB50X	-	A48AAB30X, A48AAB35X
A48AAB60X	-	A48AAB30X, A48AAB35X
A48AAD10X	-	-
A48AAG01X	-	A48AAN01X
A48AAG02X	-	A48AAN01X
A48AAJ00X	-	A48AAN01X
A48AAJ10X	-	A48AAN01X
A48AAJ33X	-	A48AAN01X
A48AAK03X	-	-
A48AAK04X	-	-
A48AAK05X	-	A48AAN01X
A48AAN01X	-	-
A48AAN02X	-	A48AAN01X
A48AAP02X	-	A48AAK04X
A48AAR00X	-	A48AAK04X
A48AAR21X	-	A48AAK03X
A48AAR25X	-	A48AAN01X
A48AAR28X	-	A48AAN01X
A48AAR29X	-	A48AAK03X
A48AAR30X	-	A48AAN01X
A48AAR50X	-	A48AAN01X
A48AAR60X	-	-
A48AAX01X	-	-
A48AAX02X	-	-
A48AAZ06X	-	A48AAZ26X
A48AAZ26X	-	-
A48AAZ40X	-	A48AAZ26X
A48ABB00X	-	-
A48ABE10X	-	-
A48ABE60X	-	-
A48ABE65X	-	-
A48ABH20X	-	A48AAN01X
A48ABH26X	-	A48AAN01X
A48ABK00X	-	A48ABK03X, A48ABK04X
A48ABK01X	-	A48ABK05X
A48ABK02X	-	A48ABK01X, A48ABK05X
A48ABK03X	-	A48ABK04X
A48ABK04X	-	A48ABK03X
A48ABK05X	-	A48ABK01X
A48ABK06X	-	A48ABK03X, A48ABK04X
A48ABK07X	-	A48ABK03X, A48ABK04X
A48ABS51X	-	-
A48ABS52X	-	-
A48ABT02X	-	A48AAN01X
A48ABT51X	-	A48AAN01X
A48ABT52X	-	A48AAN01X
A48ABW20X	-	A48ABW25X
A48ABW25X	-	A48ABW20X
A48ABW60X	-	A48ABW20X, A48ABW25X
A48ABY00X	-	A48ABW20X, A48ABW25X
A48ACB00X	-	A48AAB30X, A48AAB35X
A48ACB01X	-	A48AAB21X, A48AAB26X
A48ACB02X	-	A48AAB30X, A48AAB35X
A48ACB12X	-	A48AAB21X, A48AAB26X
A48ACB32X	-	A48AAB30X, A48AAB35X
A48ACB52X	-	A48AAB30X, A48AAB35X
A48ACD30X	-	-
A48ACD35X	-	-
A48ACE26X	-	-
A48ACE40X	-	A48ACE26X
A48ADE00X	-	A48ABK01X, A48ABK05X
A48ADE01X	-	A48ABK03X, A48ABK04X
A48ADE22X	-	A48ABK03X, A48ABK04X
A48ADU615X	-	-
A48ADU616X	-	-
A48EAB00X	-	-
A48EAB01X	-	-
A48JAF01X	-	-
A48JAF03X	-	A48AAN01X
A48JAH05X	-	-
A48JAK70X	-	-
A48JAN33X	-	A48ABK01X, A48ABK05X
A48JAN34X	-	A48ABK01X, A48ABK05X
A48JAN35X	-	A48ABK01X, A48ABK05X
A48JAN40X	-	-
A48JAN43X	-	A48ABK01X, A48ABK05X
A48JAN44X	-	A48ABK01X, A48ABK05X
A48JAN45X	-	A48ABK01X, A48ABK05X
A48JAN46X	-	A48ABK01X, A48ABK05X
A48JAN50X	-	A48ABK01X, A48ABK05X
A48JAN83X	-	A48ABK01X, A48ABK05X
A48JAN90X	-	A48ABK03X, A48ABK04X
A48JAN92X	-	-
A48JAU00X	-	-
A48JAZ70X	-	-
A48JAZ90X	-	-
A48JBH00X	-	A48ABK03X, A48ABK04X
A48JBH40X	-	A48ABK03X, A48ABK04X
A48JBK00X	-	-
A48JBM05X	-	-
A48JBM90X	-	-
A48JBS01X	-	-
A48JBV00X	-	-
A48JCL00X	-	-
A48JDD00X	-	A48ABK01X, A48ABK05X
A48JDD01X	-	A48ABK01X, A48ABK05X
A48JDD02X	-	A48ABK01X, A48ABK05X
A48JDK00X	-	-
A48JDK01X	-	-
A48JDS00X	-	-
A48JEH05X	-	-
A48JEX01X	-	-
A48JEX03X	-	-
A48JFH00X	-	-
A48JGR36X	-	-
A48JGR40X	-	-
A48JJB90X	-	-
A48JJY01X	-	-
A48JJY51X	-	-
A48JJY61X	-	-
A48JJY81X	-	-
A48JKL00X	-	-
A48JKL08X	-	A48AAB21X, A48AAB26X
A48JKL60X	-	-

Tube Type	Basing	Replacement
A48JKL68X	-	A48AAB21X, A48AAB26X
A48JLL20X	-	A48ABK01X, A48ABK05X
A48JMX20X	-	-
A48JNW00X	-	-
A48JNW20X	-	-
A48JPZ60X	-	-
A48JQA60X	-	-
A48JQG50X	-	A48ABK03X, A48ABK04X
A48JQG60X	-	A48ABK03X, A48ABK04X
A48JQG90X	-	A48ABK03X, A48ABK04X
A48JQG93X	-	A48ABK03X, A48ABK04X
A48JQH00X	-	A48ABK03X, A48ABK04X
A48JQH30X	-	A48ABK03X, A48ABK04X
A48JQH40X	-	A48ABK03X, A48ABK04X
A48JQH50X	-	A48ABK03X, A48ABK04X
A48JQH60X	-	A48ABK03X, A48ABK04X
A48JQH90X	-	A48ABK03X, A48ABK04X
A48JQH93X	-	A48ABK03X, A48ABK04X
A48JQL60X	-	-
A48JRV20X	-	-
A48JSN20X	-	-
A48JSZ60X	-	-
A48JTN31X	-	-
A48JUJ20X	-	-
A48KMW02XX	-	-
A49-12X	14BE	13VBFP22, 18VANP22, 19HCP22, 19GWP22
A49-14X	-	18VAHP22, 18VASP22, 19HCP22, 19EYP22
A49JLV50X	-	-
A51-572X	-	19VJXP22
A51-582X	-	19VLXP22
A51-592X	-	19VMFP22
A51ABU10X	-	-
A51ABU14X	-	-
A51ABU25X	-	-
A51ABU55X	-	-
A51ACB02X	-	-
A51ACB12X	-	A51ACG20X, A51ACG25X
A51ACC10X	-	-
A51ACG10X	-	-
A51ACG14X	-	A51ACG20X, A51ACG25X
A51ACG20X	-	A51ACG25X
A51ACG21X	-	-
A51ACG25X	-	A51ACG20X
A51ACG26X	-	A51ACG20X, A51ACG25X
A51ACG27X	-	A51ACG20X, A51ACG25X
A51ACG30X	-	A51ACG31X
A51ACG31X	-	A51ACG30X
A51ACG50X	-	-
A51ACG51X	-	-
A51ACG55X	-	-
A51ACG65X	-	A51ACG30X, A51ACG31X
A51ACT00X	-	A51ABU25X
A51ADL00X	-	-
A51ADM20X	-	-
A51ADM25X	-	-
A51ADM35X	-	-
A51ADR26X	-	A51ACG30X, A51ACG31X
A51ADR40X	-	A51ACG30X, A51ACG31X
A51ADU95X	-	-
A51ADU96X	-	-
A51ADU115X	-	-
A51ADU116X	-	-
A51ADY00X	-	-
A51ADY01X	-	-
A51ADY02X	-	-
A51EAF00X	-	-
A51EAT00X	-	-
A51EBD00X	-	A51ADL00X
A51JAR00X	-	A51JAR90X
A51JAR03X	-	A51JAR90X
A51JAR10X	-	A5JAR90X
A51JAR30X	-	A51JAR90X
A51JAR31X	-	A51JAR90X
A51JAR33X	-	A51JAR90X
A51JAR34X	-	A51JAR90X
A51JAR35X	-	A51JAR90X
A51JAR44X	-	A51JAR90X
A51JAR45X	-	A51JAR90X
A51JAR50X	-	A51JAR90X
A51JAR59X	-	-
A51JAR60X02	-	-
A51JAR60X	-	A51JAR90X
A51JAR61X	-	A51JAR90X
A51JAR62X	-	A51JAR90X
A51JAR65X	-	A51JAR96X
A51JAR66X	-	A51JAR96X
A51JAR67X	-	A51JAR96X
A51JAR83X	-	A51JAR90X
A51JAR90X	-	-
A51JAR91X	-	A51JAR90X
A51JAR92X	-	A51JAR90X
A51JAR93X	-	A51JAR96X
A51JAR95X	-	A51JAR96X
A51JAR96X	-	-
A51JAR97X	-	A51JAR96X
A51JAR98X	-	-
A51JAR99X	-	-
A51JCC01X	-	A51JCC11X
A51JCC03X	-	A51JCC11X
A51JCC09X	-	A51JCC11X
A51JCC11X	-	-
A51JCC21X	-	A51JCC11X
A51JCC23X	-	A51JCC11X
A51JCC51X	-	-
A51JCC61X	-	-
A51JCC90X	-	-
A51JCC91X	-	A51JCC90X
A51JEW00X	-	-
A51JFC00X	-	A51JFC60X
A51JFC01X	-	A51JFC60X
A51JFC09X	-	A51JFC60X
A51JFC30X	-	A51JFC60X
A51JFC60X	-	-
A51JFC61X	-	A51JFC60X
A51JFC80X	-	-
A51JJF00X	-	-
A51JJF01X	-	-
A51JJF60X	-	-
A51JJL00X	-	A51JJL90X
A51JJL30X	-	A51JJL90X
A51JJL90X	-	-
A51JKQ50X	-	-
A51JKY04X	-	-
A51JKY34X	-	-
A51JML03X	-	A51JML11X
A51JML11X	-	-
A51JML81X	-	-
A51JPD60X	-	-
A51JPE30X	-	-
A51JPE43X	-	A51JPE90X
A51JPE60X	-	-
A51JPE90X	-	-

PICTURE TUBES

Tube Type	Basing	Replacement
A51JPE96X	-	-
A51JPE99X	-	-
A51JQE01X	-	-
A51JRR60X	-	-
A51JSV11X	-	-
A51JSY01X	-	-
A51JTA60X	-	-
A51JUH50X	-	-
A51JZE23X	-	-
A51KKS42XX	-	-
A53EAB00X	-	-
A53JAQ36	-	-
A53JAQ36X01	-	-
A53JBM05X	-	-
A53JBM35X	-	-
A53JBM95X	-	-
A53JBW00X	-	-
A53JDE50X	-	-
A53JDE60X	-	-
A53JDH00X	-	-
A53JHA00X	-	-
A53JHK00X	-	-
A55-13X	14BE	20VAGP22, 22JP22, 22UP22
A56-120X	-	-
A56JKX00X	-	-
A56JKX30X	-	-
A56JKZ30X	-	-
A56JKZ60X	-	-
A56JMQ30X	-	-
A56JMQ60X	-	-
A59EAF00X	-	-
A59EAS00X	-	-
A59JJZ01X	-	-
A63-11X	-	25UP22
A63-16X	-	25UP22
A63-17X	-	25UP22
A63-18X	-	25UP22
A63-19X	-	25UP22
A63AAD10X	-	-
A63AAM00X	-	A63AAM05X
A63AAM01X	-	A63AAM05X
A63AAM02X	-	A63AAM10X
A63AAM03X	-	A63AAM11X
A63AAM04X	-	A63AAM11X
A63AAM05X	-	-
A63AAM06X	-	A63AAM11X
A63AAM07X	-	A63AAM05X
A63AAM08X	-	A63AAM10X
A63AAM09X	-	A63AAM10X
A63AAM10X	-	-
A63AAM11X	-	-
A63AAM12X	-	A63AAM10X
A63AAM13X	-	A63AAM05X
A63AAMM00X	-	A63AAM05X
A63AAT30X	-	-
A63AAT31X	-	A63ABZ26X
A63AAT32X	-	-
A63AAT36X	-	A63ABZ26X
A63AAX00X	-	A63ABZ26X
A63AAX01X	-	-
A63AAX02X	-	-
A63AAX12X	-	A63AAX02X
A63AAX42X	-	A63AAX02X
A63AAY00X	-	-
A63AAY01X	-	A63AAY00X
A63AAY02X	-	A63AAY00X
A63AAY11X	-	-
A63AAY12X	-	A63AAY11X
A63AAY71X	-	A63AAY11X
A63ABF00X	-	-
A63ABG05X	-	A63AAM11X
A63ABG20X	-	A63AAM10X
A63ABG25X	-	A63AAM10X
A63ABG27X	-	A63AAM05X
A63ABG35X	-	-
A63ABL15X	-	-
A63ABP00X	-	A63ABP05X
A63ABP05X	-	A63ABP00X
A63ABP10X	-	A63ABP12X
A63ABP11X	-	A63ABP13X
A63ABP12X	-	A63ABP10X
A63ABP13X	-	A63ABP11X
A63ABP20X	-	A63ABP25X
A63ABP25X	-	A63ABP20X
A63ABP26X	-	-
A63ABP50X	-	-
A63ABP51X	-	-
A63ABP55X	-	-
A63ABZ06X	-	A63ABZ26X
A63ABZ26X	-	-
A63ABZ40X	-	A63ABZ26X
A63ABZ60X	-	A63ABZ26X
A63ACM71X	-	-
A63ACM75X	-	-
A63ACP00X	-	A63ABG35X
A63ACP26X	-	A63AAM05X
A63ACP27X	-	A63AAM05X
A63ACQ00X	-	A63AAM10X
A63ACQ01X	-	A63AAM10X
A63ADG25X	-	A63AAM10X
A63ADG27X	-	A63AAM05X
A63ADG31X	-	-
A63ADG35X	-	A63AAM10X
A63ADN26X	-	A63AAM05X
A63ADN27X	-	A63AAM05X
A63ADN40X	-	A63AAM05X
A63AEH20X	-	A63AAM05X
A63JAV00X	-	-
A63JBG00X	-	-
A63JBG20X	-	-
A63JBG50X02	-	-
A63JBG50X	-	-
A63JDC05X01	-	-
A63JDF00X	-	-
A63JEC00X	-	-
A63JEC30X	-	-
A63JEZ00X	-	-
A63JGC01X	-	A63JGY93X
A63JGY33X	-	A63JGY93X
A63JGY93X	-	-
A63JKC00X	-	-
A63LQ90X	-	-
A66-501X	14BQ	24VABP22
A66AAM00X	-	-
A66AAM01X	-	-
A66AAM02X	-	-
A66AAM03X	-	-
A66AAM04X	-	-
A66ABU00X	-	A66ABU05X
A66ABU05X	-	A66ABU00X
A66ABU10X	-	-
A66ABU11X01	-	-
A66ABU11X	-	A66ABU10X
A66ABU14X01	-	-

Tube Type	Basing	Replacement
A66ABU14X	-	-
A66ABU30X	-	A66ABU35X
A66ABU32X	-	A66ABU37X
A66ABU35X	-	A66ABU30X
A66ABU36X	-	A66ABU39X
A66ABU37X	-	A66ABU32X
A66ABU38X	-	-
A66ABU39X	-	A66ABU36X
A66ABU40X	-	A66ABU51X
A66ABU50X	-	A66ACZ26X
A66ABU51X	-	-
A66ABU52X	-	A66ABU38X
A66ABU55X	-	-
A66ABU56X	-	A66ABU55X
A66ACZ26X	-	-
A66ACZ40X	-	A66ACZ26X
A66ADQ01X	-	A66ABU30X, A66ABU35X
A66ADS26X	-	A66ABU38X
A66ADS40X	-	A66ABU38X
A66ADT14X01	-	-
A66ADT14X	-	-
A66ADT16X01	-	-
A66ADT18X02	-	-
A66ADT20X01	-	-
A66ADU45X	-	-
A66ADU46X	-	-
A66ADU95X	-	-
A66ADU96X	-	-
A66AEK20X01	-	-
A66EAF00X	-	-
A66EAK50X01	-	-
A66EAK50X02	-	-
A66EAK50X03	-	-
A66EAK50X32	-	-
A66EAK50X	-	-
A66EAK51X01	-	-
A66EAK51X02	-	-
A66EAK51X03	-	-
A66EAK51X32	-	-
A66EAK51X	-	-
A66EAS00X	-	-
A66ECF00X	-	-
A66JKA01X	-	-
A66JLH50X	-	-
A66JLH70X03	-	-
A66JLH70X06	-	-
A66JLH70X	-	A66JLH50X
A66JLH86X	-	-
A66JMZ00X	-	-
A66JPP01X	-	A66ABU36X, A66ABU39X
A66JPP03X	-	A66ABU51X
A66JPQ01X	-	-
A67-100X	-	25VDMP22
A67-110X	-	-
A67JHG00X	-	-
A67JHG01X	-	-
A68ACC10X	-	-
A68ACC11X	-	A68ACC10X
A68ACC12X01	-	-
A68ACC12X	-	A68ACC10X
A68ACC14X01	-	-
A68ACC14X	-	A68ACC35X
A68ACC30X	-	A68ACC35X
A68ACC35X	-	-
A68ACT00X	-	-
A68ACT11X	-	-
A68ACT30X	-	A68ACT11X
A68ACT63X	-	A68ACT11X
A68ADZ00X	-	A68AEC00X
A68AEC00X	-	-
A68AEC01X	-	-
A68AEG10X01	-	-
A68AEG20X01	-	-
A68AEG20X02	-	-
A68AEG20X05	-	-
A68JMT10X	-	-
A68JMT50X	-	-
A76JTS90X	-	-
A76JTS99X	-	A76JTS90X
A76JYQ90X	-	-
A76JYQ96X	-	A76JYQ90X
A76JYQ98X	-	A76JYQ90X
A78JVB30X	-	-
A78JVB60X	-	-
A78JVB61X	-	A78JVB60X
A78JVB63X	-	A78JVB60X
A80EBP221X53	-	-
A80JTS96X01	-	-
A80JYV10X	-	-
A89JKA31X	-	-
A89JKA61X	-	A89JKA31X
A89JKA81X	-	-
A89JKA91X	-	A89JKA81X
A89JVU11X	-	-
A89JVU31X	-	-
A89JVU60X	-	-
A89JVU61X	-	-
A89JVU80X	-	-
A89JVU81X	-	-
A89JVU91X	-	-
AT1419L	-	-
AT1429RA	-	-
AT1429ZC	-	-
AW43-80	12AJ	17BTP4
AW43-88	8HR	17CVP4
AW53-80	12AJ	21CLP4
AW53-88	8HR	21DKP4
CPJ370BVBK1U	-	-
DA0515	-	-
E75Z12P39VRZ	-	-
J370BVBK1VST	-	-
M23JBZ00X	-	-
M29AEL10X	-	-
M29AEL60X	-	-
M29AEM10X	-	-
M29AEM60X	-	-
M29EAA	-	-
M29EAB	-	-
M29JAW40X	-	-
M29JHM00X02	-	-
M29JHM00X	-	-
M30JFX00X	-	-
M33AAW56XD02	-	-
M33ACD42Y	-	-
M33ACN40X	-	-
M34CRA61X	-	-
M34CRA70X	-	-
M34CRA71X	-	-
M34JAW43W	-	-
M34JBL09X	-	-
M34JBL60X01	-	-
M34JBL60X	-	-
M34JCA13X04	-	-
M34JCA13X37	-	-

Tube Type	Basing	Replacement
M35JFP8WW	-	-
M36AES60X	-	-
M41ECG0GH	-	-
M41ECG0WD	-	-
M46JGK00X	-	-
M48JFJ58X	-	-
M48JFK59X	-	-
M50-102W	-	-
M51JGK30X	-	-
M51JTZ91X	-	-
M66JHX00X	-	-
M66JHX30X	-	-
M68JUA21X	-	-
M68JUA98X	-	-
M68JUA99X	-	-
M78JXR25X	-	-
M79JWW11X	-	-
M79JWW31X	-	-
M79JWW61X	-	-
M79JWW81X	-	-
MV25VHMP22	-	-

Basing Diagrams

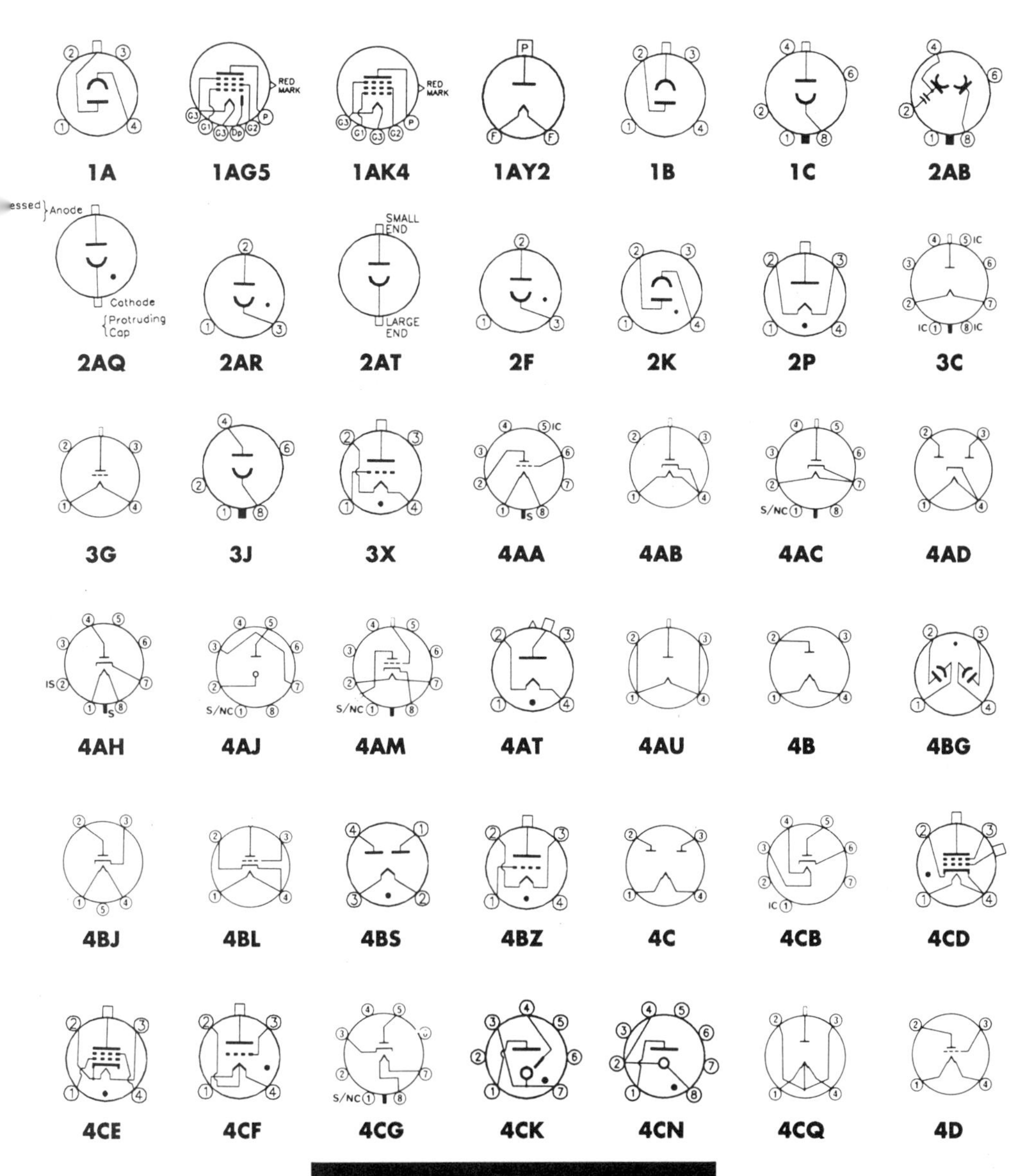

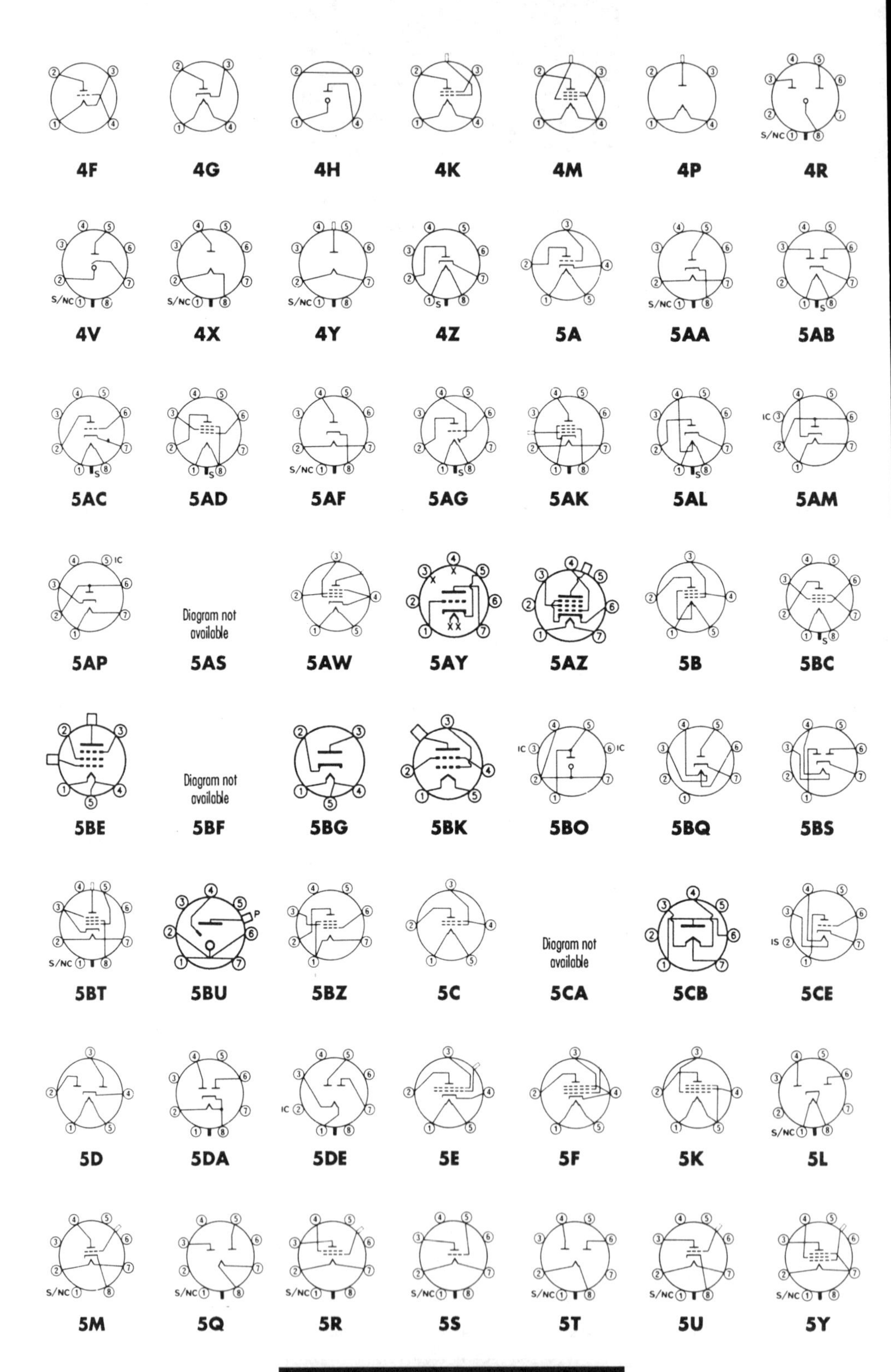
4F
4G
4H
4K
4M
4P
4R
4V
4X
4Y
4Z
5A
5AA
5AB
5AC
5AD
5AF
5AG
5AK
5AL
5AM
5AP
Diagram not available
5AS
5AW
5AY
5AZ
5B
5BC
5BE
Diagram not available
5BF
5BG
5BK
5BO
5BQ
5BS
5BT
5BU
5BZ
5C
Diagram not available
5CA
5CB
5CE
5D
5DA
5DE
5E
5F
5K
5L
5M
5Q
5R
5S
5T
5U
5Y

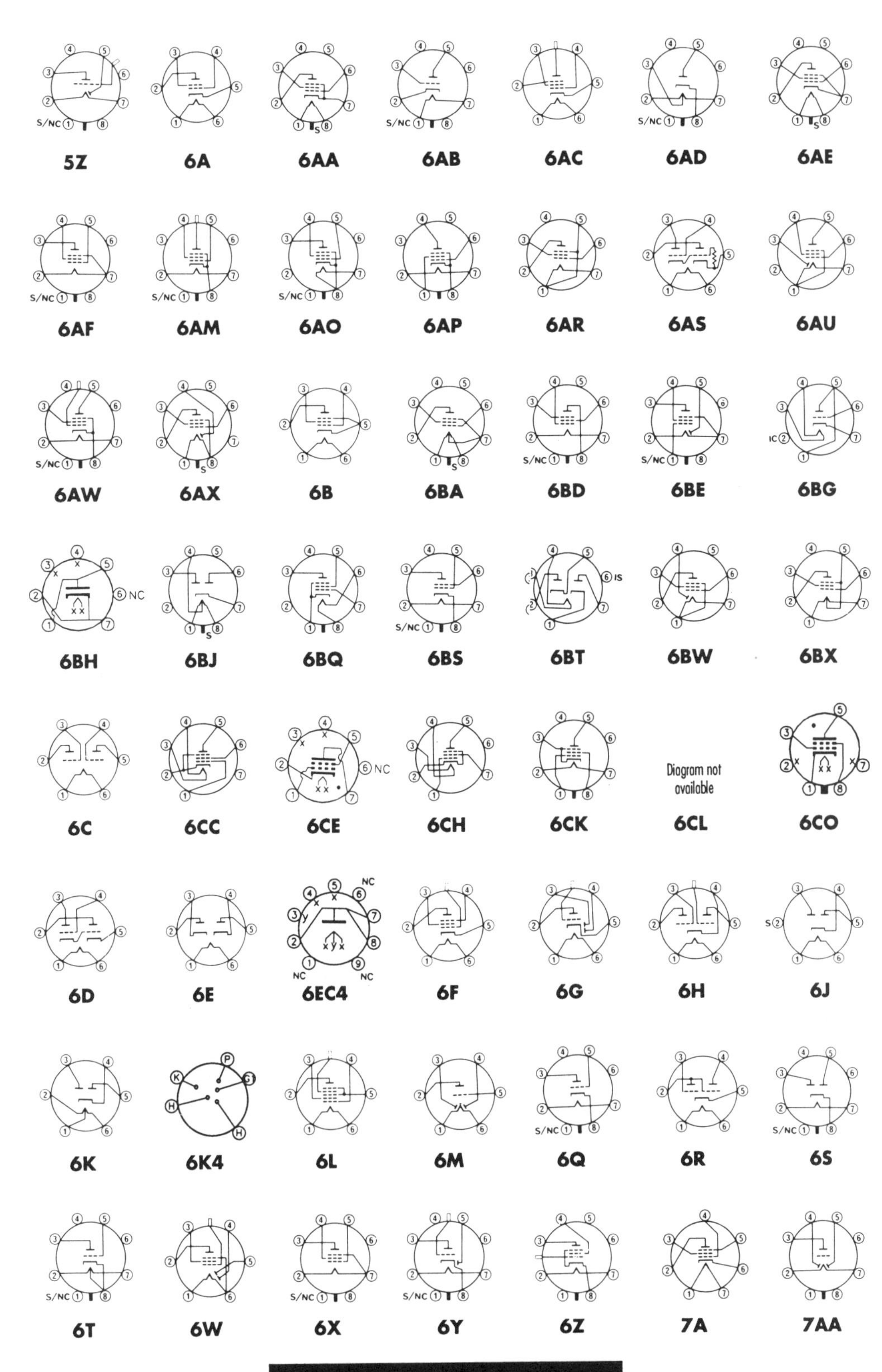
5Z
6A
6AA
6AB
6AC
6AD
6AE
6AF
6AM
6AO
6AP
6AR
6AS
6AU
6AW
6AX
6B
6BA
6BD
6BE
6BG
6BH
6BJ
6BQ
6BS
6BT
6BW
6BX
6C
6CC
6CE
6CH
6CK
Diagram not available
6CL
6CO
6D
6E
6EC4
6F
6G
6H
6J
6K
6K4
6L
6M
6Q
6R
6S
6T
6W
6X
6Y
6Z
7A
7AA

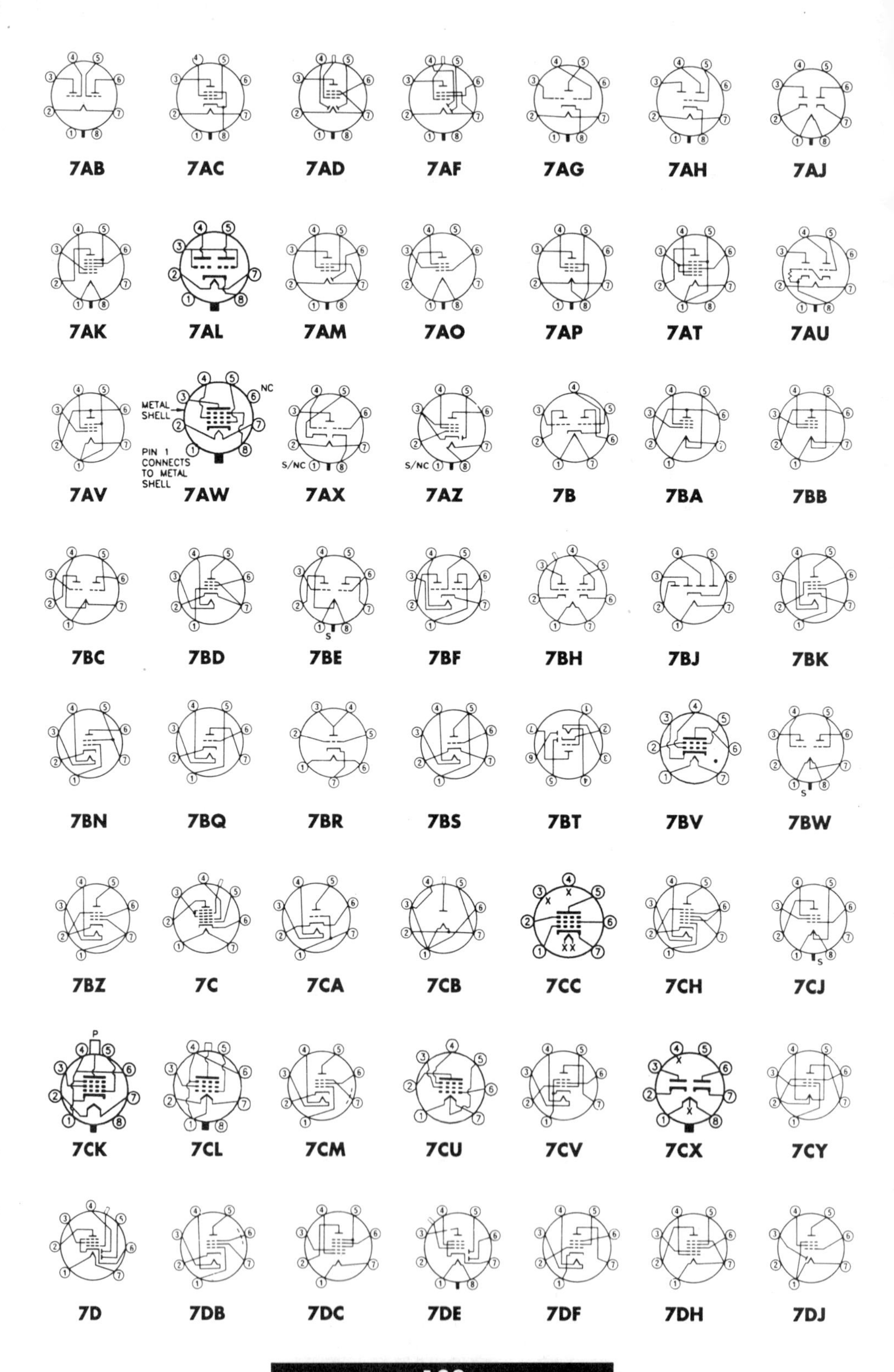
7AB
7AC
7AD
7AF
7AG
7AH
7AJ
7AK
7AL
7AM
7AO
7AP
7AT
7AU
7AV
METAL SHELL
NC
PIN 1 CONNECTS TO METAL SHELL
7AW
S/NC
7AX
S/NC
7AZ
7B
7BA
7BB
7BC
7BD
S
7BE
7BF
7BH
7BJ
7BK
7BN
7BQ
7BR
7BS
7BT
7BV
S
7BW
7BZ
7C
7CA
7CB
7CC
7CH
S
7CJ
P
7CK
7CL
7CM
7CU
7CV
7CX
7CY
7D
7DB
7DC
7DE
7DF
7DH
7DJ

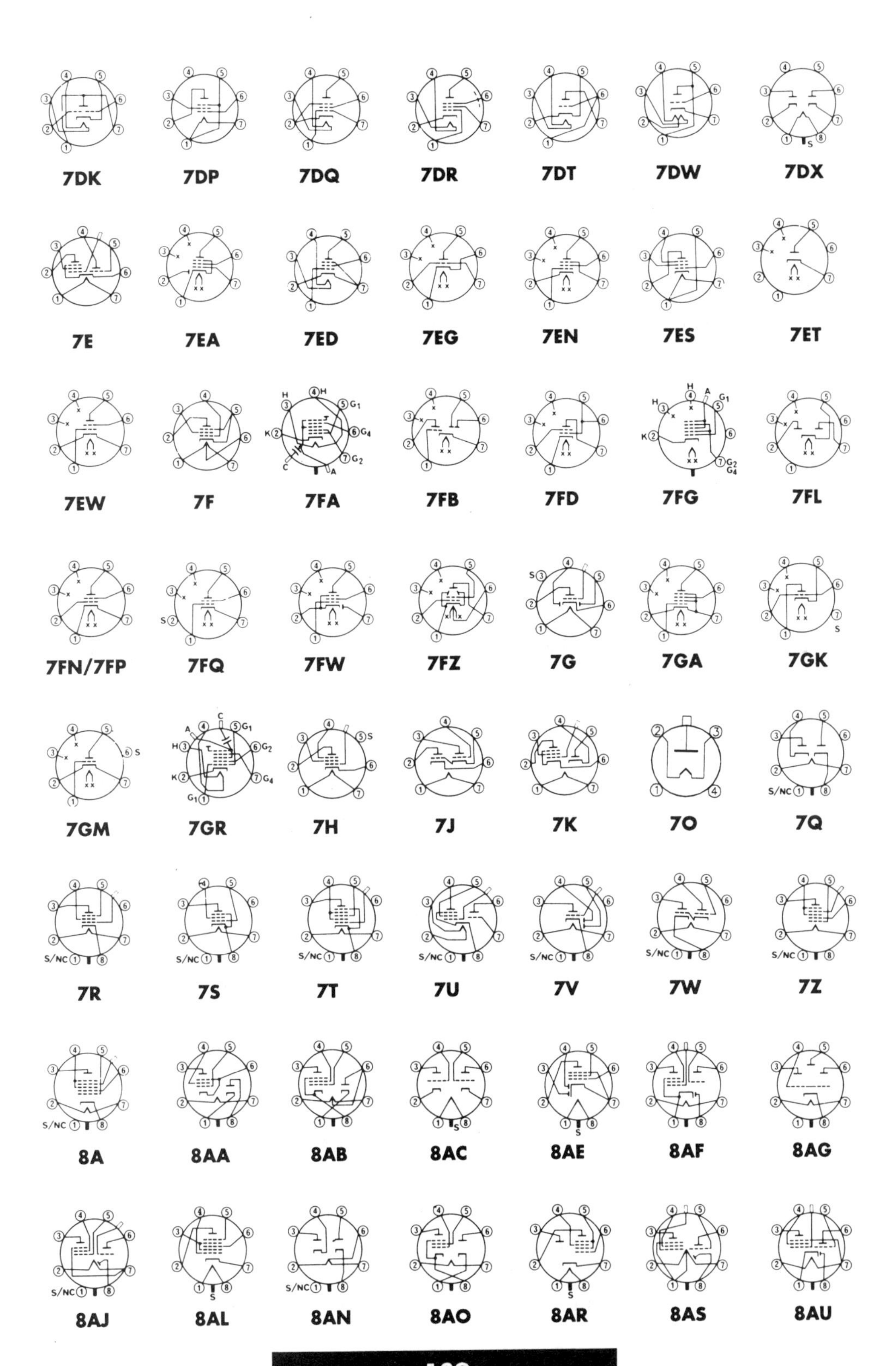
7DK
7DP
7DQ
7DR
7DT
7DW
7DX
7E
7EA
7ED
7EG
7EN
7ES
7ET
7EW
7F
7FA
7FB
7FD
7FG
7FL
7FN/7FP
7FQ
7FW
7FZ
7G
7GA
7GK
7GM
7GR
7H
7J
7K
7O
7Q
7R
7S
7T
7U
7V
7W
7Z
8A
8AA
8AB
8AC
8AE
8AF
8AG
8AJ
8AL
8AN
8AO
8AR
8AS
8AU

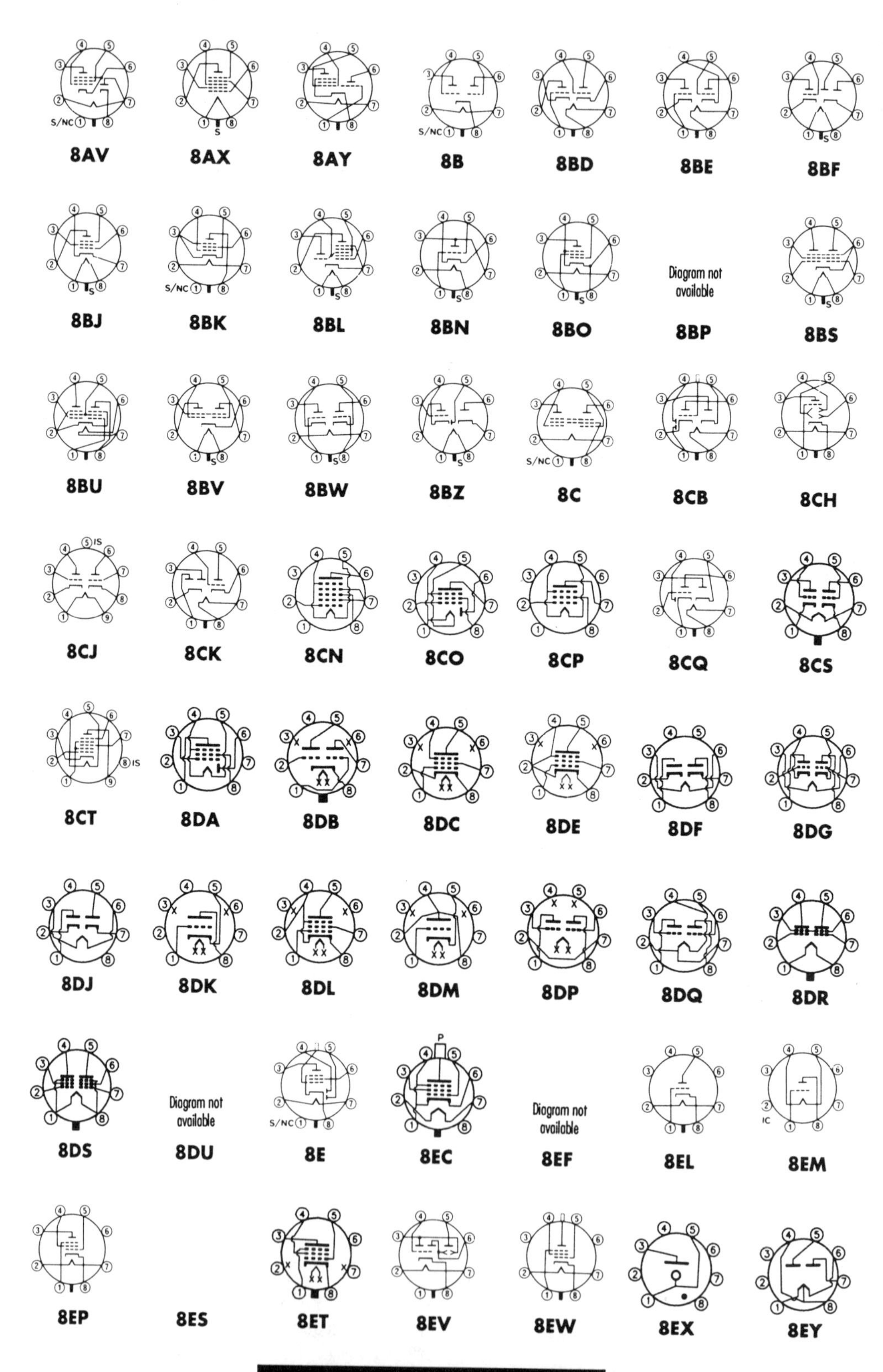
8AV
8AX
8AY
8B
8BD
8BE
8BF
8BJ
8BK
8BL
8BN
8BO
Diagram not available
8BP
8BS
8BU
8BV
8BW
8BZ
8C
8CB
8CH
8CJ
8CK
8CN
8CO
8CP
8CQ
8CS
8CT
8DA
8DB
8DC
8DE
8DF
8DG
8DJ
8DK
8DL
8DM
8DP
8DQ
8DR
8DS
Diagram not available
8DU
8E
8EC
Diagram not available
8EF
8EL
8EM
8EP
8ES
8ET
8EV
8EW
8EX
8EY

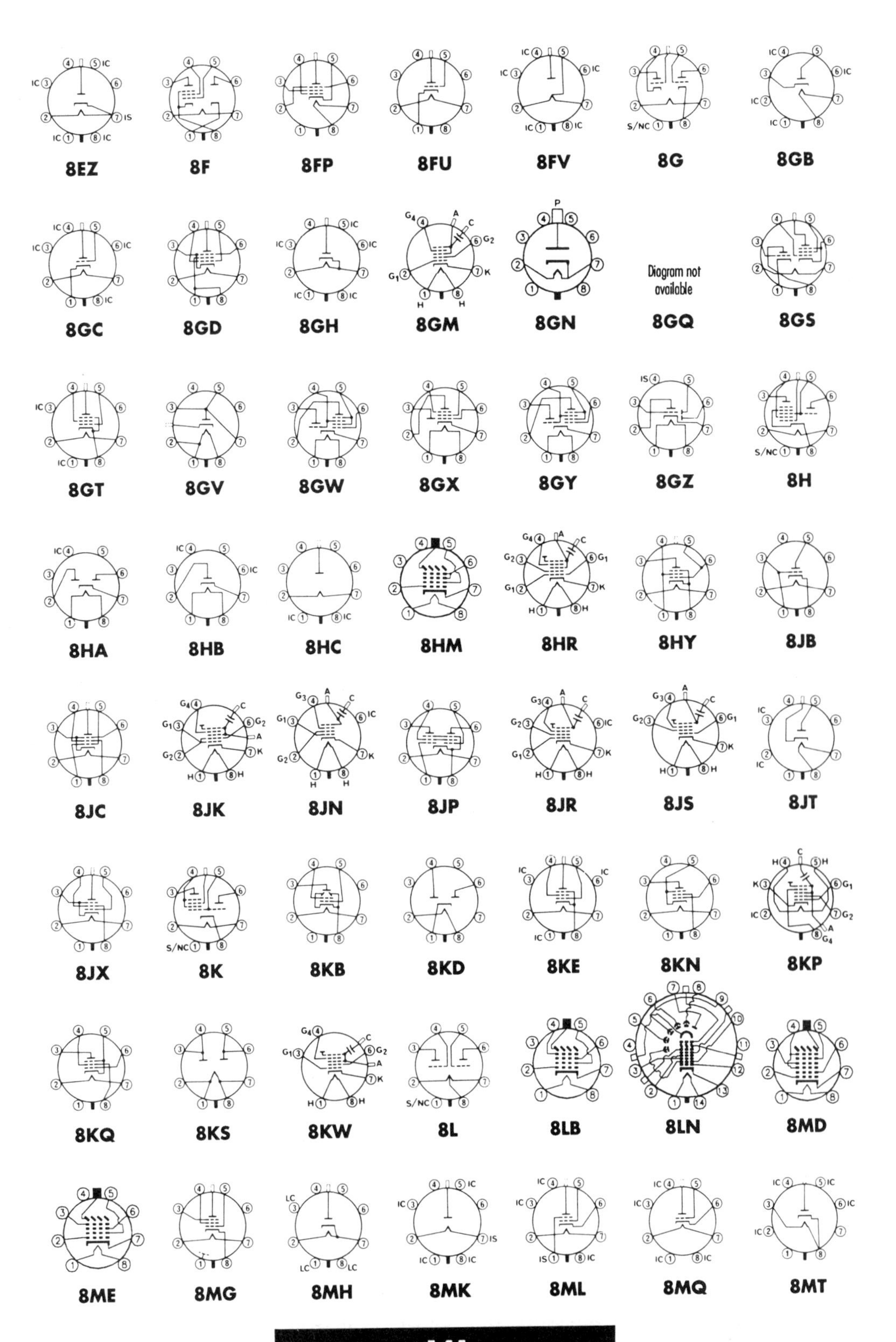
8EZ
8F
8FP
8FU
8FV
8G
8GB
8GC
8GD
8GH
8GM
8GN
Diagram not available
8GQ
8GS
8GT
8GV
8GW
8GX
8GY
8GZ
8H
8HA
8HB
8HC
8HM
8HR
8HY
8JB
8JC
8JK
8JN
8JP
8JR
8JS
8JT
8JX
8K
8KB
8KD
8KE
8KN
8KP
8KQ
8KS
8KW
8L
8LB
8LN
8MD
8ME
8MG
8MH
8MK
8ML
8MQ
8MT

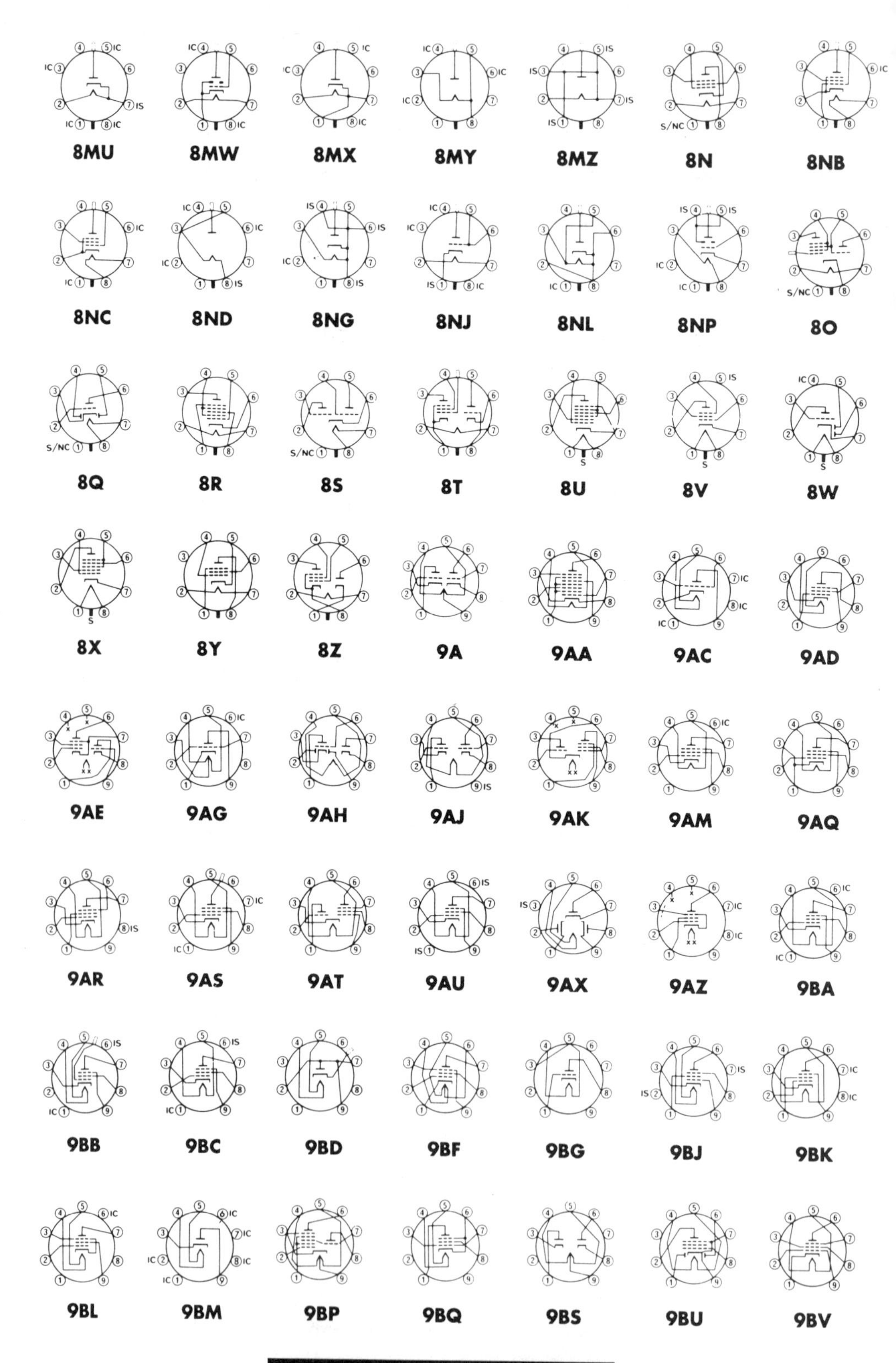
8MU
8MW
8MX
8MY
8MZ
8N
8NB
8NC
8ND
8NG
8NJ
8NL
8NP
8O
8Q
8R
8S
8T
8U
8V
8W
8X
8Y
8Z
9A
9AA
9AC
9AD
9AE
9AG
9AH
9AJ
9AK
9AM
9AQ
9AR
9AS
9AT
9AU
9AX
9AZ
9BA
9BB
9BC
9BD
9BF
9BG
9BJ
9BK
9BL
9BM
9BP
9BQ
9BS
9BU
9BV

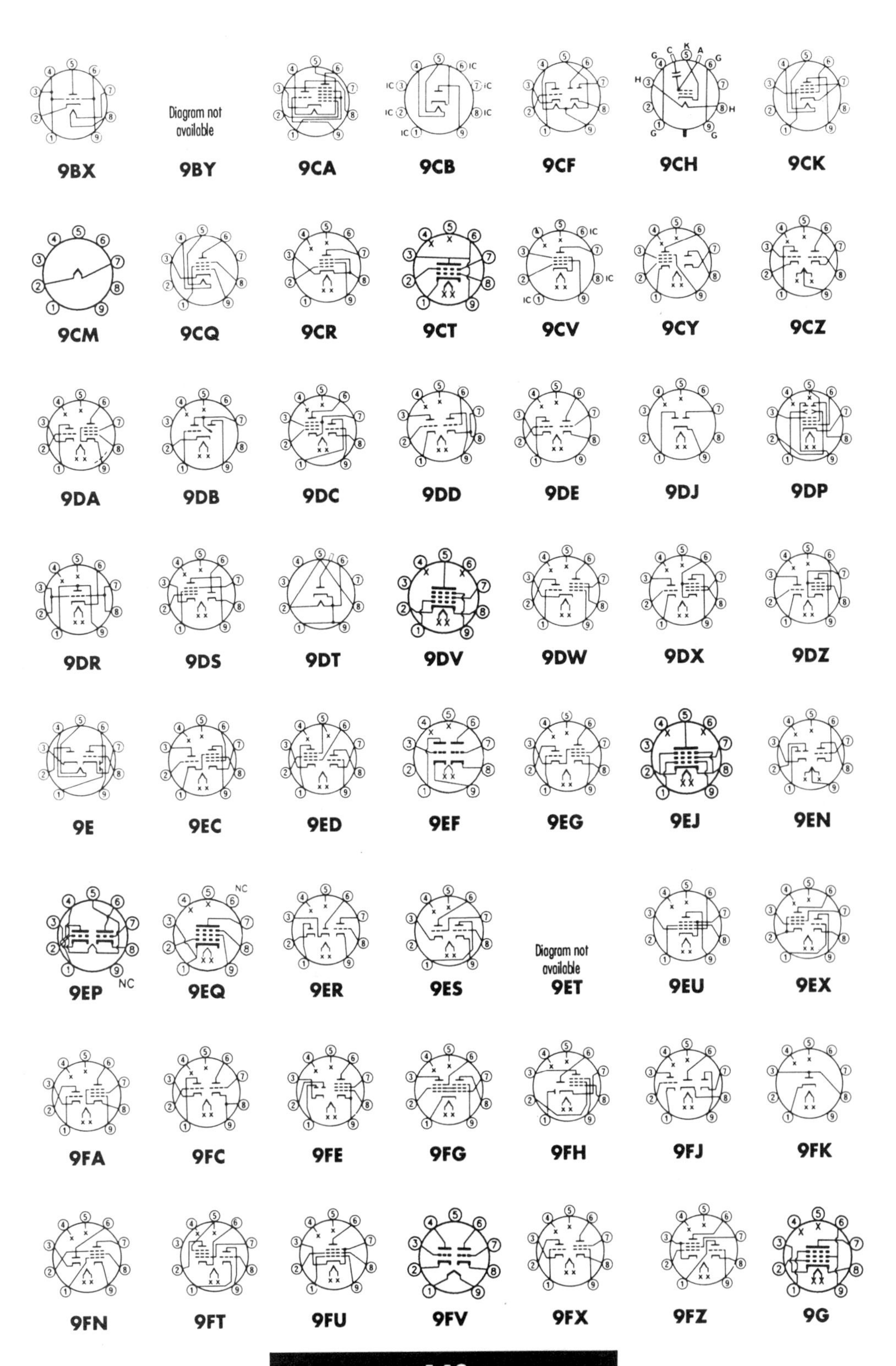
9BX
Diagram not available
9BY
9CA
9CB
9CF
9CH
9CK
9CM
9CQ
9CR
9CT
9CV
9CY
9CZ
9DA
9DB
9DC
9DD
9DE
9DJ
9DP
9DR
9DS
9DT
9DV
9DW
9DX
9DZ
9E
9EC
9ED
9EF
9EG
9EJ
9EN
9EP
9EQ
9ER
9ES
Diagram not available
9ET
9EU
9EX
9FA
9FC
9FE
9FG
9FH
9FJ
9FK
9FN
9FT
9FU
9FV
9FX
9FZ
9G

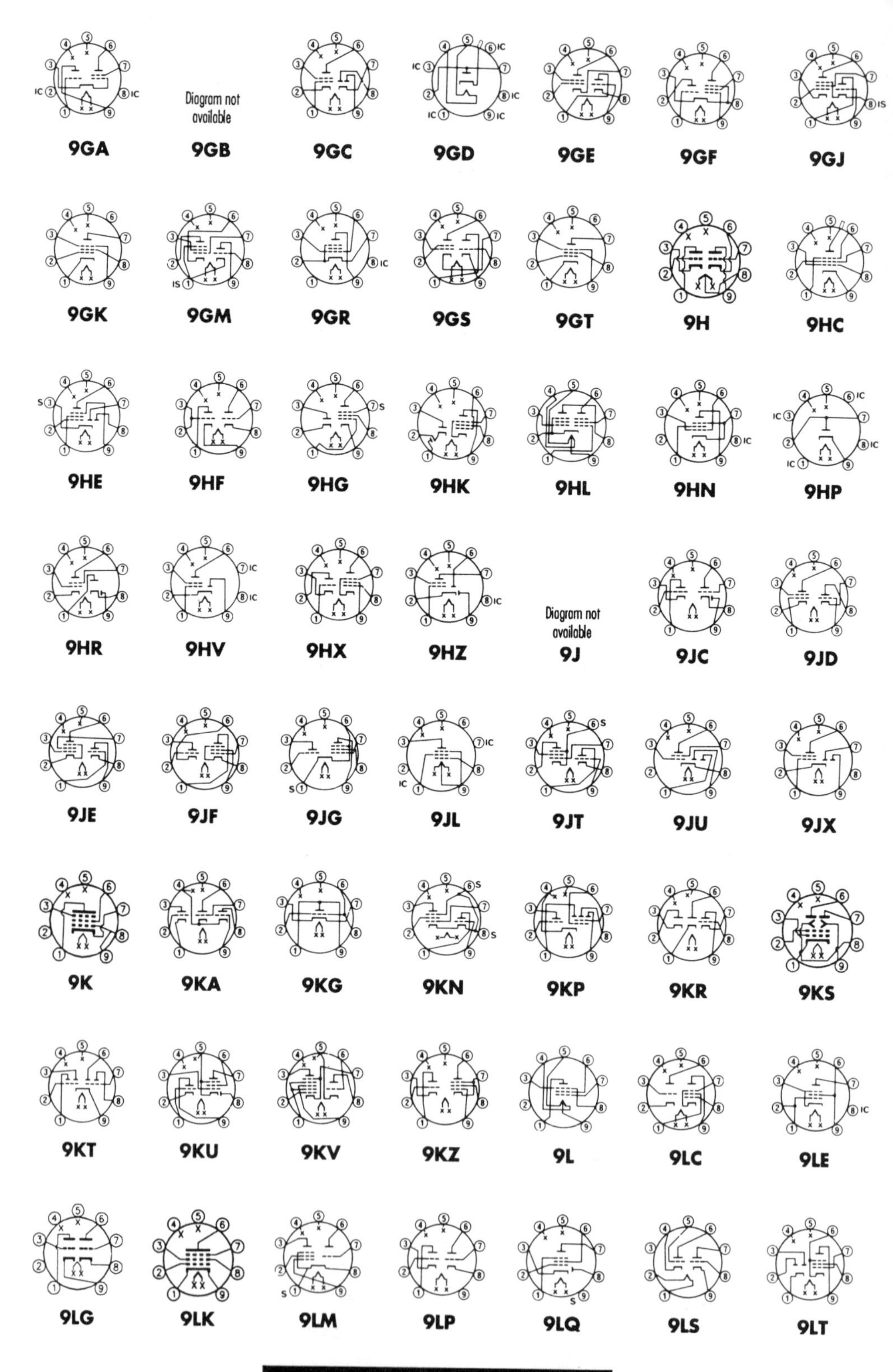
Diagram not available
9GA
9GB
9GC
9GD
9GE
9GF
9GJ
9GK
9GM
9GR
9GS
9GT
9H
9HC
9HE
9HF
9HG
9HK
9HL
9HN
9HP
9HR
9HV
9HX
9HZ
Diagram not available
9J
9JC
9JD
9JE
9JF
9JG
9JL
9JT
9JU
9JX
9K
9KA
9KG
9KN
9KP
9KR
9KS
9KT
9KU
9KV
9KZ
9L
9LC
9LE
9LG
9LK
9LM
9LP
9LQ
9LS
9LT

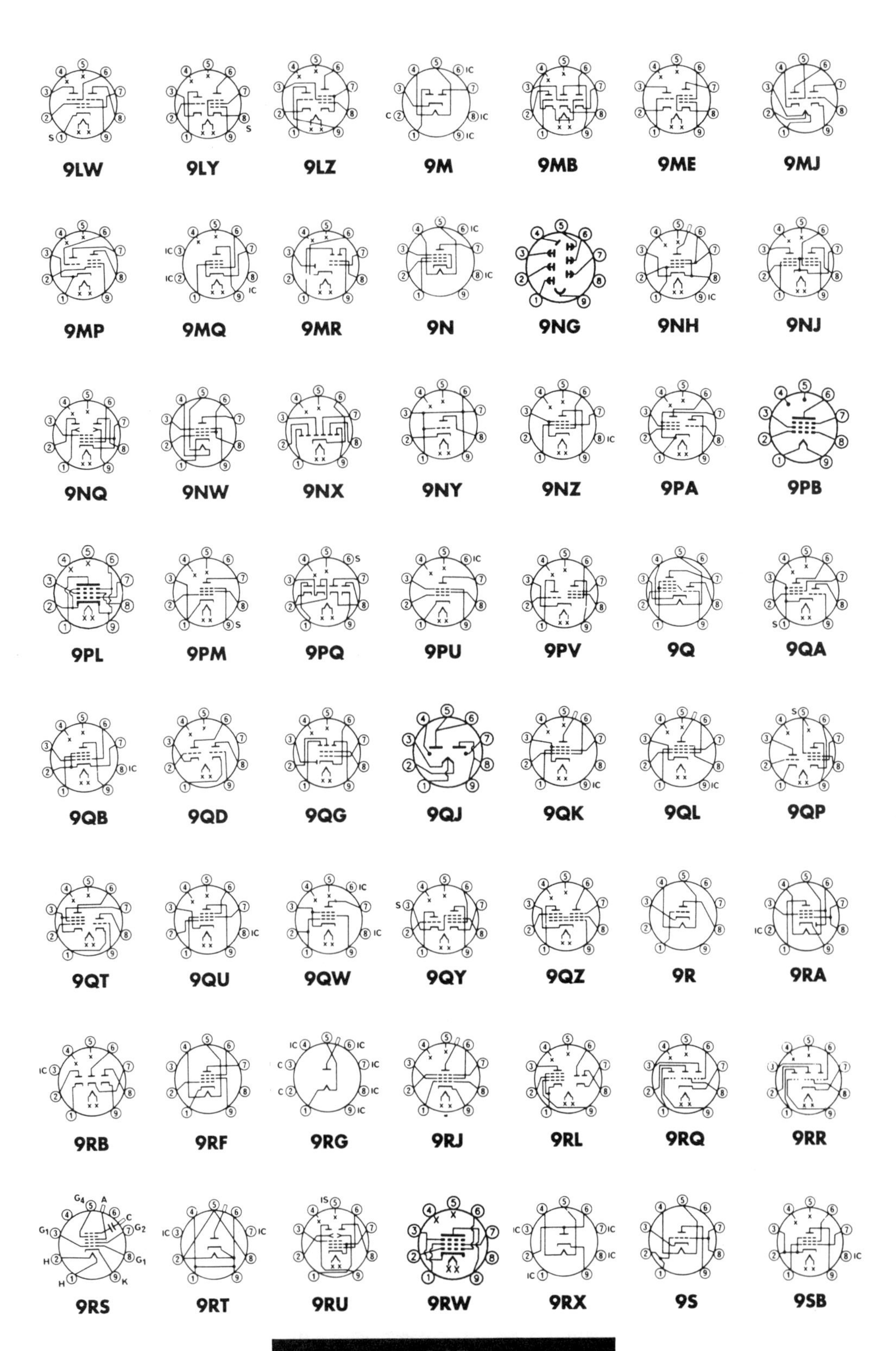
9LW
9LY
9LZ
9M
9MB
9ME
9MJ
9MP
9MQ
9MR
9N
9NG
9NH
9NJ
9NQ
9NW
9NX
9NY
9NZ
9PA
9PB
9PL
9PM
9PQ
9PU
9PV
9Q
9QA
9QB
9QD
9QG
9QJ
9QK
9QL
9QP
9QT
9QU
9QW
9QY
9QZ
9R
9RA
9RB
9RF
9RG
9RJ
9RL
9RQ
9RR
9RS
9RT
9RU
9RW
9RX
9S
9SB

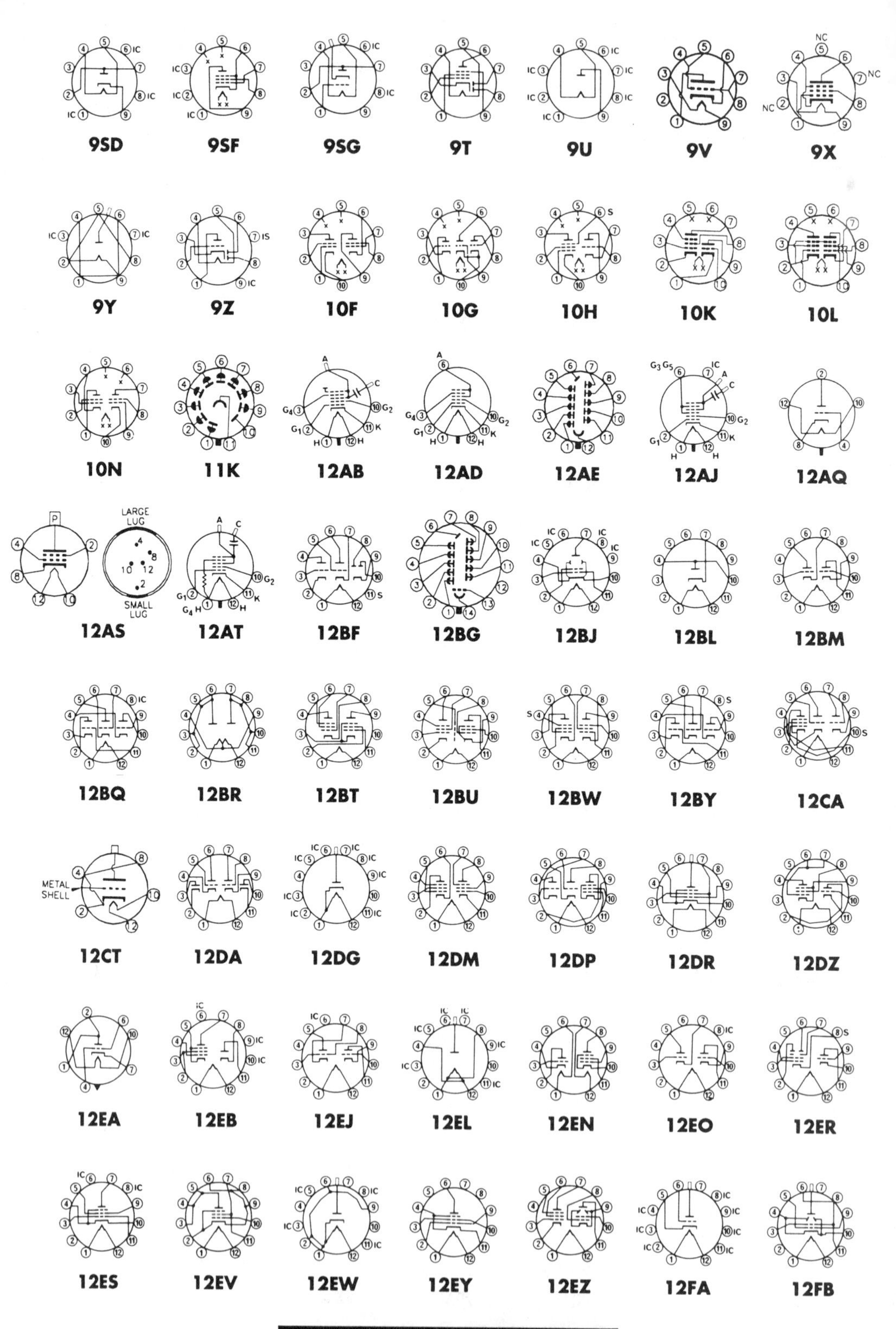
9SD
9SF
9SG
9T
9U
9V
9X
9Y
9Z
10F
10G
10H
10K
10L
10N
11K
12AB
12AD
12AE
12AJ
12AQ
12AS
LARGE LUG
SMALL LUG
12AT
12BF
12BG
12BJ
12BL
12BM
12BQ
12BR
12BT
12BU
12BW
12BY
12CA
12CT
METAL SHELL
12DA
12DG
12DM
12DP
12DR
12DZ
12EA
12EB
12EJ
12EL
12EN
12EO
12ER
12ES
12EV
12EW
12EY
12EZ
12FA
12FB

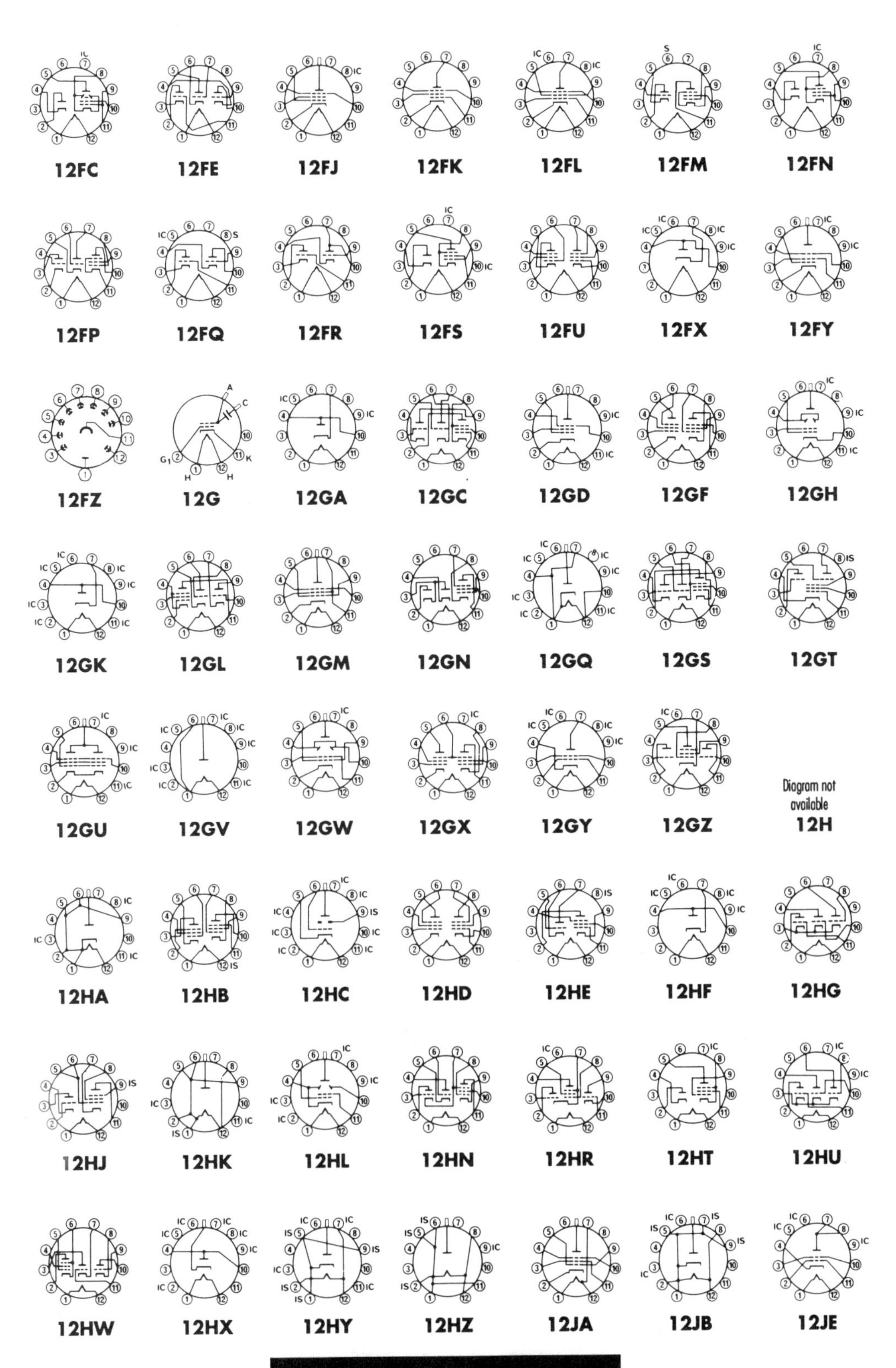
12FC
12FE
12FJ
12FK
12FL
12FM
12FN
12FP
12FQ
12FR
12FS
12FU
12FX
12FY
12FZ
12G
12GA
12GC
12GD
12GF
12GH
12GK
12GL
12GM
12GN
12GQ
12GS
12GT
12GU
12GV
12GW
12GX
12GY
12GZ
Diagram not available
12H
12HA
12HB
12HC
12HD
12HE
12HF
12HG
12HJ
12HK
12HL
12HN
12HR
12HT
12HU
12HW
12HX
12HY
12HZ
12JA
12JB
12JE

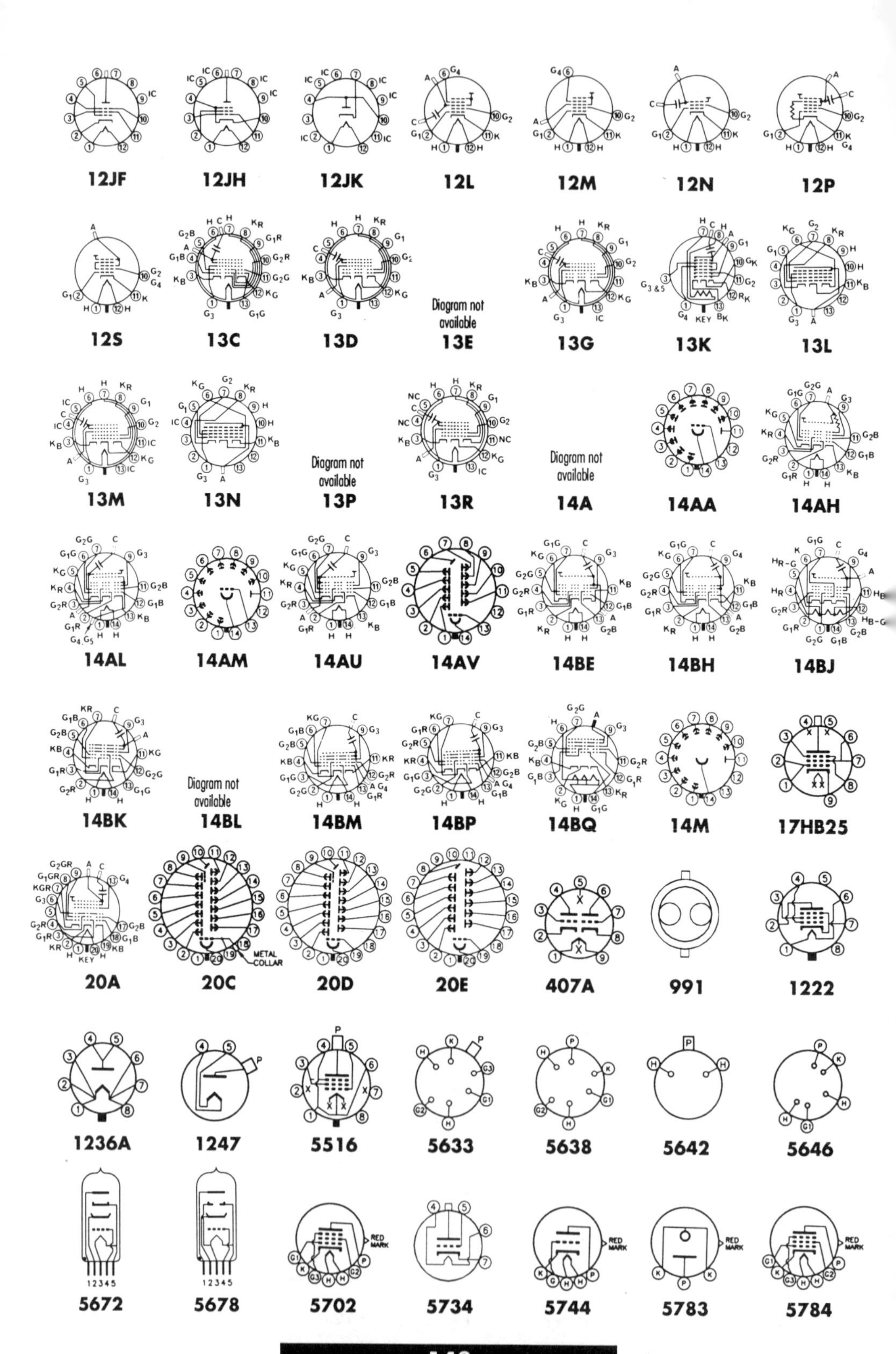
12JF
12JH
12JK
12L
12M
12N
12P
12S
13C
13D
Diagram not available
13E
13G
13K
13L
13M
13N
Diagram not available
13P
13R
Diagram not available
14A
14AA
14AH
14AL
14AM
14AU
14AV
14BE
14BH
14BJ
14BK
Diagram not available
14BL
14BM
14BP
14BQ
14M
17HB25
20A
20C
METAL COLLAR
20D
20E
407A
991
1222
1236A
1247
5516
5633
5638
5642
5646
5672
5678
5702
RED MARK
5734
5744
RED MARK
5783
RED MARK
5784
RED MARK

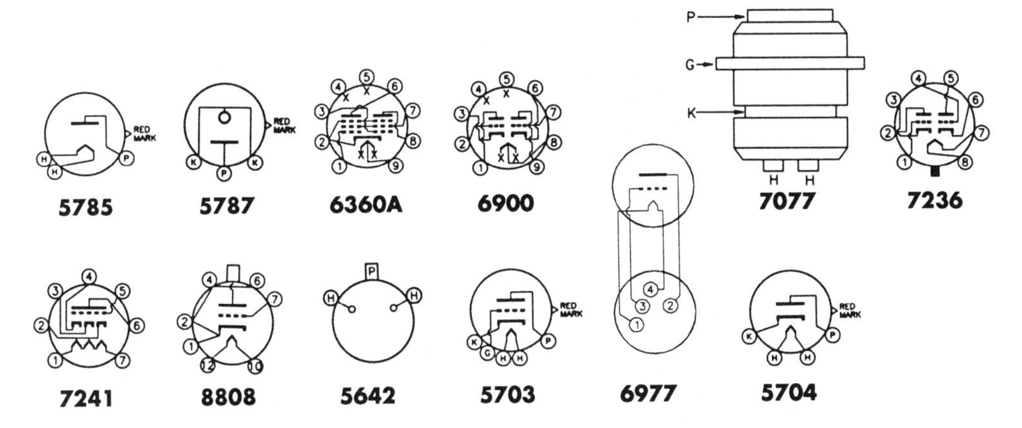
RED MARK
H
H
P
5785
RED MARK
K
P
K
5787
6360A
6900
P
G
K
H H
7077
7236
7241
8808
P
H
H
5642
RED MARK
K
G
H
H
P
5703
6977
RED MARK
K
H
H
P
5704